Shock capturing and high-order methods for hyperbolic conservation laws

Von der
Carl-Friedrich-Gauß-Fakultät
der Technischen Universität Carolo-Wilhelmina zu Braunschweig

zur Erlangung des Grades eines
Doktors der Naturwissenschaften (Dr. rer. nat.)

genehmigte Dissertation

von
Jan Glaubitz
geboren am 08.12.1990
in Braunschweig

Eingereicht am: 06.11.2019
Disputation am: 16.12.2019
1. Referent: Prof. Dr. Thomas Sonar
2. Referent: Prof. Dr. Klaus-Jürgen Förster
3. Referent: Prof. Dr. Anne Gelb

2019

Bibliografische Information der Deutschen Nationalbibliothek

Die Deutsche Nationalbibliothek verzeichnet diese Publikation in der Deutschen Nationalbibliografie; detaillierte bibliografische Daten sind im Internet über http://dnb.d-nb.de abrufbar.

ISBN 978-3-8325-5084-4

Logos Verlag Berlin GmbH
Comeniushof, Gubener Str. 47,
10243 Berlin
Tel.: +49 (0)30 42 85 10 90
Fax: +49 (0)30 42 85 10 92
INTERNET: https://www.logos-verlag.de

Acknowledgments

This thesis resulted from my doctoral position at the institute of *Computational Mathematics* at the *Technische Universität Braunschweig* and was partially funded by the German Reserach Foundation (DFG, Deutsche Forschungsgemeinschaft) under grant SO 363/15-1.

Firstly, I would like to express my sincere gratitude to my advisor Prof. Thomas Sonar for his continuous support, not only during my Ph.D. studies but already during my early undergraduate studies. His guidance helped me in all the time of research and writing of this thesis. At the same time, he provided me with much freedom and always allowed me to explore my own ideas and interests.

Besides my advisor, I would also like to express my thanks to the rest of my thesis committee: Prof. Klaus-Jürgen Förster and Prof. Anne Gelb, for their insightful comments, but also for many stimulating question which incented me to widen my research from various perspectives. I also would like to thank Prof. Anne Gelb for several joint research projects, inspiring visits and discussions, and for providing me with the opportunity to join her working group at *Dartmouth College* (New Hampshire, US).

In fact, some parts of this thesis result from joint work with many colleagues. My sincere thanks also go to Dr. Philipp Öffner (University of Zurich, Switzerland), Dr. Hendrik Ranocha (King Abdullah University of Science and Technology, Saudi Arabia), Prof. Alberto Costa Nogueira Junior and Prof. Claudio Alessandro de Carvalho Silva (State University of Campinas, Brazil), and Prof. Renato Fernandes Cantão (Universidade Federal de São Carlos, Brazil). Moreover, I would like to thank Dorian Hillebrand and Simon-Christian Klein for proofreading this thesis.

My beloved wife Alina, I would like to express my gratitude, not just for proofreading several chapters, but especially for her personal support and warm-hearted care. Last but not the least, I would like to thank my family: my parents, my sisters, and my grandparents for supporting me spiritually throughout writing this thesis and my life in general.

ABSTRACT

This thesis is concerned with the numerical treatment of hyperbolic conservation laws. These play an important role in describing many natural phenomena. Challenges in their theoretical as well as numerical study stem from the fact that spontaneous shock discontinuities can arise in their solutions, even in finite time and smooth initial states. Moreover, the numerical treatment of hyperbolic conservations laws involves many different fields from mathematics, physics, and computer science. As a consequence, this thesis also provides contributions to several different fields of research — which are still connected by numerical conservation laws, however. These contributions include, but are not limited to, the construction of stable high order quadrature rules for experimental data, the development of new stable numerical methods for conservation laws, and the investigation and design of shock capturing procedures as a means to stabilize high order numerical methods in the presence of (shock) discontinuities.

Zusammenfassung

Gegenstand der vorliegenden Arbeit ist die numerische Behandlung hyperbolischer Erhaltungsgleichungen. Diese Unterklasse von partiellen Differentialgleichungen ist von nicht zu unterschätzender Bedeutung bei der Beschreibung und Modellierung zahlreicher Phänomene in der Natur. Gleichzeitig wird ihre theoretische sowie numerische Behandlung dadurch erheblich erschwert, dass Lösungen hyperbolischer Erhaltungsgleichungen Unstetigkeiten (Stöße) entwickeln können. Auch aufgrund dessen berührt die Numerik von hyperbolischen Erhaltungsgleichungen eine Vielzahl verschiedener Gebiete aus Mathematik, Physik und Informatik. Dies hat zur Folge, dass die vorliegende Arbeit ebenso Beiträge zu verschiedenen aktuellen Fragestellungen liefert. Diese Beiträge umfassen, sind aber nicht beschränkt auf, die Entwicklung stabiler Quadraturverfahren hoher Ordnung für experimentelle Daten, die Behandlung neuartiger numerischer Methoden für Erhaltungsgleichungen und die eingehende Untersuchung und Konstruktion von Verfahren zur Stoßerfassung.

Table of Contents

CHAPTER 1

INTRODUCTION

This thesis contains most — though not all — scientific findings during my doctoral position at the institute of *Computational Mathematics* at the *Technische Universität Braunschweig* starting in April 2016. Its focus lies on the numerical treatment of hyperbolic conservation laws (CLs). Among all partial differential equations (PDEs), CLs are a particularly important, yet challenging subclass. Their importance results from the fact that they can be used to model many phenomena in the natural and engineering sciences. In fact, CLs in the form of the Euler equations were among the first PDEs to be written down [Eul57]. Since then, they have had broad application in physics as well as in other fields, such as chemistry, biology, geology, and engineering. Their challenging nature, on the other hand, stems from the well-known observation that solutions of CLs can develop spontaneous shock discontinuities, even in finite time and for smooth initial states. This observation was first made by Riemann in his ground breaking work on the propagation of waves of finite amplitude in air [Rie60]. As a consequence, the numerical treatment of CLs involves many different fields from mathematics as well as physics, which can all be quite challenging themselves. At the same time, the study of CLs has therefore always yielded new and fruitful connections between originally separated scientific fields. Most notable for this thesis are the many synergies between the numerical treatment of CLs and the field of image processing, where the treatment of image edges closely resembles (shock) discontinuities in the context of CLs. Here, we ourselves, will adapt certain objects from image processing (so-called edge sensors) to develop novel shock capturing procedures for the numerical treatment of CLs. Including these novel procedures, the contributions of this thesis can be roughly divided up into three parts:

1. Construction and investigation of stable high order quadrature rules (QRs) for the numerical integration of experimental data.

2. Development of conservative and (energy/entropy) stable high order methods for CLs.

3. Design and analysis of shock capturing procedures to stabilize high order methods for CLs in the presence of shock discontinuities.

A complete list of all publications related to this thesis can be found at the end of this chapter. In what follows, we shortly outline these contributions and their role in the current state of science, while providing an overview for the subsequent material of this thesis.

Fundamentals

The fundamentals regarding CLs as well as numerical preliminaries needed for their numerical treatment can be found in chapters 2 and 3. Readers which are already familiar with numerical methods for hyperbolic CLs and the related fields of numerical analysis, e. g., orthogonal polynomials and numerical integration, should be able to skip these two chapters.

Numerical integration of experimental data

The first novel contribution of this thesis can be found in chapters 4 and 5. In these chapters, we investigate and develop new stable high order QRs for the numerical integration of experimental data. The problem of measuring areas and volumes has always been present in everyday life. Thus, one can argue that numerical integration is as old as mathematics itself and dates back at least to the ancient Babylonians and Egyptians [BM11, Son16]. Since then, the mathematical field of numerical integration has matured considerably. Today, numerical integration describes the problem of recovering a continuous integral over a function f by using only a finite number of function evaluations $f(x_n)$, $n = 1, \ldots, N$. At least in one dimension — as considered here — there exist many excellent QRs by now, e. g., Gaussian QRs. So one might ask, what there is left for us to do in this thesis. However, it should be stressed that most of these QRs, especially the highly efficient ones, assume very specific distributions for the quadrature points x_n. Yet, experimental measurements are often performed at equidistant or even scattered points. In many applications, it is therefore impractical — if not even impossible — to obtain data to fit known QRs. In this thesis, we tackle this problem by investigating and further developing a class of QRs first proposed by Wilson [Wil70b, Wil70a] in 1970. The idea behind this approach is to formulate the well-known exactness conditions as an underdetermined least squares (LS) problem. Then, the remaining degrees of freedom can be used to also ensure stability of the QR. In principle, this approach is independent of the distribution of the quadrature points x_n and can therefore be used to construct stable high order QRs for experimental data. So far, stability of the resulting QRs has been proven by Wilson for the simple case of an unweighted integral $\int f(x)\,\mathrm{d}x$. In Chapter 4, we extend this result to weighted integrals $\int f(x)\omega(x)\,\mathrm{d}x$ including a positive weight function ω. This result can also be found in the publication [Gla19a] which resulted from this thesis. An extension to general weight functions is provided in Chapter 5. In the process of developing stable high order QRs for this case, we also discuss different stability concepts which arise when considering general weight functions. To the best of our knowledge, some of these concepts (e. g. sign-consistency) have not been considered before. Hence, the findings of Chapter 5 resulted in the publication [Gla19d]. Besides yielding to new insight regarding the stability of QRs (possibly including nonpositive weight functions), the resulting stable high order QRs are used in the subsequent chapters (and [GÖ20]) to construct stable high order methods for numerically solving CLs.

Stable high order methods for hyperbolic conservation laws

Actual numerical methods for hyperbolic CLs $\partial_t u + \partial_x f(u) = 0$ are first addressed Chapter 6. There, we present two state-of-the-art high order methods, namely the discontinuous Galerkin (DG) and the flux reconstruction (FR) method. While the DG method is derived from a weak (integral) form of the underlying CL, the FR method directly emerges from the differential form of the CL. Yet, both schemes essentially build up on a piecewise polynomial approximation of the solution u in space and can be classified as discontinuous spectral element (SE) methods. It should be stressed that in both schemes, the piecewise polynomial approximations are usually obtained by polynomial interpolation on Gauss–Legendre (GLe) or Gauss-Lobatto (GLo) points. In Chapter 7, we discuss two high order methods which, to the best of our knowledge, have not been proposed before.

The first method, presented in Chapter 7.1, builds up on the usual DG method but extends its stable discretization from GLe and GLo points to equidistant and even scattered points. The basic idea in our construction of this method is to replace usual interpolation polynomials by more general discrete least squares (DLS) approximations (resulting from orthogonal projections with respect to a discrete inner product). Moreover, the corresponding discrete inner product as well as the QR used to replace exact integrals occurring in the weak form of the CL are derived from the LS-QRs developed in the previous Chapter 4. The material presented in Chapter 7.1 resulted in the publication [GÖ20].

In Chapter 7.2, we start by investigating so-called radial basis function (RBF) methods. These methods use a global approximation of the solution u by RBFs and can therefore be classified as spectral methods. RBFs have become a powerful tool in multivariate interpolation and approximation theory since they are easy to implement, allow arbitrary scattered data, and can be spectrally accurate. As a consequence they are also often used to numerically solve PDEs. Unfortunately, Platte and Driscoll [PD06] have demonstrated that their application to CLs can yield unstable schemes in the presence of boundary conditions (BCs). Here, we show that this only holds true if RBFs are applied too naively to CLs. By carefully investigating their stability properties, we are able to develop new conservative and stable RBF methods for CLs by constructing them from a weak integral form of the CL and incorporating BCs via numerical fluxes. The material presented in this chapter resulted the publication [GG19b].

Shock capturing procedures

Probably the most outstanding challenge for numerically solving CLs lies in the famous Gibbs–Wilbraham phenomenon [HH79, Ric91, GS97]. This phenomenon was first discovered by Wilbraham [Wil48] in 1848 and rediscovered by Gibbs [Gib98, Gib99] in 1899 and essentially describes the inability of high order (polynomial) approximations to represent discontinuous functions, due to spurious oscillations around the discontinuities. In the context of high order methods for CLs, these spurious oscillations arise in the presence of shock discontinuities and often yield the numerical method to break down. As a consequence, many researchers have proposed different techniques to 'smooth out' these oscillations and to stabilize high oder numerical methods in the presence of shock discontinuities. In chapters 8, 9, and 10 we investigate and propose three different shock capturing procedures to do so for (discontinuous) SE methods. Furthermore, in Chapter 11 we build up on some of our findings and develop new promising artificial viscosity (AV) operators to stabilize (modern) finite difference (FD) methods.

We start our investigation of shock capturing methods for SE methods in Chapter 8 by revisiting the well-known AV method. This method has first been proposed in the context of classical FD schemes by von Neumann and Richtmyer [vNR50] during the *Manhattan project* at *Los Alamos National Laboratory* during the 1940's. It utilizes the well-known effect of dissipative mechanisms on (shock) discontinuities; when viscosity is incorporated, discontinuities in the solution are smeared out, yielding surfaces of discontinuities to be replaced by thin layers in which the solution varies (possibly rapidly but) continuously. Since then, AV methods have been adapted and refined by many researchers, especially in the context of discontinuous Galerkin spectral element methods (DGSEMs). In Chapter 8, we revisit the most commonly used variants of the AV method in DG methods [PP06, KWH11]. We also provide an investigation with respect to conservation and entropy stability which allows us to formulate clear criteria for the AV method which have to be satisfied in order for the AV method to preserve certain physical properties of the underlying CL. This material, presented in Chapter 8.3, resulted in the publication [GNJA+19]. Moreover, we address a strong connection to the technique of modal filtering as well as the discretization of AV methods by so-called summation by parts (SBP) operators. SBP operators allow to mimic integration by parts on a discrete level and can therefore be used to preserve many properties of the exact solution also for the (discrete) numerical solution. Even though they are an active field of research in the FD community since the work [KS77] of Kreiss and Scherer in 1977, their application to SE methods has only been initiated in 2013 by Gassner [Gas13]. In Chapter 8.6, we show how these can be utilized to construct conservative and stable discretizations of the AV method for SE methods. This analysis resulted in the publication [RGÖS18]. Additional publication which are connected to the material presented in Chapter 8 are [GÖRS16, GÖS18, ÖGR18, ÖGR19].

In Chapter 9, we then propose a new type of shock capturing in SE methods by ℓ^1 regularization. In many application, the solutions u of CLs might be discontinuous, but still piecewise smooth. In one dimension, this means that the exact solution only contains a finite number of jump discontinuities which are connected by smooth profiles. Thus, considering the jump function $[u]$ of the solution, $[u](x) \neq 0$ only holds for this finite set of jump discontinuities and $[u]$ is said to be sparse. Our idea in this chapter is to mimic this behavior of the exact solution also for the numerical solution. This is achieved by first approximating $[u]$ by certain high order edge sensors (HOES), originally proposed in the field of image processing [AGY05], and to enhance sparsity in this approximation by applying ℓ^1 regularization. It is shown in Chapter 9 that this technique, in fact, can enhance usual DG methods. It should be noted that similar investigations have been performed in [SGP17b] and [GHL19]. Yet, this thesis and the related publication [GG19a] are the first works to develop this idea in the context of SE methods. Here, we demonstrate that these methods allow some distinct advantages, such as element-to-element variations in the corresponding ℓ^1 optimization problem, yielding increased efficiency and accuracy of the method.

Another new type of shock capturing in SE methods is proposed and carefully investigated in Chapter 10. This time, the procedure is derived from some classical results in approximation theory and consists of going over from the original (polluted) approximation of u to a convex combination of the original approximation and its so-called Bernstein reconstruction. Our idea builds up on classical Bernstein operators — first introduced by Bernstein [Ber12a] in 1912 to provide a constructive proof of the famous Weierstrass approximation theorem [Wei85] — and we are able to prove that the resulting procedure is total variation diminishing (TVD) and preserves monotone (shock) profiles. Furthermore, the procedure can be modified to not just preserve but also to enforce certain bounds for the solution, such as positivity. Numerical tests demonstrate that the proposed shock capturing procedure is able to stabilize and enhance SE approximations in the presence of shocks. To the best of our knowledge, a similar approach has not been proposed before and the material presented in Chapter 10 resulted in the publication [Gla19c].

Finally, in Chapter 11, we come back to AV methods, this time in the context of modern high order FD methods though. Usual AV methods for FD methods essentially distribute viscosity equally over the whole computational domain [MSN04] or element — if a multi element/block structure is used. Here, we adapt the HOES introduced in Chapter 9 for ℓ^1 regularization and use them to construct novel AV operators which adapt themselves to the smoothness of the numerical solution. In particular, the resulting AV operators are able to calibrate the amount of viscosity and its distribution to the location of possible (shock) discontinuities. Moreover, we discuss their discretization by SBP operators. As a result, these operators are shown to preserve conservation, stability, and accuracy (in smooth regions) of the underlying method. The material presented in this final chapter resulted in the publication [Gla19b].

List of related publications

1. J. Glaubitz:
   High order edge sensor steered artificial viscosity operators.
   Submitted, 2019.

2. J. Glaubitz, A. Gelb:
Stability of radial basis function methods for one dimensional scalar conservation laws in weak form.
Submitted, 2019.

3. J. Glaubitz:
Stable high order quadrature rules for scattered data and general weight functions.
Submitted, 2019.

4. J. Glaubitz:
Discrete least squares quadrature rules on equidistant and scattered points.
Submitted, 2018.

5. J. Glaubitz, P. Öffner:
Stable discretisations of high-order discontinuous Galerkin methods on equidistant and scattered points.
Applied Numerical Mathematics, 2020. (DOI:10.1016/j.apnum.2019.12.020)

6. J. Glaubitz, P. Öffner, H. Ranocha:
Analysis of artificial dissipation of explicit and implicit time-integration methods.
Accepted in International Journal of Numerical Analysis and Modeling, 2019.
(https://arxiv.org/abs/1609.02393)

7. J. Glaubitz:
Shock capturing by Bernstein polynomials for scalar conservation laws.
Applied Mathematics and Computation 363 (2019): 124593. (DOI:10.1016/j.amc.2019.124593)

8. J. Glaubitz, A. Gelb:
High order edge sensors with ℓ^1 regularization for enhanced discontinuous Galerkin methods.
SIAM Journal of Scientific Computing, 41(2) (2019): A1304-A1330. (DOI:10.1137/18M1195280)

9. J. Glaubitz, A.C. Nogueira Jr., J.L.S. Almeida, R.F. Cantão, C.A.C. Silva:
Smooth and compactly supported viscous sub-cell shock capturing for discontinuous Galerkin methods.
Journal of Scientific Computing, 79 (2019): 249-272. (DOI:10.1007/s10915-018-0850-3)

10. P. Öffner, J. Glaubitz, H. Ranocha:
Stability of correction procedure via reconstruction with summation-by-parts operators for Burgers' equation using a polynomial chaos approach.
ESAIM: Mathematical Modelling and Numerical Analysis, 52.6 (2018): 2215-2245.
(DOI:10.1051/m2an/2018072)

11. H. Ranocha, J. Glaubitz, P. Öffner, T. Sonar:
Stability of artificial dissipation and modal filtering for flux reconstruction schemes using summation-by-parts operators.
Applied Numerical Mathematics, 128 (2018): 1-23. (DOI:10.1016/j.apnum.2018.01.019)

12. J. Glaubitz, P. Öffner, T. Sonar:
Application of modal filtering to a spectral difference method.
Mathematics of Computation, 87.309 (2018): 175-207. (DOI:10.1090/mcom/3257)

CHAPTER 2

Hyperbolic conservation laws

This chapter briefly addresses some basic concepts and results in the theory of hyperbolic CLs. To a large extend, the material is self contained, but there are many excellent books for a more detailed presentation: Some of the basic theory of CLs is, for instance, presented in the short monograph of Lax [Lax73]. The books of Whitham [Whi11] as well as Courant and Friedrichs [CF99] also present this material with a particular discussion of many applications including the Euler equations. More recent results can be found in the book of Smoller [Smo12]. Another recommendation is the wonderful monograph of Dafermos [Daf00]. In spite of explosive growth of research in the area in the last decades, this text describes, at least briefly, most topics that deal with hyperbolic CLs. Here, we roughly follow the presentation in the books of LeVeque [LeV92] and Godlewski & Raviart [GR91].

Outline

This chapter is organized as follows: We start by a general introduction of CLs in Chapter 2.1. Chapter 2.2 then addresses the (local) existence of classical solutions, while Chapter 2.3 introduces the method of characteristics and demonstrates the breakdown of (global) classical solutions for CLs. Hence, we revise the concept of weak solutions in Chapter 2.4. Unfortunately, by permitting the fairly general class of weak solutions, we lose uniqueness for solutions of CLs. This is explicitly shown in Chapter 2.5 for a simple Riemann problem. As a means to restore uniqueness as well as to motivate subsequent concepts and techniques to stabilize numerical methods for CLs, Chapter 2.6 provides a brief presentation of viscosity and its role in the theory of CLs. Chapter 2.7 then introduces the concept of (mathematical) entropy functions and the resulting entropy solutions. These are often considered as the physically reasonable weak solutions, since they are in accordance with the second law of thermodynamics. Finally, we close this chapter with some distinguished results for scalar CLs and their entropy solutions in Chapter 2.8. These results will provide us with design criteria for the later discussed numerical methods and shock capturing procedures, which are often derived by the wish to mimic certain properties of exact entropy solutions.

2.1 General form and examples

A *CL* in its *strong form* (also called *differential form*) is a system of nonlinear first order PDEs of the form

$$\partial_t \boldsymbol{u} + \nabla \cdot \boldsymbol{f}(\boldsymbol{u}) = 0 \tag{2.1}$$

with time variable $t > 0$ and d spatial variables $\boldsymbol{x} = (x_1, \ldots, x_d)^T \in \mathbb{R}^d$. Here, the m-dimensional vector of unknowns $\boldsymbol{u} = (u_1(\boldsymbol{x},t), \ldots, u_m(\boldsymbol{x},t))^T$ and the vector-valued function

$\boldsymbol{f} : \mathbb{R}^m \to \mathbb{R}^{m\times d}$ respectively are referred to as the *conserved variables* and the *flux function*. Hence, d is the number of spatial dimensions and m is the number of conserved quantities in the system (2.1). Further, $\nabla = (\partial_{x_1}, \dots, \partial_{x_d})^T$ denotes the usual *nabla operator*. In this work, we will mostly consider one dimensional CLs, i. e. $d = 1$. In this case, (2.1) reduces to

$$\partial_t \boldsymbol{u} + \partial_x \boldsymbol{f}(\boldsymbol{u}) = 0 \tag{2.2}$$

with $t > 0$ and $x \in \mathbb{R}$. Note that, assuming a differentiable solution $\boldsymbol{u}$ and flux function $\boldsymbol{f}$, the CL can be rewritten as a quasilinear system of PDEs

$$\partial_t \boldsymbol{u} + A(\boldsymbol{u})\, \partial_x \boldsymbol{u} = 0, \tag{2.3}$$

where $A(\boldsymbol{u}) = \boldsymbol{f}'(\boldsymbol{u})$ is the Jacobian matrix of the flux function. As we will see in Chapter 2.3, in general, the assumption of a differentiable solution $\boldsymbol{u}$ is not valid and solutions of the CL (2.2) and the quasilinear PDE (2.3) might differ from each other. Yet, the quasilinear form (2.3) of the CL (2.2) motivates the following definition.

Definition 2.1 (Hyperbolic CLs)
We call a system of CLs (2.1) *hyperbolic* if the matrix $A(\boldsymbol{u}) = f'(\boldsymbol{u})$ is diagonalizable for every $\boldsymbol{u}$.

Thus, for a hyperbolic system of CLs, the matrix $A(\boldsymbol{u}) = f'(\boldsymbol{u})$ has m real eigenvalues $\lambda_1(\boldsymbol{u}) \leq \dots \leq \lambda_m(\boldsymbol{u})$ and a set of corresponding (right) eigenvectors $\{\mathbf{r}_k(\boldsymbol{u})\}_{k=1}^m$ which form a basis of $\mathbb{R}^m$. In case the real eigenvalues are distinct, i. e. $\lambda_1(\boldsymbol{u}) < \dots < \lambda_m(\boldsymbol{u})$, we call the system of CLs *strictly hyperbolic*.

Remark 2.2. Hyperbolicity ensures that a system of CLs allows wavelike solutions. Let us look for solutions in the form of a traveling wave,

$$\boldsymbol{u}(x,t) = \boldsymbol{v}(x - \sigma t), \tag{2.4}$$

where $\boldsymbol{v} : \mathbb{R} \to \mathbb{R}^m$ is the profile and $\sigma \in \mathbb{R}$ is the velocity of the wave. Then, if we insert (2.4) into the quasilinear system (2.3), we get

$$-\sigma \boldsymbol{v}'(x - \sigma t) + A(\boldsymbol{v}(x - \sigma t))\boldsymbol{v}'(x - \sigma t) = 0.$$

This equation is satisfied when σ is an eigenvalue of the matrix $A(\boldsymbol{v})$ and if $\boldsymbol{v}'$ is the corresponding (right) eigenvector. This allows us to decompose solutions of hyperbolic systems into traveling waves. In particular, this provides insight into the direction in which and the speed with which information is transported. This will be important, for instance, in the study of the Riemann problem, determining appropriate time step restrictions, and to scale the later proposed viscosity operators.

Example 2.3 (The linear advection equation)
The simplest example of a scalar CL is the *linear advection equation*

$$\partial_t u + \lambda\, \partial_x u = 0$$

with flux function $f(u) = \lambda u$ and constant velocity $\lambda \in \mathbb{R}$. The linear advection equation can be used to model the transport of, e. g., a fluid by bulk motion.

Example 2.4 (The Burgers' equation)
A first example of a nonlinear CL is the *inviscid Burgers' equation*

$$\partial_t u + \partial_x \left(\frac{u^2}{2} \right) = 0$$

with flux function $f(u) = u^2/2$. This equation has been studied first by the English mathematician Bateman in 1915 [Bat15] (also see [Whi11]) and only in 1948 by the Dutch physicist Burgers [Bur48]. The inviscid Burgers' equation arises from the *viscous Burgers' equation*

$$\partial_t u + \partial_x \left(\frac{u^2}{2}\right) = \varepsilon\, \partial_{xx} u$$

when the diffusion term is neglected, i. e. $\varepsilon = 0$.

Example 2.5 (The Euler equations)
Next, let us consider the system of the *Euler equations*:

$$\partial_t \boldsymbol{u} + \partial_x \boldsymbol{f}(\boldsymbol{u}) = 0 \tag{2.5}$$

Here, $\boldsymbol{u}$ and $\boldsymbol{f}(\boldsymbol{u})$ are given by

$$\boldsymbol{u} = \begin{pmatrix} u_1 \\ u_2 \\ u_3 \end{pmatrix} = \begin{pmatrix} \rho \\ \rho u \\ E \end{pmatrix}, \quad \boldsymbol{f} = \begin{pmatrix} f_1 \\ f_2 \\ f_3 \end{pmatrix} = \begin{pmatrix} \rho u \\ \rho u^2 + p \\ u(E+p) \end{pmatrix}, \tag{2.6}$$

where, ρ is the *density*, u is the *velocity*, p is the *pressure*, and E is the *total energy per unit volume*. The Euler equations are completed by addition of an *equation of state (EOS)* with general form

$$p = p(\rho, e),$$

where $e = E/\rho - u^2/2$ is the *specific internal energy*. In this work, we only consider ideal gases for which the equation of state is given by

$$p = (\gamma - 1)\rho e \tag{2.7}$$

with γ denoting the *ratio of specific heats*. The Euler equations describe the conservation of mass, momentum, and energy. They can be derived from the Navier–Stokes equations by neglecting viscosity and thermal conductivity [Tor13, Chapter 1]. Euler [Eul57] published them in 1797 and they were among the first PDEs to be written down.

Regarding the Euler equations, we note

Lemma 2.6
The Euler equations (2.5) *are hyperbolic. In particular, the Jacobian matrix* $A(\boldsymbol{u}) = \boldsymbol{f}'(\boldsymbol{u})$ *is given by*

$$A(\boldsymbol{u}) = \begin{pmatrix} 0 & 1 & 0 \\ -\frac{1}{2}(\gamma-3)\left(\frac{u_2}{u_1}\right)^2 & (3-\gamma)\left(\frac{u_2}{u_1}\right) & \gamma - 1 \\ -\gamma\frac{u_2 u_3}{u_1^2} + (\gamma-1)\left(\frac{u_2}{u_1}\right)^3 & \gamma\frac{u_3}{u_1} - \frac{3}{2}(\gamma-1)\left(\frac{u_2}{u_1}\right)^2 & \gamma\frac{u_2}{u_1} \end{pmatrix}$$

and the real eigenvalues are

$$\lambda_1 = u - a, \quad \lambda_2 = u, \quad \lambda_3 = u + a,$$

where $a := \sqrt{\gamma p/\rho}$ *is called the* sound speed.

Proof. In (2.6), the flux vector $\boldsymbol{f}$ is expressed with respect to the *primitive variables* $\boldsymbol{w} = (\rho, u, p)^T$ and the total energy per unit volume E. Yet, the flux vector $\boldsymbol{f}$ can be rewritten with respect to the conserved variable $\boldsymbol{u}$ as

$$\boldsymbol{f}(\boldsymbol{u}) = \begin{pmatrix} u_2 \\ \frac{1}{2}(3-\gamma)\frac{u_2^2}{u_1} + (\gamma-1)u_3 \\ \gamma\frac{u_2}{u_1}u_3 - \frac{1}{2}(\gamma-1)\frac{u_2^3}{u_1^2} \end{pmatrix},$$

where we have utilized the equation of state (2.7) to rewrite the pressure as

$$p = (\gamma - 1)\left(u_3 - \frac{u_2^2}{2u_1}\right).$$

Now, the assertion follows from some simple computations. □

There are many more examples of CLs with physical meaning. Some of them are the *traffic flow equation* [LeV92, Chapter 4.1], the *Buckley–Leverett equation* [Whi11, Chapter 3], and the *shallow water equations* [LeV92, Chapter 5.4]. Yet, as we will see in the following sections, the above examples already demonstrate the many crucial phenomena that can occur for (hyperbolic) CLs.

2.2 Classical solutions

Usually, the CL (2.2) is equipped with an *initial condition (IC)* $\boldsymbol{u}(x,0) = \boldsymbol{u}_0(x)$, for $x \in \mathbb{R}$, yielding the *initial value problem (IVP)*

$$\begin{aligned} \partial_t \boldsymbol{u} + \partial_x \boldsymbol{f} &= 0, & x \in \mathbb{R},\ t > 0, \\ \boldsymbol{u}(x,0) &= \boldsymbol{u}_0(x), & x \in \mathbb{R} \end{aligned} \tag{2.8}$$

for the conserved variable $\boldsymbol{u}$. The function $\boldsymbol{u}_0 : \mathbb{R} \to \mathbb{R}^m$ is referred to as the *initial data (ID)*. First, let us address uniqueness and existence of *classical solutions*, that is of solutions $\boldsymbol{u}$ in C^1. Henceforth, $C^k(X,Y)$ denotes the linear space of all k times continuously differentiable functions mapping from X to Y. If the context clarifies what the domain X and the codomain Y are, we simply denote the space by C^k.

Lemma 2.7 (Uniqueness of classical solutions)
Let $\boldsymbol{f}$ be C^2. The IVP (2.8) has at most one classical solution.

Proof. Let $\boldsymbol{u}$ and $\boldsymbol{v}$ be classical solutions of the IVP (2.8), i. e.

$$\begin{aligned} \partial_t \boldsymbol{u} + \partial_x \boldsymbol{f}(\boldsymbol{u}) &= 0, & x \in \mathbb{R},\ t > 0, \\ \partial_t \boldsymbol{v} + \partial_x \boldsymbol{f}(\boldsymbol{v}) &= 0, & x \in \mathbb{R},\ t > 0, \\ \boldsymbol{u}(x,0) &= \boldsymbol{v}(x,0), & x \in \mathbb{R}. \end{aligned}$$

Since $\boldsymbol{u}$ and $\boldsymbol{v}$ are classical solutions and $\boldsymbol{f}$ is a C^1 function, the above IVP is equivalent to

$$\begin{aligned} \partial_t \boldsymbol{u} + A(\boldsymbol{u})\,\partial_x \boldsymbol{u} &= 0, & x \in \mathbb{R},\ t > 0, \\ \partial_t \boldsymbol{v} + A(\boldsymbol{v})\,\partial_x \boldsymbol{v} &= 0, & x \in \mathbb{R},\ t > 0, \\ \boldsymbol{u}(x,0) &= \boldsymbol{v}(x,0), & x \in \mathbb{R}. \end{aligned}$$

with $A(\boldsymbol{u}) = \boldsymbol{f}'(\boldsymbol{u})$. Then, the difference $\boldsymbol{d} = \boldsymbol{u} - \boldsymbol{v}$ satisfies

$$\begin{aligned} \partial_t \boldsymbol{d} + A(\boldsymbol{u})\,\partial_x \boldsymbol{d} + [A(\boldsymbol{u}) - A(\boldsymbol{v})]\,\partial_x \boldsymbol{v} &= 0, & x \in \mathbb{R},\ t > 0, \\ \boldsymbol{d}(x,0) &= 0, & x \in \mathbb{R}. \end{aligned}$$

Moreover, since A is C^1, which corresponds to $\boldsymbol{f}$ being C^2, there is a constant $C > 0$ such that $|A(\boldsymbol{u}) - A(\boldsymbol{v})| \leq C|\boldsymbol{d}|$ holds. It follows that $\boldsymbol{d} = 0$ and therefore the assertion. □

Next, we are interested in the existence of classical solutions. Existence of classical solutions is — as usual — harder to investigate and does not hold in general. In fact, we can guarantee the existence of classical solutions only for short periods of time and if the IC is sufficiently smooth.

Lemma 2.8 (Local existence of classical solutions)
Let $\boldsymbol{u}_0$ be in C^N with $N > 3/2$ and let $A(\boldsymbol{v}) = \boldsymbol{f}'(\boldsymbol{v})$ be symmetric for all $\boldsymbol{v} \in \mathbb{R}^m$. Then, there exists a $T > 0$ such that the IVP (2.8) has a solution $\boldsymbol{u}$ in C^1 for $t \in [0, T]$.

Proof. Here, we only sketch a proof which can be found in [Lax73, Chapter 1] and is based on the theory of linear hyperbolic systems. Let us consider the operator S, which maps a function $\boldsymbol{v}$ to the solution $\boldsymbol{u}$ of the linear IVP

$$\begin{aligned} \partial_t \boldsymbol{u} + A(\boldsymbol{v})\, \partial_x \boldsymbol{u} &= 0, && x \in \mathbb{R},\ t > 0, \\ \boldsymbol{u}(x, 0) &= \boldsymbol{u}_0(x), && x \in \mathbb{R}, \end{aligned}$$

with constant coefficients. For more details on linear hyperbolic systems of first order (with constant coefficients) see [Eva10, Chapter 7.3]. Next, we define the norm $\|\cdot\|_{N,T}$ by

$$\|\boldsymbol{u}\|_{N,T}^2 = \sup_{0 \le t \le T} \int_{\mathbb{R}} \sum_{n=0}^{N} |\partial_x^n \boldsymbol{u}|^2 \, \mathrm{d}x$$

and assume that A is a symmetric matrix. Then, by utilizing energy estimates for linear symmetric hyperbolic systems of first order and the Sobolev inequalities, the following two properties can be shown for the operator S:

1. For sufficiently large $R > 0$ and sufficiently small $T > 0$, the operator S maps the ball $B_R[\|\cdot\|_{N,T}]$ of radius R with respect to the norm $\|\cdot\|_{N,T}$ into itself, i. e.

$$\|\boldsymbol{v}\|_{N,T} \le R \implies \|\boldsymbol{u}\|_{N,T} \le R$$

   for $\boldsymbol{u} = S\boldsymbol{v}$.

2. S is a contraction on $B_R[\|\cdot\|_{N,T}]$ with respect to the norm $\|\cdot\|_{0,T}$, that is

$$\|S\boldsymbol{v}_1 - S\boldsymbol{v}_2\|_{0,T} \le \frac{1}{2} \|\boldsymbol{v}_1 - \boldsymbol{v}_2\|_{0,T}.$$

Finally, note that the ball $B_R[\|\cdot\|_{N,T}]$ is closed with respect to the norm $\|\cdot\|_{0,T}$. Hence, the operator S has a unique fixed point in C^1, which solves the original IVP (2.8) for $t \in [0, T]$. □

2.3 Breakdown of classical solutions

Lemmas 2.7 and 2.8 tell us that for sufficiently smooth ID the IVP (2.8) is ensured to have a unique classical solution, at least for a finite period of time. Naturally, the question arises, whether this classical solution remains to exists outside of a finite period of time. Unfortunately, this is not the case and can already be demonstrated for the inviscid Burgers equation (Example 2.4) by utilizing the so-called *method of characteristics*. Let us consider the IVP for a scalar CL,

$$\begin{aligned} \partial_t u + \partial_x f(u) &= 0, && x \in \mathbb{R},\ t > 0, \\ u(x, 0) &= u_0(x), && x \in \mathbb{R}. \end{aligned} \tag{2.9}$$

The idea behind the method of characteristics is to identify curves $t \mapsto (x(t), t)^T$ in the t-x-plane along which the solution u of (2.9) is constant. A curve satisfying this property is called a *characteristic curve* (or short *characteristic*). Let u be a solution of (2.9) and define $z(t) := u(x(t), t)$. Then, z describes the values of u along the curve $t \mapsto (x(t), t)^T$ and the task is to find $x(t)$ such that $\dot{z}(t) = 0$ holds. Note that if $\dot{z}(t) = 0$ holds, the chain rule yields

$$\dot{x}(t)(\partial_x u)(x(t), t) + (\partial_t u)(x(t), t) = 0. \tag{2.10}$$

Let us assume that u is a classical solution of (2.9) and therefore satisfies the quasilinear equation $\partial_t u + a(u)\, \partial_x u = 0$ with $a(u) = f'(u)$. In this case, (2.10) becomes

$$\dot{x}(t)(\partial_x u)(x(t), t) - a(u(x(t), t))(\partial_x u)(x(t), t) = 0. \tag{2.11}$$

Since u is assumed to be constant along the characteristic curve, $u(x(t), t)$ can be replaced by $u_0(x_0)$ with $x_0 = x(0)$ and the above equation is equivalent to

$$[\dot{x}(t) - a(u_0(x_0))]\, (\partial_x u)(x(t), t) = 0.$$

This motivates us to look for solutions of the ordinary differential equation

$$\dot{x}(t) = a(u_0(x_0))$$

with IC $x(0) = x_0$. The solution of this simple equation is given by

$$x(t) = a(u_0(x_0))t + x_0. \tag{2.12}$$

Inserting (2.12) into (2.11), in fact, we see that the classical solution u is constant along $(x(t), t)$ with $x(t)$ given by (2.12), making it a characteristic curve. Note that characteristic curves of CLs are straight lines, where the slope only depends on the values of the ID u_0. Finally, classical solutions $u = u(x, t)$ of the scalar IVP (2.9) are implicitly given by

$$u = u_0(x - a(u)t), \tag{2.13}$$

assuming a sufficiently small period of time.

Example 2.9 (Characteristics of the linear advection equation)
For the linear advection equation, we have $a(u) = f'(u) = \lambda$, which corresponds to a constant propagation speed of $\lambda \in \mathbb{R}$. This is also reflected in the characteristics for the linear advection equation. These are given by

$$x(t) = \lambda t + x_0. \tag{2.14}$$

Thus, the solution of the linear advection equation with ID u_0 is given by

$$u(x, t) = u_0(x - \lambda t)$$

and corresponds to the ID u_0 to be transported in time with speed λ. This is illustrated in Figure 2.1(a) for $\lambda = 1$.

Example 2.10 (Characteristics of the inviscid Burgers' equation)
For the inviscid Burgers' equation, on the other hand, we have $a(u) = u$. This yields characteristics of different slopes, given by

$$x(t) = u_0(x_0)t + x_0$$

and deepening on the ID u_0. As long as the characteristics do not intersect, the solution $u = u(x, t)$ is implicitly given by

$$u = u_0(x - ut).$$

Naturally, the question arises what happens when the characteristics *do intersect*.

The above example for the Burgers' equation already demonstrates that, beyond a finite and possibly small period of time, we can not hope for a classical solution anymore. Let us consider the Burgers' equation with ID u_0 given by

$$u_0(x) = \cos x.$$

Let us assume that u is a classical solution. Then, u would be constant along the characteristics given by (2.14). In particular, u would a constant value of 1 along the characteristic $x_1(t)$ with origin $x_0 = 0$, given by

$$x_1(t) := u_0(0)t + 0 = t,$$

and a constant value of -1 along the characteristic $x_2(t)$ with origin $x_0 = \pi$, given by

$$x_2(t) := u_0(\pi)t + \pi = \pi - t.$$

Yet, it is evident that both characteristics will intersect at $x = \pi/2$ for time $t = \pi/2$. This is illustrated in Figure 2.1(b). As a result, the solution will have two opposing values for the same set of arguments, resulting in the *breakdown of classical solutions* for CLs.

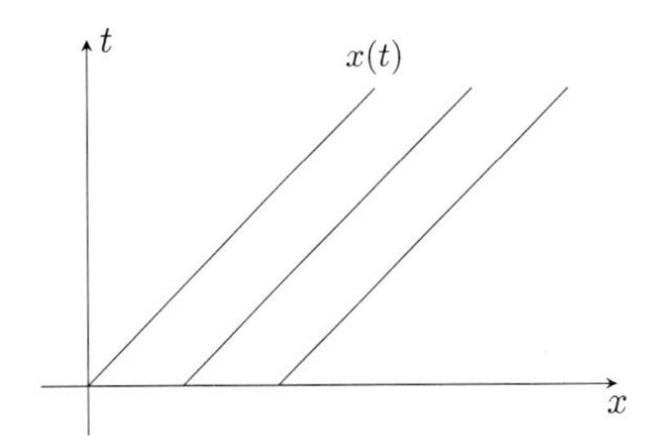

(a) Characteristics for the linear advection equation.

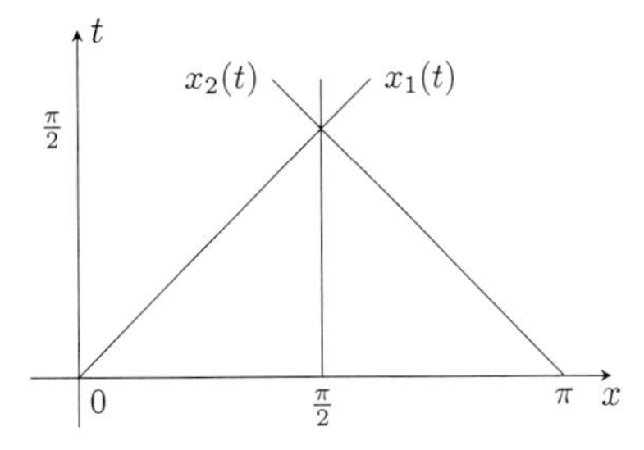

(b) Characteristics for the inviscid Burgers' equation with $u_0(x) = \cos x$.

Figure 2.1: Characteristics for the linear advection equation (left) and the inviscid Burgers' equation with ID $u_0(x) = \cos x$ (right).

2.4 Weak solutions

In the last section, we have observed that, in general, classical solutions do not exist for CLs. Hence, we need a framework of generalized solutions that do not have to be differentiable or even continuous. Such a framework is provided by a mathematical technique which can also be applied more generally to rewrite a PDEs in a form where less smoothness is required. The basic idea is to multiply the whole equation by a smooth test function, to integrate over some domain, and to use integration by parts to move derivatives off the function $\boldsymbol{u}$ onto the smooth test function then. This results in an equation which includes fewer derivatives on the solution $\boldsymbol{u}$ and therefore requires less smoothness. For systems of CLs (2.2) in one spatial dimension, it is sufficient to use test functions $\varphi \in C_0^1(\mathbb{R} \times \mathbb{R}_0^+, \mathbb{R}^m)$, where $C_0^1(\mathbb{R} \times \mathbb{R}_0^+, \mathbb{R}^m)$ denotes the space of functions that are continuously differentiable and have compact support. The latter requirement means that φ is zero outside of some bounded set, making its support lie in a compact set. Let us now multiply every component of (2.2) with a test function φ and perform integration over space and time. This results in

$$\int_0^\infty \int_{-\infty}^\infty \varphi\, \partial_t \boldsymbol{u} + \varphi\, \partial_x \boldsymbol{f}(\boldsymbol{u}) \, \mathrm{d}x \, \mathrm{d}t = 0,$$

where integration is performed component-wise. Thus, applying integration by parts yields

$$\int_0^\infty \int_{-\infty}^\infty (\partial_t \varphi)\boldsymbol{u} + (\partial_x \varphi)\boldsymbol{f}(\boldsymbol{u}) \, \mathrm{d}x \, \mathrm{d}t = - \int_{-\infty}^\infty \varphi(x,0)\boldsymbol{u}(x,0) \, \mathrm{d}x,$$

where most boundary terms drop out due to φ having compact support. The remaining boundary term can be used to incorporate the IC:

$$\int_0^\infty \int_{-\infty}^\infty (\partial_t \varphi)\boldsymbol{u} + (\partial_x \varphi)\boldsymbol{f}(\boldsymbol{u}) \, \mathrm{d}x \, \mathrm{d}t = - \int_{-\infty}^\infty \varphi(x,0)\boldsymbol{u}_0(x) \, \mathrm{d}x \tag{2.15}$$

This yields the following definition of weak solutions.

Definition 2.11 (Weak solutions)
A function $\boldsymbol{u} = \boldsymbol{u}(x,t)$ is called a *weak solution* of the IVP (2.8) if the integral form (2.15) holds for all test functions $\varphi \in C_0^1(\mathbb{R} \times \mathbb{R}_0^+)$.

Remark 2.12 (Physical interpretation of CLs)**.** From (2.15) another integral equation can be derived, which might better describe the physical interpretation behind CLs. For $t_2 > t_1 > 0$, let us consider special test functions φ with

$$\varphi(x,t) = \begin{cases} 1 & \text{if } (x,t) \in [x_1, x_2] \times [t_1, t_2], \\ 0 & \text{if } (x,t) \notin [x_1 - \varepsilon, x_2 + \varepsilon] \times [t_1 - \varepsilon, t_2 + \varepsilon], \end{cases}$$

and a smooth transition in the intermediate strip of width $\varepsilon > 0$. Then, for $\varepsilon \to 0$, the integral form (2.15) becomes

$$\int_{x_1}^{x_2} \boldsymbol{u}(x,t_2) \, \mathrm{d}x = \int_{x_1}^{x_2} \boldsymbol{u}(x,t_1) \, \mathrm{d}x + \int_{t_1}^{t_2} \boldsymbol{f}(\boldsymbol{u}(x_1,t)) \, \mathrm{d}t - \int_{t_1}^{t_2} \boldsymbol{f}(\boldsymbol{u}(x_2,t)) \, \mathrm{d}t$$

or equivalently

$$\frac{\mathrm{d}}{\mathrm{d}t} \int_{x_1}^{x_2} \boldsymbol{u}(x,t) \, \mathrm{d}x = \boldsymbol{f}(\boldsymbol{u}(x_1,t)) - \boldsymbol{f}(\boldsymbol{u}(x_2,t)). \tag{2.16}$$

The above integral forms allow the following physical interpretation: Assume that $\boldsymbol{u}$ describes, for instance, the density of a fluid. Then, the corresponding CL models a physical process in which mass — corresponding to the integral on the left hand side of (2.16) — is neither created nor destroyed and can only change because of the fluid flowing across the endpoints x_1 or x_2.

The above derivation of weak solutions shows that every classical solution $\boldsymbol{u} \in C^1$ is also a weak solution. Obviously, the reverse is not true, since weak solutions do not even have to be continuous. Still, we can note a strong connection for piecewise C^1 functions.

Definition 2.13 (Piecewise C^k functions)
We say that a function $\boldsymbol{u} : \mathbb{R} \times \mathbb{R}_0^+ \to \mathbb{R}^m$ is *piecewise* C^k — we denote this as $\boldsymbol{u} \in C_p^k(\mathbb{R} \times \mathbb{R}_0^+, \mathbb{R}^m)$ or short $\boldsymbol{u} \in C_p^k$ — when there is a finite number of curves in $\mathbb{R} \times \mathbb{R}_0^+$ outside of which $\boldsymbol{u}$ is C^k and across which $\boldsymbol{u}$ has jump discontinuities. The latter requirement means that if γ is one of these curves with normal vector $\mathbf{n} = (n_t, n_x)$ then the two one-sided limits

$$\boldsymbol{u}_\pm(x,t) := \lim_{\varepsilon \to 0} \boldsymbol{u}\left((x,t) \pm \varepsilon \mathbf{n}\right)$$

exist.

Theorem 2.14 (Characterization of weak solutions for C_p^1 functions)
Let $\boldsymbol{u}$ be a piecewise C^1 function (in the sense of Definition 2.13). Then, $\boldsymbol{u}$ is a weak solution of (2.8) if and only if the following two conditions are satisfied:

(i) $\boldsymbol{u}$ is a classical solution of (2.8) in regions where $\boldsymbol{u}$ is C^1.

(ii) $\boldsymbol{u}$ satisfies the jump condition

$$(\boldsymbol{u}_+ - \boldsymbol{u}_-)\, n_t + (\boldsymbol{f}(\boldsymbol{u}_+) - \boldsymbol{f}(\boldsymbol{u}_-))\, n_x = 0 \tag{2.17}$$

along the curves of discontinuity.

Proof. See Chapter 2 in [GR91]. □

The jump condition (2.17) is typically referred to as the *Rankine–Hugoniot jump condition*. Denoting $[\boldsymbol{u}] = \boldsymbol{u}_+ - \boldsymbol{u}_-$ as well as $[\boldsymbol{f}] = \boldsymbol{f}(\boldsymbol{u}_+) - \boldsymbol{f}(\boldsymbol{u}_-)$, the Rankine–Hugoniot jump condition can be rewritten as

$$[\boldsymbol{u}]n_t + [\boldsymbol{f}]n_x = 0.$$

If $n_x \neq 0$, let us set $\mathbf{n} = (-s, 1)$ with a suitable $s \in \mathbb{R}$. Then, the Rankine–Hugoniot jump condition is equivalent to

$$s[\boldsymbol{u}] = [\boldsymbol{f}]. \tag{2.18}$$

The value s is referred to as the *shock speed.*

Remark 2.15. For continuous $\boldsymbol{u}$, the Rankine–Hugoniot jump condition is automatically satisfied. In this case, it is sufficient to check that $\boldsymbol{u}$ is a classical solution of (2.8) in region where $\boldsymbol{u}$ is C^1 to verify that $\boldsymbol{u}$ is a weak solution of (2.8).

Remark 2.16. For scalar CLs, the Rankine–Hugoniot jump condition becomes scalar as well and the shock speed s is simply given by

$$s = [f]/[u], \tag{2.19}$$

assuming $[u] \neq 0$. For systems of CLs, on the other hand, $[\boldsymbol{u}]$ and $[\boldsymbol{f}]$ are vectors, while the shock speed s is still scalar. Hence, we can not always solve for s. Instead, the Rankine–Hugoniot jump condition only allows jump discontinuities for which the vectors $[\boldsymbol{u}]$ and $[\boldsymbol{f}]$ are linearly dependent.

It can be fairly exhausting to check if the integral form (2.15) is satisfied for all test functions $\varphi \in C_0^1$. For a piecewise C^1 function, however, Theorem 2.14 provides a useful characterization of weak solutions. This is demonstrated in the following examples.

Example 2.17 (The linear advection equation)
Let us consider the scalar linear advection equation $\partial_t u + \partial_x(\lambda u) = 0$ with constant velocity $\lambda \in \mathbb{R}$. For C^1 ID u_0, we have already derived an exact classical solution by the method of characteristics, given by

$$u(x,t) = u_0(x - \lambda t).$$

But even for a more general ID u_0, the above equation provides a solution, though a weak solution for $u_0 \notin C^1$. As an example let us consider the discontinuous ID

$$u_0(x) = \begin{cases} 1 & \text{if } x < 0, \\ 0 & \text{if } x > 0. \end{cases}$$

Then, the function

$$u(x,t) = u_0(x - \lambda t) = \begin{cases} 1 & \text{if } x < \lambda t, \\ 0 & \text{if } x > \lambda t, \end{cases} \tag{2.20}$$

is piecewise smooth and only discontinuous across the curve parametrized by $(\gamma(t), t)$ with $\gamma(t) = \lambda t$. Outside of this curve (2.20) obviously is a classical solution and across this curve we have $[u] = 1$ and $[f] = \lambda$. Hence, the Rankin–Hugoniot jump condition $s[u] = [f]$ is only satisfied for $s = \lambda$, which is in accordance with (2.20). Thus, (2.20) is a weak solution of the linear advection equation. Note that, for instance,

$$u(x,t) = \begin{cases} 1 & \text{if } x < 0, \\ 0 & \text{if } x > 0, \end{cases}$$

also satisfies the first condition of Theorem 2.14. Yet, for $\lambda \neq 0$, the Rankine–Hugoniot jump condition is violated, ruling the function out as a weak solution.

Example 2.18 (The Burgers' equation)
Let us consider the inviscid Burgers' equation $\partial_t u_t + \partial_x(u^2/2) = 0$ with ID

$$u_0(x) = \begin{cases} 1 & \text{if } x < 0, \\ 0 & \text{if } x > 0. \end{cases}$$

In this case a discontinuity is present already at $t = 0$. To identify a weak solution, we have to determine at which speed the discontinuity will propagate over time. Note that we have $[u] = 1$ and $[f] = 1/2$. Hence, the Ranking-Hugoniot jump condition yields a shock speed of $s = 1/2$ and we expect

$$u(x,t) = \begin{cases} 1 & \text{if } x < t/2, \\ 0 & \text{if } x > t/2, \end{cases} \tag{2.21}$$

to provide a weak solution. Here, the curve of discontinuity is parametrized by $(\gamma(t), t)$ with $\gamma(t) = \frac{t}{2}$. Across this curve, the second condition of Theorem 2.14 is satisfied by construction of u. Finally, note that outside of the curve of discontinuity, u is (locally) constant and obviously satisfies the Burgers' equation in the classical sense. Thus, the first condition of Theorem 2.14 is fulfilled as well and (2.21), in fact, is a weak solution. Again it should be stressed that the first condition of Theorem 2.14 also allows other choices for the shock speed, e. g. $s = 0$. Yet, only $s = 1/2$ satisfies the Rankine–Hugoniot jump condition and therefore yields a weak solution. A wrong shock speed of $s = 0$ as well as the right shock speed of $s = 1/2$ are illustrated in Figure 2.2.

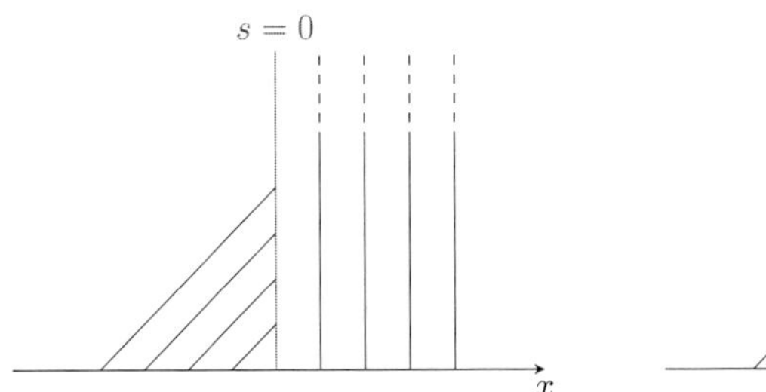

(a) Characteristics corresponding to a wrong shock speed for the Burgers' equation.

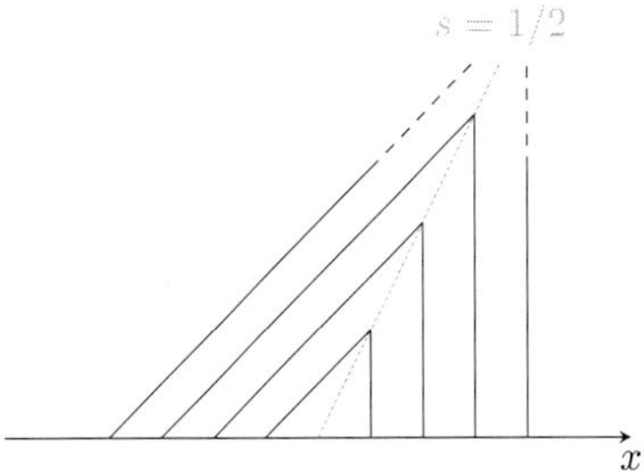

(b) Characteristics corresponding to the right shock speed for the Burgers' equation.

Figure 2.2: Characteristics corresponding to the inviscid Burgers' equation and a piecewise constant ID u_0.

2.5 The Riemann problem

In this section, we generalize the prior example to demonstrate missing uniqueness for weak solutions. Let us consider a system of m CLs with piecewise constant ID,

$$\boldsymbol{u}_t + \boldsymbol{f}(\boldsymbol{u})_x = 0,$$
$$\boldsymbol{u}(x,0) = \begin{cases} \boldsymbol{u}_L & \text{if } x < 0, \\ \boldsymbol{u}_R & \text{if } x > 0, \end{cases}$$

where $\boldsymbol{u}_L, \boldsymbol{u}_R \in \mathbb{R}^m$ denote constant left and right states. To such IVPs, where the ID is piecewise constant, we refer to as *the Riemann problem.* The form of the solution $\boldsymbol{u}$ depends on the relation between $\boldsymbol{u}_L$ and $\boldsymbol{u}_R$. To demonstrate that weak solutions of CLs do not have to be unique, it is sufficient to consider the inviscid Burgers' equation. Here, two cases can occur for the relation between u_L and u_R, namely $u_L > u_R$ and $u_L < u_R$. The case $u_L = u_R$ corresponds to constant ID and is not of interest. For the first case, $u_L > u_R$, Example 2.18 can be considered as a representative. In this case, a unique weak solution is given by

$$u(x,t) = \begin{cases} u_L & \text{if } x < st, \\ u_R & \text{if } x > st, \end{cases}$$

with shock speed

$$s = \frac{[f]}{[u]} = \frac{u_R^2 - u_L^2}{2(u_R - u_L)} = \frac{u_R + u_L}{2}.$$

It is the second case, $u_L < u_R$, that will demonstrate nonuniqueness of weak solutions.

Example 2.19 (Nonuniqueness of weak solutions)
Let us consider the Burgers' equation with piecewise constant ID

$$u_0(x) = \begin{cases} 0 & \text{if } x < 0, \\ 1 & \text{if } x > 0. \end{cases}$$

This corresponds to the Riemann problem with $u_L < u_R$. The characteristics for this Riemann problem are illustrated in Figure 2.3 and, by the method of characteristics, we can immediately note the values of the solution to the left of $x = 0$ and to the right of $x = t$:

$$u(x,t) = \begin{cases} 0 & \text{if } x < 0, \\ 1 & \text{if } x > t. \end{cases}$$

Yet, there seems to be a void of information for $0 < x < t$. One option to fill this void is a discontinuous transition, yielding

$$u_1(x,t) = \begin{cases} 0 & \text{if } x < st, \\ 1 & \text{if } x > st, \end{cases}$$

where the shock speed s is given by the Rankine–Hugoniot jump condition as

$$s = \frac{[f]}{[u]} = \frac{u_R + u_L}{2} = \frac{1}{2}.$$

It is clear that u_1 with $s = 1/2$ is a weak solution. Another option, however, is a linear transition, yielding a continuous function

$$u_2(x,t) = \begin{cases} 0 & \text{if } x < 0, \\ x/t & \text{if } 0 \le x \le t, \\ 1 & \text{if } x > t. \end{cases}$$

Note that for $0 \le x \le t$, we have $\partial_t u_2 = -\partial_x(u_2^2/2) = -x/t^2$. Hence, u_2 is a weak solution as well. Weak solutions of such form, where an initial discontinuity becomes continuous, are typically referred to as *rarefaction waves*.

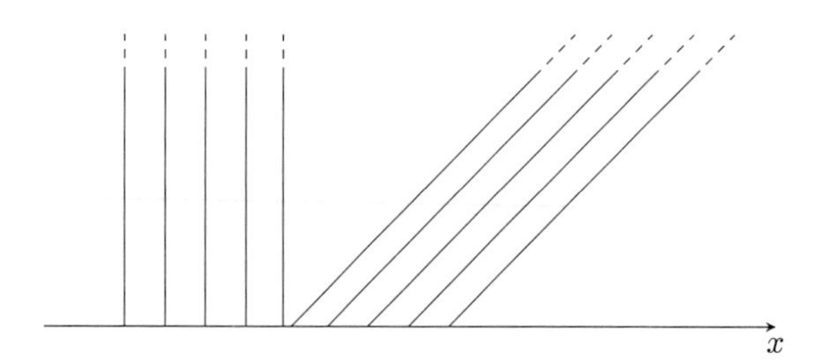

Figure 2.3: Characteristics corresponding to the Riemann problem for the inviscid Burgers' equation with $u_L = 0$ and $u_R = 1$.

In the above example, we have found two different weak solutions for the same IVP. This demonstrates that by permitting the fairly large class of weak solutions, we have lost uniqueness of solutions. In what follows, we are concerned with distinguishing physically relevant weak solutions and thereby reestablish uniqueness.

2.6 The role of viscosity

Example 2.19 shows that there can be different weak solutions for the same IVP. We want to pick out the physically relevant weak solution. An approach to do so is the so-called *vanishing viscosity method*. The basic idea is to add a *viscosity term* to the CL and consider the resulting parabolic system

$$\partial_t \boldsymbol{u}_\varepsilon + \partial_x \boldsymbol{f}(\boldsymbol{u}_\varepsilon) = \varepsilon \partial_{xx} \boldsymbol{u}_\varepsilon \tag{2.22}$$

for $\varepsilon > 0$, which is sometimes referred to as the *viscosity extension*. Assume that the parabolic system (2.22) provides a unique (smooth) solution for every $\varepsilon > 0$. Then, we are interested in the *vanishing viscosity limit*

$$\boldsymbol{u} := \lim_{\varepsilon \to 0} \boldsymbol{u}_\varepsilon, \tag{2.23}$$

where $\boldsymbol{u}_\varepsilon$ solves (2.22) for $\varepsilon > 0$. This approach is motivated by physics. Remember that many CLs are derived from more complex models by neglecting certain effects, such as viscosity. The inviscid Burgers' equation arises from the viscous Burgers' equation when viscosity is neglected; see Example 2.4. The Euler equations can be derived from the Navier–Stokes equations by neglecting viscosity and thermal conductivity. Hence, it seems reasonable to consider the weak solution which arises from the vanishing viscosity limit (2.23) as the physically relevant one.

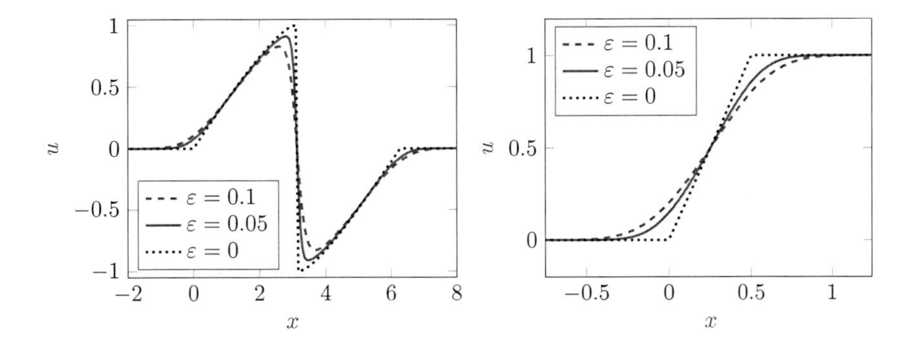

(a) Vanishing viscosity solution for a shock discontinuity.

(b) Vanishing viscosity solution for a rarefaction wave.

Figure 2.4: Vanishing viscosity limits for the inviscid Burgers' equation.

Figure 2.4 illustrates the vanishing viscosity method for the Burgers' equation with two different IVPs. The first IVP with ID

$$u_0(x) = \begin{cases} \sin x & \text{if } 0 \leq x \leq 2\pi, \\ 0 & \text{otherwise}, \end{cases}$$

demonstrates the vanishing viscosity method for a (shock) discontinuity in the vanishing viscosity limit at time $t = \pi/2$; see Figure 2.4(a). The second ID,

$$u_0(x) = \begin{cases} 0 & \text{if } x < 0, \\ 1 & \text{if } x > 0, \end{cases}$$

illustrate the vanishing viscosity method for a rarefaction wave in the vanishing viscosity limit at time $t = 0.5$; see Figure 2.4(b).

2.7 Entropy solutions

For practical considerations, it seems unreasonable to determine all weak solutions and to additionally build the vanishing viscosity limit of the parabolic systems (2.22) to pick out the physically relevant weak solution. A more direct approach to identify physically relevant weak solutions is based on the concept of *entropy*.

Definition 2.20 (Entropy functions and fluxes)
A convex function $\eta : \mathbb{R}^m \to \mathbb{R}$ with $\boldsymbol{u} \mapsto \eta(\boldsymbol{u})$ is called an *entropy function* for the system of CLs (2.2) if there exists a function $\psi : \mathbb{R}^m \to \mathbb{R}$ with $\boldsymbol{u} \mapsto \psi(\boldsymbol{u})$, called *entropy flux*, such that

$$\eta'(\boldsymbol{u})\boldsymbol{f}'(\boldsymbol{u}) = \psi'(\boldsymbol{u}) \tag{2.24}$$

holds, where $\eta'(\boldsymbol{u})$ and $\psi'(\boldsymbol{u})$ denote

$$\eta' = \nabla\eta^T = (\partial_{u_1}\eta, \ldots, \partial_{u_m}\eta), \quad \psi' = \nabla\psi^T = (\partial_{u_1}\psi, \ldots, \partial_{u_m}\psi).$$

Remark 2.21. The concept of entropy functions and fluxes as well as the subsequent definition of entropy solutions can be motivated by the following observation: Let $\boldsymbol{u}$ be a classical solution of the system of CLs (2.2). Moreover, let η and ψ be a pair of an associated entropy function and flux. Then, $\boldsymbol{u}$ satisfies an additional CL of the form

$$\partial_t\eta(\boldsymbol{u}) + \partial_x\psi(\boldsymbol{u}) = 0. \tag{2.25}$$

The above equation arises from noting

$$\begin{aligned}
\partial_t\eta(\boldsymbol{u}) &= \eta'(\boldsymbol{u})\partial_t\boldsymbol{u} \\
&= -\eta'(\boldsymbol{u})\partial_x\boldsymbol{f}(\boldsymbol{u}) \\
&= -\eta'(\boldsymbol{u})\boldsymbol{f}'(\boldsymbol{u})\partial_x\boldsymbol{u} \\
&= -\psi'(\boldsymbol{u})\partial_x\boldsymbol{u} \\
&= -\partial_x\psi(\boldsymbol{u}).
\end{aligned}$$

Here, the first and last equation follow from the chain rule, the second last equation is a consequence of (2.24), and all other equations hold since $\boldsymbol{u}$ is a classical solution of (2.2). For a weak solution of (2.2), on the other hand, the additional CL (2.25) does not hold in general. Instead, a weak solution $\boldsymbol{u}$ should satisfy the additional *entropy inequality*

$$\partial_t\eta(\boldsymbol{u}) + \partial_x\psi(\boldsymbol{u}) \leq 0 \tag{2.26}$$

in the weak sense; that is

$$\int_0^\infty \int_{-\infty}^\infty (\partial_t\varphi)\eta(\boldsymbol{u}) + (\partial_x\varphi)\psi(\boldsymbol{u})\,\mathrm{d}x\,\mathrm{d}t \geq 0 \tag{2.27}$$

should hold for all test functions $\varphi \in C_0^1(\mathbb{R} \times \mathbb{R}_0^+)$ with $\varphi \geq 0$. Note that the weak form of the entropy inequality (2.27) only involves nonnegative test functions, since we have an inequality now. The entropy inequality is motivated by the vanishing viscosity method and will become clearer when consulting Theorem 2.26, which ensures the existence of entropy solutions by the vanishing viscosity method.

After introducing the entropy inequality (2.26) as an additional condition to single out physically relevant solutions among the different weak solutions, let us define the resulting entropy solutions.

Definition 2.22 (Entropy solutions)
A weak solution $\boldsymbol{u}$ of the IVP (2.8) is called an *entropy solution* if $\boldsymbol{u}$ satisfies the entropy inequality (2.26) for all entropies η.

For scalar CLs it is easy to find entropy functions. In this case, in fact, any convex function η is an entropy function. A particular entropy function, which will be of interest later, is given by the *square entropy* $\eta(u) = u^2/2$. Yet, finding entropy functions for general systems of CLs can be highly nontrivial. In fact, the existence of entropy functions is a specific property of a system of CLs. In all practical examples from Mechanics and Physics, however, entropy functions with a physical meaning are known. A classification for hyperbolic systems of CLs regarding their entropy functions has been given by Serre [Ser91].

Example 2.23 (Entropy functions for the Euler equations)
For the Euler equations (2.5), a pair of an entropy function and flux is given by

$$\eta = -\rho s, \quad \psi = -\rho s \boldsymbol{u},$$

where s is the specific (physical) entropy $s = \ln(p\rho^{-\gamma}) = \ln(p) - \gamma \ln(\rho)$. There are many other possible choices for the (mathematical) entropy function, e. g. $\eta = \rho g(s)$ for any convex function g; see [Har83]. Yet, the above choice is the only one which is consistent with the entropy condition from thermodynamics [HFM86] in the presence of heat transfer.

Similar to the characterization of weak solutions, we can also characterize entropy solutions for piecewise C^1 functions.

Theorem 2.24 (Characterization of entropy solutions for C_p^1 functions)
Let $\boldsymbol{u}$ be a piecewise C^1 function (in the sense of Definition 2.13) which is a weak solution of (2.8). *Then, $\boldsymbol{u}$ is an entropy solution of* (2.8) *if and only if the following two conditions are satisfied:*

(i) $\boldsymbol{u}$ is a classical solution of (2.8) *in regions where $\boldsymbol{u}$ is C^1.*

(ii) $\boldsymbol{u}$ satisfies the jump inequality

$$s[\eta(\boldsymbol{u})] \geq [\psi(\boldsymbol{u})] \tag{2.28}$$

across the curves of discontinuity and for all pairs of entropy functions and fluxes. Here, s is given by the Rankine–Hugoniot jump condition (2.19).

Proof. The proof follows by arguing as in the proof of Theorem 2.14 and can be found in [GR91, Chapter 2]. □

For scalar CLs with a strictly convex flux function, the situation can be further simplified.

Theorem 2.25
Let $f : \mathbb{R} \to \mathbb{R}$ be a strictly convex function. Further, let $u : \mathbb{R} \times \mathbb{R}_0^+ \to \mathbb{R}$ be a piecewise C^1 weak solution that satisfies the entropy inequality (2.26) *for* one *strictly convex entropy function $\eta = \eta_0$. Then, u satisfies* (2.26) *for all entropy functions and therefore is an entropy solution.*

Proof. Let η and ψ be an arbitrary pair of an entropy function and corresponding entropy flux. To show that the entropy inequality (2.26) holds for this pair, it is sufficient to check that the jump condition (2.28) is satisfied across curves of discontinuity. Note that since we consider a scalar CL, the Rankine–Hugoniot jump condition yields $s = [f]/[u]$ and (2.28) can be rewritten as

$$\frac{f(u_+) - f(u_-)}{u_+ - u_-} [\eta(u_+) - \eta(u_-)] - [\psi(u_+) - \psi(u_-)] \geq 0. \tag{2.29}$$

For fixed u_-, let us define $E_\eta : \mathbb{R} \to \mathbb{R}$ by

$$E_\eta(v) := \frac{f(v) - f(u_-)}{v - u_-} [\eta(v) - \eta(u_-)] - [\psi(v) - \psi(u_-)]$$

for $v \neq u_-$ and $E_\eta(u_-) := 0$. Then, the jump condition (2.29) becomes

$$E_\eta(u_+) \geq 0. \tag{2.30}$$

Next, we show that E_η is monotonically decreasing if η is a (convex) entropy and strictly monotonically decreasing if η is a strictly convex entropy. Let $v \neq u_-$. Then, we get

$$E_\eta'(v) = \frac{f'(v)(v - u_-) - f(v) + f(u_-)}{(v - u_-)^2} [\eta(v) - \eta(u_-)] + \frac{f(v) - f(u_-)}{v - u_-} \eta'(v) - \psi'(v),$$

and since $\eta'(v) f'(v) = \psi'(v)$, this yields

$$\begin{aligned}
E_\eta'(v) &= [f'(v)(v - u_-) - f(v) + f(u_-)] \frac{\eta(v) - \eta(u_-)}{(v - u_-)^2} \\
&\quad + [f(v) - f(u_-) - f'(v)(v - u_-)] \frac{\eta'(v)}{v - u_-} \\
&= - [f(v) - f(u_-) - f'(v)(v - u_-)] \frac{\eta(v) - \eta(u_-)}{(v - u_-)^2} \\
&\quad + [f(v) - f(u_-) - f'(v)(v - u_-)] \frac{\eta'(v)(v - u_-)}{(v - u_-)^2} \\
&= [f(v) - f(u_-) - f'(v)(v - u_-)] \frac{\eta(u_-) - \eta(v) - \eta'(v)(u_- - v)}{(v - u_-)^2}.
\end{aligned}$$

Note that for a strictly convex flux function f, we have

$$f(v) - f(u_-) - f'(v)(v - u_-) < 0.$$

Further, the inequality

$$\eta(u_-) - \eta(v) - \eta'(v)(u_- - v) = - [\eta(v) - \eta(u_-) - \eta'(v)(v - u_-)) \geq 0$$

holds for an entropy η, where the inequality is strict for a strictly convex entropy. Thus, E_η is (strictly) decreasing when η is a (strictly) convex entropy function. In particular, (2.30) holds for the strictly convex entropy function η_0, resulting in

$$E_{\eta_0}(u_+) \geq E_{\eta_0}(u_-)$$

and therefore $u_+ \leq u_-$. Finally, since E_η is decreasing for every other entropy function η as well, we have

$$E_\eta(u_+) \geq E_\eta(u_-) = 0,$$

which yields the assertion. □

2.8 Some results for scalar conservation laws

In this final subsection, we gather some important results for the IVP for scalar CLs (2.9). Most of these results will come in handy later for the development of numerical methods. We start by proving the existence and uniqueness of entropy solutions for scalar CLs. The uniqueness result will follow from Kruzkov's theorem, which states that entropy solutions continuously depend on the ID. The existence result, on the other hand, essentially results from the vanishing viscosity method.

Theorem 2.26 (Existence of entropy solutions)
Let u_0 be in $L^1(\mathbb{R}) \cap L^\infty(\mathbb{R}) \cap \mathrm{BV}(\mathbb{R})$ and f be in $C^1(\mathbb{R})$. Then, the IVP (2.9) has an entropy solution u. For this entropy solution, we have

$$\begin{aligned} u &\in L^\infty(\mathbb{R} \times \mathbb{R}^+) \cap \mathcal{B}([0,T]; L^1(\mathbb{R})) \text{ for any } T > 0, \\ u(\cdot, t) &\in BV(\mathbb{R}) \text{ for all } t \geq 0, \end{aligned}$$

where $\mathcal{B}([0,T]; X)$ denotes the space of continuous and bounded functions from $[0,T]$ to X. Furthermore, the following estimates hold for the entropy solution u:

$$\begin{aligned} \|u(\cdot,t)\|_{L^\infty(\mathbb{R})} &\leq \|u_0\|_{L^\infty(\mathbb{R})} \text{ for almost every } t \geq 0 \\ \mathrm{TV}(u(\cdot,t)) &\leq \mathrm{TV}(u_0(\cdot)) \text{ for all } t \geq 0 \\ \|u(\cdot,t_2) - u(\cdot,t_1)\|_{L^1(\mathbb{R})} &\leq C\, \mathrm{TV}(u_0)\, |t_2 - t_1| \text{ for all } t_1, t_2 \geq 0 \end{aligned}$$

In the above theorem, $L^1(\mathbb{R})$ and $L^\infty(\mathbb{R})$ respectively denote the spaces of integrable and essentially bounded functions. The latter means that $v \in L^\infty(\mathbb{R})$ if and only if there exists a constant $a \geq 0$ such that $|v(x)| \leq a$ for almost all $x \in \mathbb{R}$. Finally, $\mathrm{BV}(\mathbb{R})$ denotes the space of functions with bounded total variation (TV), where the TV of a function v is defined by

$$\mathrm{TV}(v) := \sup_{\substack{x_0 < \cdots < x_J \\ J \in \mathbb{N}}} \sum_{j=0}^{J-1} |v(x_{j+1}) - v(x_j)| \,.$$

In the above definition, the supremum essentially runs over the set of all partitions of $\mathbb{R}$.

Proof. Here, we only outline how the entropy solution u is constructed. A more detailed proof can be found, for instance, in [GR91, Chapter 2.3]. Let $u_{0,\varepsilon}$ and f_ε be suitable regularizations of the ID u_0 and the flux function f, obtained by convolution with a mollifier. The basic idea is to use the vanishing viscosity method to construct a sequence of approximate solutions $(u_\varepsilon)_{\varepsilon>0}$, where u_ε solves the parabolic regularization

$$\begin{aligned} \partial_t u_\varepsilon + \partial_x f_\varepsilon(u_\varepsilon) &= \varepsilon \partial_{xx} u_\varepsilon, & x \in \mathbb{R},\ t > 0, \\ u_\varepsilon(x,0) &= u_{0,\varepsilon}(x), & x \in \mathbb{R}, \end{aligned} \tag{2.31}$$

of the IVP (2.9). It can be shown that there exists a subsequence, still denoted by $(u_\varepsilon)_{\varepsilon>0}$, such that

$$\|u_\varepsilon\|_{L^\infty(\mathbb{R}\times\mathbb{R}^+)} \leq \|u_0\|_{L^\infty(\mathbb{R})}$$

for all $\varepsilon > 0$ and there is an $u \in L^\infty(\mathbb{R} \times \mathbb{R}_0^+)$ with

$$\begin{aligned} &u_\varepsilon \to u \text{ in } L^1_{\mathrm{loc}}(\mathbb{R} \times \mathbb{R}^+), \\ &u_\varepsilon \to u \text{ almost everywhere in } \mathbb{R} \times \mathbb{R}^+ \end{aligned}$$

for $\varepsilon \to 0$. Here, $L^1_{loc}(\mathbb{R})$ denotes the space of all local integrable functions, that is of all functions which are integrable on compact sets $A \subset \mathbb{R}$. Then, in fact, the resulting vanishing viscosity limit u can be shown to provide an entropy solution of (2.9). □

Theorem 2.27 (Kruzkov's theorem)
Let u and v be two entropy solutions of (2.9) with ID u_0 and v_0 in $L^\infty(\mathbb{R})$ such that

$$u, v \in L^\infty(\mathbb{R} \times \mathbb{R}_0^+) \cap \mathcal{B}([0,T]; L^1_{loc}(\mathbb{R})) \text{ for all } T > 0.$$

Then, for

$$M := \max\left\{|f'(\xi)| \;\middle|\; |\xi| \le \max\{\|u\|_{L^\infty(\mathbb{R}\times\mathbb{R}_0^+)}, \|v\|_{L^\infty(\mathbb{R}\times\mathbb{R}_0^+)}\}\right\},$$

we have

$$\int_{|x|\le R} |u(x,t) - v(x,t)| \,\mathrm{d}x \le \int_{|x|\le R+Mt} |u_0(x) - v_0(x)| \,\mathrm{d}x \tag{2.32}$$

for all $R > 0$ and almost all $t > 0$.

Proof. The basic idea of the proof is to choose

$$\eta(u) = |u-k|, \quad \psi(u) = \operatorname{sign}(u-k)\,[f(u) - f(k)], \quad k \in \mathbb{R},$$

as an entropy function and entropy flux associated to (2.9). Then, the assertion essentially follows from a clever choice of the test functions in the weak formulation of the corresponding entropy inequality. Again, we refer to [GR91, Chapter 2.4] for a detailed proof. □

Theorem 2.28 (Uniqueness of entropy solutions)
Let u_0 be in $L^1(\mathbb{R}) \cap L^\infty(\mathbb{R}) \cap \mathrm{BV}(\mathbb{R})$ and let f be in $C^1(\mathbb{R})$. Then, the entropy solution u of the IVP (2.9), given by Theorem 2.26, is unique.

Proof. Kruzkov's theorem has already established that entropy solutions continuously depend on their ID. In fact, if we choose $u_0 = v_0$ in (2.32), we immediately get uniqueness of entropy solutions. □

Remark 2.29. Another consequence of Kruzkov's theorem is that it bounds the speed with which information can propagate. Let us note from Theorem 2.26 that

$$\|u(\cdot,t)\|_{L^\infty(\mathbb{R})} \le \|u_0\|_{L^\infty(\mathbb{R})}$$

for almost every $t \ge 0$. Hence, we have

$$M \le M_0 := \max\left\{|f'(\xi)| \;\middle|\; |\xi| \le \max\{\|u_0\|_{L^\infty(\mathbb{R})}, \|v_0\|_{L^\infty(\mathbb{R})}\}\right\}$$

and by choosing $u_0 = v_0$ in B_{R+M_0t}, we obtain that $u(x,t) = v(x,t)$ for $x \in B_R$. Thus, the entropy solution u has a finite domain of dependence. The function value $u(x,t)$ only depends on the values of u_0 in the set $\{y \in \mathbb{R} \mid |y-x| \le M_0 t\}$.

The main result of the section is the following Theorem, which ensures existence and uniqueness of entropy solutions of the scalar IVP (2.9) in the general case of ID $u_0 \in L^\infty(\mathbb{R})$. The proof can be found in [GR91, Chapter 2.5].

Theorem 2.30 (The ultimate existence and uniqueness result)
Let u_0 be in $L^\infty(\mathbb{R})$ and let f be in $C^1(\mathbb{R})$. Then, the IVP (2.9) has a unique entropy solution

$$u \in L^\infty(\mathbb{R} \times [0,T]).$$

This solution satisfies

$$\|u(\cdot,t)\|_{L^\infty(\mathbb{R})} \le \|u_0\|_{L^\infty(\mathbb{R})}$$

for almost all $t \ge 0$. Moreover, if u and v are two entropy solutions with ID u_0 and v_0, we have

$$u_0 \ge v_0 \text{ almost everywhere} \implies u(\cdot,t) \ge v(\cdot,t) \text{ almost everywhere.}$$

Finally, if $u_0 \in L^\infty(\mathbb{R}) \cap \mathrm{BV}(\mathbb{R})$, then $u(\cdot,t) \in \mathrm{BV}(\mathbb{R})$ with

$$\mathrm{TV}(u(\cdot,t)) \le \mathrm{TV}(u_0).$$

We can summarize some interesting properties of entropy solutions from Theorem 2.30: Let $t_2 \geq t_1 \geq 0$ and $\alpha, \beta \in \mathbb{R}$, then

$$\begin{aligned}
&(\text{P1}) \quad \|u(\cdot,t_2)\|_{L^\infty(\mathbb{R})} \leq \|u(\cdot,t_1)\|_{L^\infty(\mathbb{R})}\,, \\
&(\text{P2}) \quad \text{TV}(u(\cdot,t_2)) \leq \text{TV}(u(\cdot,t_1)), \\
&(\text{P3}) \quad \alpha \leq u(\cdot,t_1) \leq \beta \implies \alpha \leq u(\cdot,t_2) \leq \beta
\end{aligned}$$

hold almost everywhere. All of these are important properties of the entropy solution and we will try to mimic some of them for numerical solutions approximating the entropy solution. Property (P1) and (P2) tell us that both, the L^∞ norm as well as the TV, decrease over time for entropy solutions. The latter property is often referred to as u being *TVD*. Property (P3), on the other hand, states that entropy solutions preserve bounds. Assuming the entropy solution u is bounded from below by α and from above by β at time t_1, the solution will be bounded in the same way at all later times as well.

CHAPTER 3

NUMERICAL PRELIMINARIES

In this chapter, we gather some basic numerical preliminaries. These are needed for the development and study of the numerical methods which are discussed in the later chapters. For all material covered in this chapter, we can recommend some excellent text books.

Outline

This chapter is organized as follows: Chapter 3.1 addresses the topic of approximation and interpolation, first in general normed linear spaces and inner product spaces, later in more specific function spaces. Most notably, the advantages of (polynomial) interpolation of univariate functions and its pitfalls in higher dimensions are discussed. For more details on these topics, we refer to the book [Tre13] of Trefethen on approximation theory as well as to the books [Fas07] of Fasshauer and [Wen04] of Wendland on multivariate function approximation by RBFs. Next, Chapter 3.2 introduces the concept of orthogonal polynomials, which can be used, for instance, to efficiently compute polynomial best approximations in inner product spaces. The book [Gau04] of Gautschi provides an extensive treatment of this topic. Chapters 3.3 and 3.4 address numerical differentiation and integration. Both topics can be found in nearly any text book on numerical analysis. Yet, it turns out that numerical integration, in fact, seems to be a more delicate problem and is therefore addressed individually in a great number of monographs. Here, we mention the books of Krylov [KS06], Davis & Rabinowitz [DR07], and Brass & Petras [BP11]. Further, numerical integration will be addressed again in Chapters 4 and 5, which present novel contributions to the numerical integration of experimental data. Finally, Chapter 3.5 is concerned with integrating systems of ordinary differential equations in time. A general recommendation for the numerical treatment of ordinary differentiation equations are the well-known books of Hairer & Wanner and co-authors [HNW91, WH96, HLW06]. In the context of CLs, so-called strong stability preserving (SSP) Runge–Kutta (RK) methods are often preferred. More details on these methods can be found, for instance, in the wonderful book [GKS11] of Gottlieb, Ketcheson, and Shu. An almost universal recommendation for nearly all the material in this chapter — except for RBFs and SSP methods for time integration — is the book [Gau97] of Gautschi.

3.1 Approximation and interpolation

In almost all of the later investigated methods, we will need to approximate functions by some simpler objects, such as polynomials. Thus, in this section, we gather some elemental concepts and results from approximation theory. We start by briefly revisiting basic existence and uniqueness results for best approximations in general normed spaces in Chapter 3.1.1.

In particular, these directly apply to best approximations in inner product spaces, which are discussed in Chapter 3.1.2 then. The concept of orthogonality further allows us to derive simple formulas for best approximations. In Chapter 3.1.3, we start to restrict ourselves to function approximations, in particular to polynomial interpolation in one dimension. The general interpolation problem is addressed in Chapter 3.1.4. Here, the main result is the Mairhuber–Curtis theorem, which yields us to realize that many interpolation techniques from one dimension are not valid in higher dimension anymore. We close with 3.1.5 and RBF interpolation as one particularly important interpolation technique which also remains valid in higher dimensions.

3.1.1 Approximation in normed linear spaces

Let $(X, \|\cdot\|)$ be a normed linear space, $f \in X$, and $V \subset X$ be a linear subspace. An omnipresent problem in numerical mathematics and approximation theory is to find an approximation $v^* \in V$ such that

$$\|f - v^*\| \leq \|f - v\| \quad \forall v \in V. \tag{3.1}$$

Then, the approximation $v^* \in V$ is called *best approximation* of f from V with respect to $\|\cdot\|$. In many cases, X is a space of functions, say $C^0([a, b])$, and V is a finite dimensional linear subspace. Then, the idea behind the best approximation is to approximate the possibly complicated or just partially known function $f \in X$ by a simpler function $v \in V$. The most prominent choice for the subspace V might be the space of polynomials of degree at most d, denoted by $\mathbb{P}_d$. On the one hand, polynomials allow a simple representation of many important operations on function, such as differentiation and integration. On the other hand, the famous approximation theorem of Weierstrass [Wei85] states that every continuous function $f \in C^0([a, b])$ can be approximated arbitrarily accurate by a polynomials on a compact interval $[a, b]$, i. e.

$$\forall \varepsilon > 0 \;\; \exists d \in \mathbb{N}, p \in \mathbb{P}_d : \quad \|f - p\|_\infty < \varepsilon.$$

Here, we have used the *supremum norm*

$$\|f\|_\infty := \max_{x \in [a,b]} |f(x)|$$

for $f \in C^0([a, b])$. The Weierstrass approximation theorem was proven 1885 by Weierstrass himself [Wei85]. The first constructive proof of the theorem was provided by Bernstein using the so-called *Bernstein polynomials*. We will introduce these polynomials in Chapter 10. Also see [Tre13, Chapter 6] or [Sch11, Chapter 2.1]. Next, we address existence and uniqueness of best approximations in a more general setting.

Theorem 3.1 (Existence of best approximations in finite dimensional subspaces)
Let $(X, \|\cdot\|)$ be a normed linear space, $f \in X$, and $V \subset X$ be a finite dimensional *linear subspace. Then, for every $f \in X$, the exists (at least one) best approximation $v^* \in V$.*

Proof. Let us define the function $\delta_V : V \to \mathbb{R}_0^+$ by

$$\delta_V(v) := \|f - v\|.$$

Then, δ_V measures the distance of an approximation v to the element f and the assertion is equivalent to δ_V possessing a global minimum. It is sufficient to look for such a global minimum in the set

$$T := \{v \in V \mid \|v\| \leq 2\|f\|\} \subset V.$$

Note that $0 \in T$ with $\delta_V(0) = \|f\|$ and for every $v \in V \setminus T$, we have

$$\delta_V(v) = \|f - v\| \geq \|v\| - \|f\| > \|f\| = \delta_V(0).$$

Hence, there can be no global minimum of δ_V outside of T. Finally, the existence of a global minimum inside of T follows by a simple argument from real analysis: The set T is bounded and closed, and since V is a finite dimensional linear space, T is compact. Thus, as a continuous function, δ_V possesses a global minimum on T. □

Thus, existence of best approximations can be guarantee, for instance, by restricting ourselves to finite dimensional linear subspaces $V \subset X$. At the same time, it should be noted that best approximations, in general, are not unique. This can already be noted from the simple example of $X = \mathbb{R}^2$, $\|\cdot\| = \|\cdot\|_\infty$, and $V = \mathbb{R} \times \{0\}$. Yet, we can establish uniqueness when we restrict ourselves to strictly convex norms.

Definition 3.2 (Strictly convex norms)
Let $(X, \|\cdot\|)$ be a normed linear space. The norm $\|\cdot\|$ is called *strictly convex* if

$$\left\| \frac{1}{2}(x + y) \right\| < 1$$

holds for all $x, y \in X$ with $x \neq y$ and $\|x\| = \|y\| = 1$.

Note that in the above example, the norm $\|\cdot\|_\infty$ is convex — as any norm — but not strictly convex. For a strictly convex norm, however, we get the following uniqueness result.

Theorem 3.3 (Uniqueness of best approximations for strictly convex norms)
Let $(X, \|\cdot\|)$ be a normed linear space, $f \in X$, and be $V \subset X$ a linear subspace. If $\|\cdot\|$ is strictly convex, there is at most one best approximation of f.

Proof. Let us denote the set of all best approximations of f in V by

$$B := \{v^* \in V \mid \|f - v^*\| \leq \|f - v\| \, \forall v \in V\}.$$

It is easy to check that B is a convex subset of V. Next, let us assume that there are two different best approximations $u^*, v^* \in V$ of f. Note that $\|f - u^*\| = \|f - v^*\|$ and without loss of generality $\|f - u^*\| > 0$. Then, also the convex combination $w^* = (u^* + v^*)/2$ would be a best approximation. Yet, since $\|\cdot\|$ is strictly convex, we observe

$$\begin{aligned} \|f - w^*\| &= \left\| \frac{1}{2} \left([f - u^*] + [f - v^*] \right) \right\| \\ &= \|f - u^*\| \left\| \frac{1}{2} \left(\frac{f - u^*}{\|f - u^*\|} + \frac{f - v^*}{\|f - v^*\|} \right) \right\| \\ &< \|f - u^*\|, \end{aligned}$$

which is a contradiction to u^* being a best approximation. □

3.1.2 Approximation in inner product spaces

Next, let us focus on best approximations in inner product spaces $(H, \langle \cdot, \cdot \rangle)$, where the norm $\|\cdot\|$ is induced by the inner product,

$$\|x\|^2 = \langle x, x \rangle, \quad x \in H. \tag{3.2}$$

Often, such spaces are also called *pre-Hilbert spaces*, since we do not demand the space to be complete. It should be stressed that every norm that is induced by an inner product is strictly convex. Thus, we can immediately state

Theorem 3.4 (Existence and uniqueness of the best approximation)
Let $(H, \langle \cdot, \cdot \rangle)$ be an inner product space and $V \subset H$ a finite dimensional linear subspace. Then, for every $f \in H$, there exists a unique best approximation $v^* \in V$ of f.

Proof. The existence of the best approximation $v^* \in V$ follows directly from Theorem 3.1 since V is finite dimensional. For uniqueness, it is sufficient to show that the induced norm $\|\cdot\|$ given by (3.2) is strictly convex (in the sense of Definition 3.2). Then, the assertion follows from Theorem 3.3. Hence, let $x, y \in H$ with $x \neq y$ and $\|x\| = \|y\| = 1$. To prove strict convexity, remember that in every inner product space the *parallelogram identity*

$$\|x+y\|^2 + \|x-y\|^2 = 2\left(\|x\|^2 + \|y\|^2\right)$$

holds for all $x, y \in H$. Thus, we get

$$\begin{aligned}\left\|\frac{1}{2}(x+y)\right\|^2 &= \frac{1}{4}\|x+y\|^2 \\ &= \frac{1}{4}\left[2\left(\|x\|^2 + \|y\|^2\right) - \|x-y\|^2\right] \\ &= 1 - \frac{1}{4}\|x-y\|^2 \\ &< 1,\end{aligned}$$

for $x \neq y$. This proves strict convexity of the induced norm $\|\cdot\|$. □

The actual advantage of considering inner product spaces is the concept of orthogonality. We call two vectors $x, y \in H$ *orthogonal* if $\langle x, y \rangle = 0$. Utilizing the concept of orthogonality, we can give the following characterization of the best approximation.

Theorem 3.5 (Characterization of the best approximation)
Let $(H, \langle \cdot, \cdot \rangle)$ be an inner product space, $f \in H$, and let $V \subset H$ be a finite dimensional linear subspace. Then, $v^* \in V$ is the best approximation of f if and only if

$$\langle f - v^*, v \rangle = 0 \quad \forall v \in V \tag{3.3}$$

holds.

Condition (3.3) means that the error $f - v^*$ is orthogonal to the space V. Thus, the best approximation v^* is given by orthogonal projection of f onto V.

Proof. Let $v^* \in V$. First, we show that (3.3) is a necessary condition for v^* being a best approximation. Then, in a second step, we prove that (3.3) also is a sufficient condition.

$\Longrightarrow$: Let us assume that there exists a $v \in V$ with $v \neq 0$ such that (3.3) is violated, i. e.

$$\alpha := \langle f - v^*, v \rangle \neq 0.$$

In this case, we have

$$\|f - (v^* + \lambda v)\|^2 = \|f - v^*\|^2 - 2\text{Re}(\lambda \langle f - v^*, v \rangle) + |\lambda|^2 \|v\|^2.$$

If we choose $\lambda = \frac{\overline{\alpha}}{\|v\|^2}$, this yields

$$\|f - (v^* + \lambda v)\|^2 = \|f - v^*\|^2 - \frac{|\alpha|^2}{\|v\|^2} < \|f - v^*\|^2.$$

Hence, v^* can not be the best approximation.

$\Longleftarrow$: Let (3.3) hold for v^* and let $v \in V$. Then, we have

$$\begin{aligned} \|f - v^*\|^2 &= \langle f - v^*,\, f - v + v - v^* \rangle \\ &= \langle f - v^*,\, f - v \rangle + \langle f - v^*,\, v - v^* \rangle . \end{aligned}$$

Since $v - v^* \in V$, condition (3.3) yields $\langle f - v^*,\, v - v^* \rangle = 0$ and we get

$$\|f - v^*\|^2 = \langle f - v^*,\, f - v \rangle \le \|f - v^*\|\, \|f - v\|$$

by the Cauchy-Schwarz inequality. Thus, $\|f - v^*\| \le \|f - v\|$ for all $v \in V$, which means that v^* is the best approximation of f in V.

□

The characterization of best approximations provided by Theorem 3.5 allows — at least formally — an easy computation of the best approximation v^*. Let $\{v_k\}_{k=1}^n$ be a basis of the finite dimensional subspace V. Then, the best approximation has a unique representation as a linear combination of the basis elements:

$$v^* = \sum_{k=1}^{n} \alpha_k v_k \tag{3.4}$$

Here, the scalar coefficients α_k are elements of the underlying field $\mathbb{F}$, when H is a linear space over $\mathbb{F}$ with $\mathbb{F} = \mathbb{R}$ or $\mathbb{F} = \mathbb{C}$. These coefficients can be determined by consulting (3.3). Note that, because of the sesquilinearity of the inner product, it is sufficient to check (3.3) only for the basis elements $v_1, \ldots, v_n$. Moreover, using the representation (3.4), we get a system of linear equations,

$$\sum_{k=1}^{n} \alpha_k \langle v_k,\, v_l \rangle = \langle f,\, v_l \rangle, \quad l = 1, \ldots, n.$$

By denoting $\boldsymbol{\alpha} = (\alpha_1, \ldots, \alpha_n)^T$, $\mathbf{b} = (\langle f,\, v_1 \rangle, \ldots, \langle f,\, v_n \rangle)^T$, and

$$G = \begin{pmatrix} \langle v_1,\, v_1 \rangle & \ldots & \langle v_n,\, v_1 \rangle \\ \vdots & & \vdots \\ \langle v_1,\, v_n \rangle & \ldots & \langle v_n,\, v_n \rangle \end{pmatrix},$$

the system of linear equations can be rewritten in matrix vector notation as

$$G\boldsymbol{\alpha} = \mathbf{b}. \tag{3.5}$$

The matrix G, called the *Gram matrix*. It is Hermitian (self-adjoint), i. e. $G = G^H := \overline{G}^T$, and positive definite since

$$\begin{aligned} \boldsymbol{\beta}^H G \boldsymbol{\beta} &= \sum_{k,l=1}^{n} \beta_k \overline{\beta_l} \langle v_k,\, v_l \rangle \\ &= \left\langle \sum_{k=1}^{n} \beta_k v_k,\, \sum_{l=1}^{n} \beta_l v_l \right\rangle \\ &= \left\| \sum_{k=1}^{n} \beta_k v_k \right\|^2 \\ &> 0 \end{aligned}$$

holds for $\beta \in \mathbb{F}^n$ with $\beta \neq 0$. The last estimate follows from the basis elements $\{v_k\}_{k=1}^n$ being linearly independent. In particular, the Gram matrix G is regular and (3.5) can be uniquely solved for the coefficients $\boldsymbol{\alpha}$ of the best approximation v^*. This can be done, for instance, by Gaussian elimination [TBI97, Lecture 20]. Of course, the condition number of G depends on the basis $\{v_k\}_{k=1}^n$ and an unclever choice might result in a poor result for the desired vector of coefficients $\boldsymbol{\alpha}$, when computed numerically. The real beauty of approximating in inner product spaces is revealed, when we choose $\{v_k\}_{k=1}^n$ to be an *orthogonal basis*; that is

$$\langle v_k, v_l \rangle = \delta_{kl} \|v_k\|^2, \quad k,l = 1,\dots,n.$$

Here, δ_{kl} denotes the usual *Kronecker delta* defined by

$$\delta_{kl} = \begin{cases} 1 & \text{if } k = l, \\ 0 & \text{if } k \neq l. \end{cases}$$

When using such an orthogonal basis, the Gram matrix G reduces to a diagonal matrix and (3.5) consists of n independent equations

$$\alpha_k \|v_k\|^2 = \langle f, v_k \rangle, \quad k = 1,\dots,n.$$

Hence, the coefficients are simply given by

$$\alpha_k = \frac{\langle f, v_k \rangle}{\|v_k\|^2}, \quad k = 1,\dots,n,$$

in this case. Sometimes, we further demand the basis elements to be *normalized*, i. e.

$$\langle v_k, v_l \rangle = \delta_{kl}, \quad k,l = 1,\dots,n.$$

To such a basis $\{v_k\}_{k=1}^n$, we refer to as an *orthonormal basis*. Note that, when using an orthonormal basis, the Gram matrix G even reduces to the identity matrix and we get

$$\alpha_k = \langle f, v_k \rangle, \quad k = 1,\dots,n,$$

for the coefficients. We summarize our above observations in

Theorem 3.6 (Best approximations for orthogonal bases)
Let $(H, \langle\cdot,\cdot\rangle)$ be an inner product space and let $V \subset H$ be a finite dimensional linear subspace. Moreover, let $\{v_k\}_{k=1}^n$ be an orthogonal basis of V. Then, the best approximation $v^* \in V$ of $f \in H$ is uniquely given by

$$v^* = \sum_{k=1}^{n} \frac{\langle f, v_k \rangle}{\|v_k\|^2} v_k.$$

Orthogonal and orthonormal bases are an recurring tool in many of the subsequent chapters. Of particular importance for us will be bases of orthogonal polynomials. Hence, these are addressed in more detail in Chapter 3.2.

3.1.3 Polynomial interpolation in one dimension

In many situation, we are faced with the problem to approximate a real valued function $f : [a,b] \to \mathbb{R}$ by matching its values at a given set of points. If we restrict ourselves to approximate f by polynomials, this yields the problem of *polynomial interpolation*.

Definition 3.7 (Polynomial interpolation)
Given $n+1$ distinct points $\{x_i\}_{i=1}^{n+1}$ in $[a,b]$, referred to as *interpolation points*, and the values $f_i = f(x_i)$, $i = 1, \dots, n+1$, of a function f. The problem of polynomial interpolation is to find a polynomial $p_n \in \mathbb{P}_n(\mathbb{R})$ such that

$$p_n(x_i) = f_i, \quad i = 1, \dots, n+1. \tag{3.6}$$

Then, p_n is called the *interpolation polynomial* of f with respect to the interpolation points $\{x_i\}_{i=1}^{n+1}$.

Remark 3.8. At least heuristically, the interpolation polynomial can be considered as a best approximation in the special case of $X = C^0([a,b])$, $V = \mathbb{P}_n(\mathbb{R})$, and

$$\|v\| = \max_{i=1,\dots,n+1} |v(x_i)|.$$

Yet, it should be noted that $\|\cdot\|$ only is a norm when restricted to V. On the whole space X, $\|\cdot\|$ is a seminorm.

Still, we can prove existence and uniqueness for the interpolation polynomial.

Theorem 3.9 (Existence and uniqueness of the interpolation polynomial)
Let $\{x_i\}_{i=1}^{n+1}$ be a set of $n+1$ distinct interpolation points and $f_i = f(x_i)$, $i = 1, \dots, n+1$, the values of a function f. Then, there exists a unique interpolation polynomial $p_n \in \mathbb{P}_n(\mathbb{R})$ satisfying (3.6).

Proof. First, let us address the existence of the interpolation polynomial by defining the *Lagrange basis polynomials* $\ell_k \in \mathbb{P}_n(\mathbb{R})$ as

$$\ell_k(x) := \prod_{\substack{i=1 \\ i \neq k}}^{n+1} \frac{x - x_i}{x_k - x_i}. \tag{3.7}$$

Obviously, the Lagrange basis polynomials satisfy $\ell_k(x_i) = \delta_{ik}$. Thus, an interpolation polynomial is given by

$$p_n(x) = \sum_{k=1}^{n+1} f_k \ell_k(x), \tag{3.8}$$

which is referred to as the *Lagrange interpolation formula*. Next, let us prove uniqueness of the interpolation polynomial. Assume there exists a second polynomial $q_n \in \mathbb{P}_n(\mathbb{R})$ that satisfies (3.6). Then, define $r \in \mathbb{P}_n(\mathbb{R})$ as the difference between the two interpolation polynomials, i.e.

$$r(x) := p_n(x) - q_n(x).$$

Since p_n and q_n both satisfy (3.6), the polynomial r satisfies $r(x_i) = 0$ for $i = 1, \dots, n+1$ and therefore has $n+1$ roots. Hence, by the fundamental theorem of algebra, $r \equiv 0$. □

Remark 3.10. The Lagrange interpolation formula (3.8) is attractive for theoretical considerations but only partially suited for practical computations. More favorable formulas — that are computationally more efficient and allow additional interpolation points to be added with more ease — are, for instance, the *Barycentric interpolation formula* [Gau97, Chapter 2.2.5] and the *Newton interpolation formula* [Gau97, Chapter 2.2.6].

Next, let us briefly address convergence of polynomial interpolation with respect to the maximum norm. We have already noted the Weierstrass approximation theorem, which states that every continuous function can be approximated arbitrarily accurate by polynomials. Unfortunately, the situation is more complicated for polynomial interpolation. Faber proved in 1914 that no sequence of interpolation polynomials, no matter how the points are distributed, will converge for all continuous function [Fab14]. Moreover, Runge showed already in 1901 that interpolation polynomials on equidistant points may diverge exponentially, even if f is analytic (holomorphic) [Run01]. This observation is today known as the *Runge phenomenon*. Keeping these results in mind, polynomial interpolation might not seem to be a suitable method for function approximation. Yet, as long as the function f at least is a little bit smooth, we can find interpolation points for which the interpolation polynomials actually converge. For instance, if we use Chebyshev points, the interpolation polynomials already converge for every Lipschitz continuous function f. And, in fact, the smoother f is, the faster the interpolation polynomials with respect to the Chebyshev points converge. If f has ν derivatives with the νth derivative being of bounded variation V, we have $\|f - p_n\|_\infty = \mathcal{O}(Vn^{-\nu})$. For more details, see the excellent work of Trefethen [Tre11] and references therein.

3.1.4 General interpolation and the Mairhuber–Curtis theorem

So far, we have been concerned with polynomial interpolation in one dimension. Generalizing this idea, let us now consider a set $\Omega \subset \mathbb{R}^d$ with at least n points in it and let $V \subset C^0(\Omega)$ be an n-dimensional linear space of continuous functions. Given n distinct points $\{x_i\}_{i=1}^n$ in Ω and corresponding function values $f_i = f(x_i)$, $i = 1, \dots, n$, the general interpolation problem is to find a function $v \in V$ such that

$$v(x_i) = f_i, \quad i = 1, \dots, n. \tag{3.9}$$

As demonstrated by polynomial interpolation, the general interpolation problem can be solved for every set of points in one dimension if a suitable subspace V is chosen. Unfortunately, it is not possible to find subspaces V for which the interpolation problem can be solved for every set of points in higher dimensions anymore. To prove this, let us define Haar spaces first.

Definition 3.11 (Haar spaces)
Let $\Omega \subset \mathbb{R}^d$ be a set with at least n points in it and let $V \subset C^0(\Omega)$ be an n-dimensional subspace of continuous functions. We call V a *Haar space* if for any set of n distinct points $\{x_i\}_{i=1}^n \subset \Omega$ and corresponding values $\{f_i\}_{i=1}^n$, we can find a unique function $v \in V$ such that

$$v(x_i) = f_i, \quad i = 1, \dots, n. \tag{3.10}$$

Thus, Haar spaces are exactly the n-dimensional subspaces of continuous functions for which the interpolation problem can always be solved uniquely. We note some more characterizations of Haar spaces, which will be useful to prove the later Mairhuber–Curtis theorem.

Lemma 3.12 (Characterization of Haar spaces)
Let $\Omega \subset \mathbb{R}^d$ be a set with at least n points in it and let $V \subset C^0(\Omega)$ be an n-dimensional subspace of continuous functions. Then, the following are equivalent:

(i) V is a Haar space

(ii) Every $v \in V \setminus \{0\}$ has at most $n-1$ zeros

(iii) For any set of distinct points $\{x_i\}_{i=1}^n \subset \Omega$ and any basis $\{v_k\}_{k=1}^n$ of V, we have

$$\det \underbrace{\begin{pmatrix} v_1(x_1) & \dots & v_n(x_1) \\ \vdots & & \vdots \\ v_1(x_n) & \dots & v_n(x_n) \end{pmatrix}}_{=:A} \neq 0$$

Proof. We start by proving that (i) implies (ii). Therefore, let V be a Haar space and let $v \in V \setminus \{0\}$. Assume that v had n zeros $x_1, \dots, x_n$. Then, the interpolation problem with points $\{x_i\}_{i=1}^n$ and corresponding function values $f_i = 0$, $i = 1, \dots, n$, would have two different solutions, $0 \in V$ and $v \in V$.

Next, we prove that (ii) implies (iii). Assume that the a set of distinct points $\{x_i\}_{i=1}^n \subset \Omega$ and a basis $\{v_k\}_{k=1}^n$ of V such that $\det A = 0$. Then, the vectors $(v_k(x_1), \dots, v_k(x_n))^T \in \mathbb{R}^n$ for $k = 1, \dots, n$ are linearly dependent and we can find a vector of coefficients $\mathbf{c} := (c_1, \dots, c_n)^T$ with $\mathbf{c} \neq 0$ such that

$$v := \sum_{k=1}^n c_k v_k(x_i) = 0, \quad i = 1, \dots, n.$$

Hence, $v \in V$ has n zeros. At the same time, we can note that $v \neq 0$, since $\{v_k\}_{k=1}^n$ is a basis of V and $\mathbf{c} \neq 0$. Yet, this is a contradiction to (ii).

Finally, we show that (iii) implies (i). It is sufficient to observe that the interpolation problem (3.10) can be reformulated as a system of linear equations $A\mathbf{c} = \mathbf{f}$ with $\mathbf{c} = (c_1, \dots, c_n)^T$, $\mathbf{f} = (f_1, \dots, f_n)^T$, and A given as in (iii). Thus, if A is regular, the interpolation problem is uniquely solved by $v := \sum_{k=1}^n c_k v_k$. This condition, of A being regular, is exactly matched by (iii). □

The following theorem states that there can be no Haar spaces in more than one dimension.

Theorem 3.13 (Mairhuber–Curtis theorem)
Let $\Omega \subset \mathbb{R}^d$, $d \geq 2$, contain an inner point. Then, there exists no Haar space $V \subset C^0(\Omega)$ of dimension $n \geq 2$.

Proof. Let us assume that there exists an n-dimensional Haar space $V \subset C^0(\Omega)$ with $n \geq 2$ and basis $\{v_k\}_{k=1}^n$. Since Ω contains an inner point, it also contains an open ball with radius $\delta > 0$, $B_\delta(x_0) \subset \Omega$. Let us fix $n-2$ distinct points $x_3, \dots, x_n$ inside of the ball $B_\delta(x_0)$. Further, we are able two find two curves $x_1, x_2 : [0,1] \to B_\delta(x_0)$ with $x_1(0) = x_2(1)$ and $x_2(0) = x_1(1)$ such that $x_1(t)$ and $x_2(t)$ neither intersect with each other nor with any of the other points. Note that this is only possible for $d \geq 2$. Then, considering

$$D(t) = \det \begin{pmatrix} v_1(x_1(t)) & \dots & v_n(x_1(t)) \\ v_1(x_2(t)) & \dots & v_n(x_2(t)) \\ v_1(x_3) & \dots & v_n(x_3) \\ \vdots & & \vdots \\ v_1(x_n) & \dots & v_n(x_n) \end{pmatrix},$$

we note $D(0) = -D(1)$. This is because $D(1)$ is obtained from $D(0)$ by switching the first two rows. Hence, since $D(t)$ is a continuous function, there exists a $t \in (0,1)$ such that $D(t) = 0$. By Lemma 3.12, this is a contradiction to V being a Haar space. □

In particular, the above Theorem by Mairhuber and Curtis tells us that in higher dimensions there can be situation in which the interpolation problem can not be solved uniquely. No matter which linear subspace $V \subset C^0(\Omega)$ we choose, there will always be a set of distinct interpolation points $\{x_i\}_{i=1}^n$ for which the interpolation problem is not well-defined. This includes polynomial interpolation in higher dimensions.

3.1.5 Radial basis function interpolation

Regarding interpolation in higher dimensions, $\Omega \subset \mathbb{R}^d$, we are in a seemingly difficult situation. As stated by the Mairhuber–Curtis theorem, it is not possible to find a finite dimensional subspace $V \subset C^0(\Omega)$ such that the interpolation problem (3.9) can be solved uniquely for every data set $\{(x_i, f_i)\}_{i=1}^n$. One way to overcome this problem is to restrict oneself to very specific combinations of spaces V and sets of interpolation points $X = \{x_i\}_{i=1}^n$. In such approaches, one usually tries to find so-called V-unisolvent point sets X. Unfortunately, were are not allowed to choose the interpolation points freely then. Thus, we investigate another approach here. The basic idea is to bypass the Mairhuber–Curtis theorem by allowing the space V to change with the interpolation points X, i. e. $V = V(X)$. This is realized by using RBFs.

Definition 3.14 (Radial basis functions)
Let $\phi : [0, \infty) \to \mathbb{R}$ and the function $g : \mathbb{R}^d \to \mathbb{R}$ be given by $g(x) = \phi(\|x\|)$. Then, g only depends on the norm (radius) of x and is therefore called *radial*. The function ϕ is referred to as an *RBF* (or *kernel*).

RBFs originate from Hardy's work [Har71] on cartography from 1971. Since then, they have been successfully used in many applications. Popular examples for RBFs are listed in the following table. The parameter $\varepsilon > 0$ in Table 3.1 is typically called the *shape parameter*.

RBF	$\phi(r)$	parameters	order
Gaussian	$\exp(-(\varepsilon r)^2)$	$\varepsilon > 0$	0
Polyharmonic splines	r^{2k-1}	$k \in \mathbb{N}$	$m = k$
	$r^{2k} \log r$	$k \in \mathbb{N}$	$m = k + 1$
Multiquadrics	$\sqrt{(\varepsilon r)^2 + 1}$	$\varepsilon > 0$	1
Inverse multiquadrics	$\frac{1}{\sqrt{(\varepsilon r)^2+1}}$	$\varepsilon > 0$	0
Inverse quadrics	$\frac{1}{(\varepsilon r)^2+1}$	$\varepsilon > 0$	0

Table 3.1: Some popular RBFs.

Let $\Omega \subset \mathbb{R}^d$ and $X = \{x_i\}_{i=1}^n$ be a set of distinct points in Ω. Then, we span a finite dimensional subspace $V \subset C^0(\Omega)$ by translating a single RBF for every point x_k. This results in n basis functions $v_k(x) := \phi(\|x - x_k\|)$, $k = 1, \ldots, n$, and a finite dimensional subspace $V = \operatorname{span}\{v_k \mid k = 1, \ldots, n\}$ which changes with the set of points X. This gives rise to RBF interpolants.

Definition 3.15 (RBF interpolant)
Let $\{(x_i, f_i)\}_{i=1}^n$ be a set of n data pairs. We call a function $s : \mathbb{R}^d \to \mathbb{R}$ given by

$$s(x) = \sum_{k=1}^{n} \alpha_k \phi\left(\|x - x_k\|\right) \tag{3.11}$$

an *RBF interpolant* if it solves the corresponding interpolation problem; that is

$$s(x_i) = f_i, \quad i = 1, \ldots, n, \tag{3.12}$$

holds. Here, the interpolation points $\{x_i\}_{i=1}^n \subset \mathbb{R}^d$ are also referred to as *centers*.

Again, the interpolation condition (3.12) yields a system of linear equations

$$\underbrace{\begin{pmatrix} \phi(\|x_1 - x_1\|) & \ldots & \phi(\|x_1 - x_n\|) \\ \vdots & & \vdots \\ \phi(\|x_n - x_1\|) & \ldots & \phi(\|x_n - x_n\|) \end{pmatrix}}_{=\Phi} \underbrace{\begin{pmatrix} \alpha_1 \\ \vdots \\ \alpha_N \end{pmatrix}}_{=\boldsymbol{\alpha}} = \underbrace{\begin{pmatrix} f_1 \\ \vdots \\ f_n \end{pmatrix}}_{=\mathbf{f}}, \tag{3.13}$$

which can be solved for the coefficients $\boldsymbol{\alpha}$ if the matrix Φ is invertible. Often, the RBF interpolants (3.11) are modified to include polynomials together with matching constraints on the expansion coefficients; see [Sch95b, Buh00, FFBB16, FBW16]. Including polynomials of degree up to $m-1$ results in RBF interpolants

$$s(x) = \sum_{k=1}^{n} \alpha_k \phi(\|x - x_k\|) + \sum_{l=1}^{P} \beta_l q_l(x) \tag{3.14}$$

with $P = \dim \mathbb{P}_{m-1}(\mathbb{R}^d)$ and constraints

$$\sum_{k=1}^{n} \alpha_k q_l(x_k) = 0, \quad l = 1, \ldots, P, \tag{3.15}$$

where $\{q_l\}_{l=1}^P$ is a basis of the space of polynomials in d variables of degree at most $m-1$, $\mathbb{P}_{m-1}(\mathbb{R}^d)$. Let us denote

$$Q = \begin{pmatrix} q_1(x_1) & \ldots & q_P(x_1) \\ \vdots & & \vdots \\ q_1(x_n) & \ldots & q_P(x_n) \end{pmatrix} \in \mathbb{R}^{n \times P}$$

as well as $\boldsymbol{\beta} = (\beta_1, \ldots, \beta_P)^T$. Then, (3.15) can be rewritten as

$$Q^T \boldsymbol{\alpha} = 0. \tag{3.16}$$

Thus, given the interpolation condition (3.13), the counterpart to (3.13) is

$$\begin{pmatrix} \Phi & Q \\ Q^T & 0 \end{pmatrix} \begin{pmatrix} \boldsymbol{\alpha} \\ \boldsymbol{\beta} \end{pmatrix} = \begin{pmatrix} \mathbf{f} \\ 0 \end{pmatrix}. \tag{3.17}$$

There are various reasons for including polynomials in RBF interpolants [Sch95b, Buh00, FFBB16, FBW16]:

1. Polynomial terms can ensure that (3.17) is uniquely solvable when working with conditionally positive definite RBFs (see Definition 3.16), assuming the set of centers $\{x_i\}_{i=1}^n$ is unisolvent on the space of polynomials of degree at most $m-1$ (see Definition 3.17). See, for instance, [Fas07, Chapter 7].

2. Numerical tests demonstrate that including a constant improves the accuracy of derivative approximations. In particular, adding a constant avoids oscillatory representations of constant functions.

3. Including polynomial terms of low order can also improve the accuracy of RBF interpolants near domain boundaries due to regularizing the far-field growth of RBF interpolants [FDWC02].

In what follows, we will outline the first reason in more detail. To investigate if (3.17) has a unique solution for every set of data pairs it is sufficient to consider the homogeneous problem

$$\begin{pmatrix} \Phi & Q \\ Q^T & 0 \end{pmatrix} \begin{pmatrix} \boldsymbol{\alpha} \\ \boldsymbol{\beta} \end{pmatrix} = 0. \tag{3.18}$$

The inhomogeneous system (3.17) has a unique solution if and only if the homogeneous system (3.18) only admits the trivial solution $\boldsymbol{\alpha} = \boldsymbol{\beta} = 0$. Note that the homogeneous system (3.18) can be rewritten as

$$\Phi\boldsymbol{\alpha} + Q\boldsymbol{\beta} = 0, \tag{3.19}$$

$$Q^T\boldsymbol{\alpha} = 0. \tag{3.20}$$

If we multiply (3.19) by $\boldsymbol{\alpha}^T$ from the left and utilize the additional constraint (3.16), we get

$$\boldsymbol{\alpha}^T\Phi\boldsymbol{\alpha} = 0, \tag{3.21}$$

while still requiring $Q^T\boldsymbol{\alpha} = 0$. This observation motivates the following definition.

Definition 3.16 (Conditionally definite RBFs)
An RBF ϕ is called *conditionally positive definite* of order m on $\mathbb{R}^d$ if

$$\boldsymbol{\alpha}^T\Phi\boldsymbol{\alpha} > 0 \tag{3.22}$$

holds for every set of distinct centers $X = \{x_i\}_{i=1}^n \subset \mathbb{R}^d$ and all nonzero vectors $\boldsymbol{\alpha} \in \mathbb{R}^n$ that satisfy the additional constraint

$$\sum_{i=1}^n \alpha_i p(x_i) = 0 \quad \forall p \in \mathbb{P}_{m-1}(\mathbb{R}^d). \tag{3.23}$$

Analogously, it is said to be a *conditionally negative definite* of order m on $\mathbb{R}^d$ if $-\phi$ is conditionally positive definite of order m on $\mathbb{R}^d$. When ϕ is either conditionally positive or conditionally negative definite, we say it is *conditionally definite.* A conditionally (positive/negative) definite ϕ of order 0 in $\mathbb{R}^d$ is said to be *(positive/negative) definite* on $\mathbb{R}^d$.

Here, the term "conditionally" is motivated by the fact that (3.22) does not have to hold for *all* nonzero vectors $\boldsymbol{\alpha}$, but only for the nonzero vectors $\boldsymbol{\alpha}$ that satisfy (3.23) and therefore lie in the nullspace of Q^T. Thus, if ϕ is conditionally definite, (3.21) implies $\boldsymbol{\alpha} = 0$ and only

$$Q\boldsymbol{\beta} = 0 \tag{3.24}$$

remains to consider. For this equation to imply $\boldsymbol{\beta} = 0$, we require the columns of Q to be linearly independent. In general, this depends on the point set X and yields the following definition of unisolvent point sets.

Definition 3.17 (Unisolvent point sets)
Let $V \subset C^0(\Omega)$ be a finite dimensional subspace of continuous functions. A set of points $X = \{x_i\}_{i=1}^n$ is called V*-unisolvent* if the only function $v \in V$ interpolating zero data on X is the zero function $v \equiv 0$, i. e.

$$v(x) = 0 \quad \forall x \in X \implies v \equiv 0$$

holds for all $v \in V$.

Regarding the remaining equation (3.24) it is therefore sufficient to require the point set X to be $\mathbb{P}_{m-1}(\mathbb{R}^d)$-unisolvent. This means that any polynomial with degree at most m can be uniquely determined by its function values on X. Moreover, note that $\{q_l\}_{l=1}^P$ is a basis of $\mathbb{P}_{m-1}(\mathbb{R}^d)$. Hence, for an $\mathbb{P}_{m-1}(\mathbb{R}^d)$-unisolvent point set X, (3.24) implies $\beta = 0$. Note that for low degrees $m-1$, unisolvency is a rather weak requirement. For instance for $m = 2$ it means that $X \subset \mathbb{R}^2$ should not lie on a straight line. We summarize our above observations in

Theorem 3.18 (Existence and uniqueness of RBF interpolants)
Let $\{(x_i, f_i)\}_{i=1}^n$ be a set of n data pairs in $\mathbb{R}^d$. Further, let ϕ be a conditionally definite RBF of order m in $\mathbb{R}^d$, let $X = \{x_i\}_{i=1}^n$ be $\mathbb{P}_{m-1}(\mathbb{R}^d)$-unisolvent, and let $\{q_l\}_{l=1}^P$ be a basis of $\mathbb{P}_{m-1}(\mathbb{R}^d)$. Then, there exists a unique RBF interpolant

$$s(x) = \sum_{k=1}^{n} \alpha_k \phi(\|x - x_k\|) + \sum_{l=1}^{P} \beta_l q_l(x)$$

with additional constraint

$$\sum_{k=1}^{n} \alpha_k q_l(x_k) = 0, \quad l = 1, \dots, P,$$

that satisfies the interpolation condition (3.12).

Finally, we address a relatively simple characterization of conditionally positive definite functions.

Definition 3.19 (Completely monotone)
An infinetly differentiable function $\psi : (0, \infty) \to \mathbb{R}$ is called *completely monotone* if

$$(-1)^k \psi^{(k)}(r) \geq 0$$

holds for all $k \in \mathbb{N}_0$ and $r \in (0, \infty)$.

Using completely monotone functions, the following characterization of conditionally positive definite functions was first conjectured by Michelli in 1986 [Mic84].

Theorem 3.20 (Characterization of conditionally positive definite functions)
Let $\phi : [0, \infty) \to \mathbb{R}$ be a continuous RBF and let $\phi_{\surd} : [0, \infty) \to \mathbb{R}$ be given by $\phi_{\surd}(r) = \phi(\sqrt{r})$. Assume $g_m := (-1)^m \phi_{\surd}$ is well-defined and is not constant. Then, the following are equivalent:

(a) ϕ is conditionally positive definite of order m on $\mathbb{R}^d$ for all $d \in \mathbb{N}$

(b) The function g_m is completely monotone on $(0, \infty)$

Proof. The implication (a) $\Longrightarrow$ (b) was shown by Michelli [Mic84] and the case $m = 1$ of the reversed implication (b) $\Longrightarrow$ (a) was already given by Schoenberg [Sch38]. A complete proof of the above theorem, including (b) $\Longrightarrow$ (a) for $m > 1$ was done by Guo, Hu, and Sun and can be found in [GHS93]. Their proof utilizes a characterization of conditionally positive definite functions as certain Laplace–Stieltjes integrals and would exceed the scope of this work. □

By Theorem 3.20 it is now easy to verify that the RBFs listed in Table 3.1 are actually conditionally definite of the listed order m. For more details on RBF approximations, we refer to the works [Sch95a, Buh03, Isk03, Wen04, Sch05, PD05, FZ07] and references therein.

3.2 Orthogonal polynomials

Univariate orthogonal polynomials are an important tool, especially for many of our later computations and methods. Here, we present a quick review of them and some of their properties. For more details, we recommend Gautschi's monograph [Gau04].

3.2.1 Continuous and discrete inner products

Let $\lambda : \mathbb{R} \to \mathbb{R}$ be a nondecreasing function with finite limits as $x \to \pm\infty$. Moreover, let us assume that the induced positive measure $\mathrm{d}\lambda$ has finite moments of all orders,

$$m_k = m_k(\mathrm{d}\lambda) := \int_{\mathbb{R}} x^k \,\mathrm{d}\lambda(x) < \infty, \quad k \in \mathbb{N}_0,$$

with $m_0 > 0$. For any pair of polynomials $u, v \in \mathbb{P}(\mathbb{R})$, we can define an inner product

$$\langle u,\, v\rangle = \int_{\mathbb{R}} u(x)v(x)\,\mathrm{d}\lambda(x) \tag{3.25}$$

then. If we want to exhibit the associated measure, we write $\langle u,\, v\rangle_{\mathrm{d}\lambda}$. In many applications, the measure $\mathrm{d}\lambda$ is *absolutely continuous* and therefore is given by $\mathrm{d}\lambda = \omega(x)\,\mathrm{d}x$, where ω is a nonnegative and integrable function, called *weight function*. In this case, we usually denote the corresponding inner product by

$$\langle u,\, v\rangle_\omega = \int_{\mathbb{R}} u(x)v(x)\omega(x)\,\mathrm{d}x. \tag{3.26}$$

A *discrete measure*, on the other hand, is a measure which support consists of a finite (or denumerable infinite) number of distinct points $\{x_n\}_{n=1}^N$ at which λ has positive jumps of strength w_n. Such a measure is sometimes denoted as $\mathrm{d}\lambda_N$. For the associated inner product, we write

$$\langle u,\, v\rangle_{\mathbf{x},\mathbf{w}} = \sum_{n=1}^{N} w_n u(x_n)v(x_n). \tag{3.27}$$

Here, $\mathbf{x}$ and $\mathbf{w}$ respectively denote the vectors $(x_1,\ldots,x_N)^T$ and $(w_1,\ldots,w_N)^T$. If $\mathbf{x}$ and $\mathbf{w}$ are clear by the context, we simply write $\langle u,\, v\rangle_N$. Note that (3.27) is positive definite on $\mathbb{P}_{N-1}(\mathbb{R})$, but not on $\mathbb{P}_m(\mathbb{R})$ with $m \geq N$.

3.2.2 Bases of orthogonal polynomials

The use of an inner product also allows us to utilize the concept of orthogonality. In particular, we are interested in orthogonal bases of polynomials. These will provide an efficient tool which can be applied to many problems from approximation theory.

Definition 3.21 (Orthogonal polynomials)
Let $\{\pi_k\}_{k\geq 0}$ be a set of real monic polynomials

$$\pi_k(x) = x^k + \ldots, \quad k = 0, 1, \ldots.$$

These are called *orthogonal polynomials* with respect to the measure $\mathrm{d}\lambda$ if they satisfy

$$\langle \pi_k, \pi_l \rangle_{\mathrm{d}\lambda} = 0 \ \text{ if } \ k \neq l,$$
$$\|\pi_k\|_{\mathrm{d}\lambda} > 0,$$

for $k, l \in \mathbb{N}_0$. We denote them by $\pi_k(\cdot) = \pi_k(\cdot; \mathrm{d}\lambda)$. Normalization yields the *orthonormal polynomials*

$$\tilde{\pi}_k = \frac{\pi_k}{\|\pi_k\|_{\mathrm{d}\lambda}}, \quad k = 0, 1, \dots,$$

which satisfy

$$\langle \tilde{\pi}_k, \tilde{\pi}_l \rangle_{\mathrm{d}\lambda} = \delta_{kl}$$

for $k, l \in \mathbb{N}_0$. Analogously, these are denoted by $\tilde{\pi}_k(\cdot) = \tilde{\pi}_k(\cdot; \mathrm{d}\lambda)$.

Theorem 3.22
If the inner product (3.25) is positive definite on $\mathbb{P}(\mathbb{R})$, there exists a unique infinite sequence $(\pi_k)_{k \in \mathbb{N}_0}$ of monic orthogonal polynomials.

Proof. The polynomials can be constructed, for instance, by applying the Gram–Schmidt orthogonalization procedure [TBI97] to the sequence of monomials $(e_k)_{k \in \mathbb{N}_0}$ with $e_k(x) := x^k$. Hence, we have $\pi_0 \equiv 1$ and for $k \geq 1$ the polynomials are recursively given by

$$\pi_k = e_k - \sum_{l=0}^{k-1} c_l \pi_l \quad \text{with} \quad c_l = \frac{\langle e_k, \pi_l \rangle_{\mathrm{d}\lambda}}{\langle \pi_l, \pi_l \rangle_{\mathrm{d}\lambda}}.$$

Finally, since the inner product is assumed to be positive definite, we have $\langle \pi_l, \pi_l \rangle_{\mathrm{d}\lambda} > 0$ and the polynomial π_k is uniquely defined. Further, by construction, π_k is orthogonal to all polynomials π_l with $l < k$. □

Theorem 3.23
If the inner product (3.25) is positive definite on $\mathbb{P}_d(\mathbb{R})$ but not on $\mathbb{P}_n(\mathbb{R})$ with $n > d$, there exists only a finite number of $d+1$ orthogonal polynomials $\pi_0, \dots, \pi_d$.

Proof. Again, we can apply the Gram–Schmidt orthogonalization procedure to the sequence of monomials $(e_k)_{k \in \mathbb{N}_0}$ to construct the polynomials $\pi_0, \dots, \pi_d$ as

$$\pi_k = e_k - \sum_{l=0}^{k-1} c_l \pi_l \quad \text{with} \quad c_l = \frac{\langle e_k, \pi_l \rangle_{\mathrm{d}\lambda}}{\langle \pi_l, \pi_l \rangle_{\mathrm{d}\lambda}}.$$

for $k \leq d$. Note that we have $\langle \pi_l, \pi_l \rangle_{\mathrm{d}\lambda} > 0$ just for $l \leq d$, because the inner product is assumed to be positive definite only on $\mathbb{P}_d(\mathbb{R})$. In fact, there can be no more orthogonal polynomials. By assumption there exists a polynomial $p \in \mathbb{P}_{d+1}(\mathbb{R}) \setminus \mathbb{P}_d(\mathbb{R})$ such that $\|p\|_{\mathrm{d}\lambda} = 0$. Thus, we can find coefficients $\alpha_0, \dots, \alpha_{d+1}$ with $\alpha_{d+1} \neq 0$ such that

$$p = \alpha_{d+1} \pi_{d+1} + \sum_{k=0}^{d} \alpha_k \pi_k.$$

This yields

$$0 = \|p\|_{\mathrm{d}\lambda}^2 = \alpha_{d+1}^2 \, \|\pi_{d+1}\|_{\mathrm{d}\lambda}^2 + \sum_{k=0}^{d} \alpha_k^2 \, \|\pi_k\|_{\mathrm{d}\lambda}^2$$

and therefore $\|\pi_{d+1}\|_{\mathrm{d}\lambda} = 0$. Hence, π_{d+1} can not be an orthogonal polynomial. □

Example 3.24 (Legendre polynomials)
The Legendre polynomials, denoted by P_k, $k \in \mathbb{N}_0$, are orthogonal with respect to the inner product

$$\langle u, v\rangle = \int_{-1}^{1} u(x)v(x)\,\mathrm{d}x$$

on $[-1,1]$. Further, the Legendre polynomial have the following useful properties [Gau04, Chapter 1.5.1]:

1. They are normalized in the sense that $P_k(1) = 1$
2. Their norms are given by $\|P_k\|^2 = 2/(2k+1)$
3. Each P_k is bounded by 1 on $[-1,1]$
4. They satisfy a three term recurrence relations:
$$\begin{aligned}(k+1)P_{k+1}(x) &= (2k+1)xP_k(x) - kP_{k-1}(x),\\ P_0(x) = 1, &\quad P_1(x) = x\end{aligned} \tag{3.28}$$
5. They satisfy the *Legendre differential equation*:
$$\frac{\mathrm{d}}{\mathrm{d}x}\left[(1-x^2)\frac{\mathrm{d}}{\mathrm{d}x}P_k(x)\right] = -k(k+1)P_k(x) \tag{3.29}$$

Remark 3.25 (Construction of orthogonal polynomials)**.** It should be stressed that — depending on the measure $\mathrm{d}\lambda$ — for many orthogonal polynomials no explicit formula is known. In this case, we utilize numerical algorithms to construct bases of orthogonal polynomials. Such algorithms include the Stieltjes procedure [Gau04] as well as the Gram–Schmidt orthogonalization [TBI97]. However, it should be noted that neither the *Stieltjes procedure* (without normalization),

$$\pi_0 \equiv 0, \quad \pi_1 \equiv 1, \quad \pi_{k+1} = (x - \alpha_k)\,\pi_k - \beta_k \pi_{k-1}, \tag{3.30}$$

where the recursion coefficients are given by

$$\alpha_k = \frac{\langle x\pi_k, \pi_k\rangle_{\mathrm{d}\lambda}}{\langle \pi_k, \pi_k\rangle_{\mathrm{d}\lambda}} \quad \text{and} \quad \beta_k = \frac{\langle \pi_k, \pi_k\rangle_{\mathrm{d}\lambda}}{\langle \pi_{k-1}, \pi_{k-1}\rangle_{\mathrm{d}\lambda}}, \tag{3.31}$$

nor the *classical Gram–Schmidt orthogonalization* applied to an initial basis $\{v_k\}_{k\geq 0}$ of (typically nonorthogonal) polynomials,

$$\begin{aligned}\pi_1 &= v_1, \quad \tilde{\pi}_1 = \frac{\tilde{\pi}_1}{\|\tilde{\pi}_1\|_{\mathrm{d}\lambda}},\\ \pi_{k+1} &= v_{k+1} - \sum_{i=1}^{k} \langle \tilde{\pi}_i, v_{k+1}\rangle_{\mathrm{d}\lambda}\,\tilde{\pi}_i, \quad \tilde{\pi}_{k+1} = \frac{\pi_{k+1}}{\|\pi_{k+1}\|_{\mathrm{d}\lambda}},\end{aligned} \tag{3.32}$$

are recommended, since they are numerically unstable. Hence, in this work, we construct bases of discrete orthonormal polynomials (DOPs) by the numerical stable *modified Gram–Schmidt orthogonalization* [TBI97], where π_{k+1} is computed by

$$\begin{aligned}\pi_{k+1}^{(1)} &= v_{k+1} - \langle \tilde{\pi}_1, v_{k+1}\rangle_{\mathrm{d}\lambda}\,\tilde{\pi}_1,\\ \pi_{k+1}^{(i)} &= \pi_{k+1}^{(i-1)} - \left\langle \tilde{\pi}_i, \pi_{k+1}^{(i-1)}\right\rangle_{\mathrm{d}\lambda}\tilde{\pi}_i, \quad i = 2,\dots,k,\\ \pi_{k+1} &= \pi_{k+1}^{(k)},\end{aligned} \tag{3.33}$$

instead of (3.32).

3.2.3 Application to least squares approximations

We shortly demonstrate how bases of orthogonal polynomials can be used to determine best approximations with respect to inner product function spaces. Let $f : [-1, 1] \to \mathbb{R}$ be a continuous function. Assume, we wish to approximate f by a polynomial of degree at most n. One option would be to construct such an approximation by looking for the best approximation of f from $\mathbb{P}_n(\mathbb{R})$ with respect to the inner product

$$\langle u, v \rangle = \int_{-1}^{1} u(x)v(x)\,\mathrm{d}x.$$

According to Theorem 3.6, this best approximation — let us denote it by p_n — exists and is uniquely given by

$$p_n(x) = \sum_{k=0}^{n} \alpha_k \pi_k(x) \quad \text{with} \quad \alpha_k = \frac{\langle f, \pi_k \rangle}{\langle \pi_k, \pi_k \rangle}. \tag{3.34}$$

Here, the π_k denote a basis of orthogonal polynomials which span $\mathbb{P}_d(\mathbb{R})$. For the above inner product, such a basis is given by the Legendre polynomials, i. e. $\pi_k = P_k$. Thus, we can summarize that the best approximation of f from $\mathbb{P}_n(\mathbb{R})$ is given by

$$p_n(x) = \sum_{k=0}^{n} \frac{2 \langle f, P_k \rangle}{2k+1} P_k(x).$$

Moreover, the Legendre polynomials can be calculated from the three term recurrence relation (3.28). Many more applications of orthogonal polynomials occur, for instance, in numerical integration, high order numerical methods for PDEs, and edge/discontinuity detection. These will be addressed in later sections.

3.3 Numerical differentiation

In many situation, we are not directly interested in approximating or reconstructing a (unknown) function $f : \mathbb{R} \to \mathbb{R}$ itself, but rather its derivative. This problem is typically called *numerical differentiation*. This is an especially important step in the numerical treatment of CLs (2.2), where the spatial derivative of the flux $f(u)$ has to be approximated among other things. In what follows, we briefly review some basic concepts of numerical differentiation.

3.3.1 Finite difference approximations

Let $f : \mathbb{R} \to \mathbb{R}$ be a differentiable function. The derivative of f at a point x_0 is defined by the limit of the *difference quotient* as x approaches x_0,

$$f'(x_0) := \lim_{x \to x_0} \frac{f(x_0) - f(x)}{x_0 - x}. \tag{3.35}$$

Let us assume that we only know the values of f at a set of distinct points $x_1 < \dots < x_N$. Then, the limit process in (3.35) can not be performed. If we still want to approximate the derivative f' at a point x_k, we can take different approaches. Perhaps the simplest idea for such an approximation is given by *FDs*. Here, the idea is to approximate $f'(x_k)$ by the difference quotient involving x_k itself and one of its neighboring points x_{k-1} or x_{k+1}. This yields

Definition 3.26 (Some basic FD approximations)
The *forward FD approximation* of $f'(x_k)$ is

$$f'(x_k) \approx \frac{f(x_{k+1}) - f(x_k)}{x_{k+1} - x_k}. \tag{3.36}$$

Analogously, the *backward FD approximation* of $f'(x_k)$ is

$$f'(x_k) \approx \frac{f(x_k) - f(x_{k-1})}{x_k - x_{k-1}}. \tag{3.37}$$

Of course, other choices are possible as well and $f'(x_k)$ could be approximated, for instance, by the *central FD approximation*

$$f'(x_k) \approx \frac{f(x_{k+1}) - f(x_{k-1})}{x_{k+1} - x_{k-1}}.$$

as well. For a sufficiently smooth function f, the Taylor series can be used to show the following convergence results for the above FD approximations.

Theorem 3.27
Let x_{k-1}, x_k, x_{k+1} be equally spaced, that is $x_{k+1} - x_k = x_k - x_{k-1} = h$. Then, we have

$$\begin{aligned} \frac{f(x_{k+1}) - f(x_k)}{x_{k+1} - x_k} = \frac{f(x_k) - f(x_{k-1})}{x_k - x_{k-1}} &= f'(x_k) + \mathcal{O}(h), \\ \frac{f(x_{k+1}) - f(x_{k-1})}{x_{k+1} - x_{k-1}} &= f'(x_k) + \mathcal{O}(h^2) \end{aligned}$$

for the forward, backward and central FD approximation.

Proof. We only address the central FD approximation. The two other FD approximations can be handled similarly. Let us consider the Taylor series of $f(x_{k+1})$ and $f(x_{k-1})$, given by

$$\begin{aligned} f(x_{k+1}) &= f(x_k) + hf'(x_k) + \frac{h^2}{2} f''(x_k) + \mathcal{O}(h^3), \\ f(x_{k-1}) &= f(x_k) - hf'(x_k) + \frac{h^2}{2} f''(x_k) + \mathcal{O}(h^3). \end{aligned}$$

Summing up the two Taylor series, we get

$$f(x_{k+1}) - f(x_{k+1}) = 2hf'(x_k) + \mathcal{O}(h^3)$$

and therefore

$$\frac{f(x_{k+1}) - f(x_{k+1})}{x_{k+1} - x_{k-1}} = f'(x_k) + \mathcal{O}(h^2).$$

□

More general FD approximations are possible and can be used to derive higher order approximations to the derivative of f. Further important problems are the construction of one-sided FD approximations, the design of FD approximations which are exact if f is a polynomial of a certain degree, and the investigation of FD approximations on nonequally spaced grid points. We do not discuss these problems in a general setting here. Yet, we can answer them easily for the subsequent formulas for numerical differentiation.

3.3.2 Beyond finite differences

Another approach to construct approximations of the derivative f' is to approximate f first and to differentiate the approximation then. Again assuming that we only know the values of f at a set of distinct points $x_0 < \cdots < x_N$, a natural idea is to determine the interpolation polynomial $f_N \in \mathbb{P}_N(\mathbb{R})$ of f with respect to the points $\{x_n\}_{n=0}^N$ and to approximate f' by f_N'. In this case, the Lagrange interpolation formula (3.8) provides us with the approximation formula

$$f'(x) \approx f_N'(x) = \sum_{k=0}^{N} f_k \ell_k'(x). \tag{3.38}$$

This approach comes with some potential advantages:

1. Theoretically, we are able to construct approximate derivatives on general grid points
2. The approximation formula (3.38) handles polynomials up to degree N exactly
3. The formula also provides us with an approximation at the end points x_0 and x_N For FD approximations, these often have to be treated separately by (noncentral) one-sided approximations
4. For the order of convergence of the approximation formula (3.38) we can draw on the corresponding approximation results for interpolation polynomials

Of course, we also have to keep in mind some pitfalls of polynomial interpolation. Regarding the first point above, general grid points might be chosen but will not always yield a reasonable approximation for f'. In particular, this is due to the famous Runge phenomenon, which has already been discussed in Chapter 3.1.3. Thus, the approximation formula (3.38) is only reasonable when we can restrict ourselves to very specific sets of grid points. Examples of suitable grid points are GLe, GLo and Chebyshev points. The later ones will be introduced in the forthcoming Chapter 3.4. For more details on interpolation based derivative approximations see [Gau97, Chapter 3.1].

3.4 Numerical integration

Another omnipresent problem in mathematics and the applied sciences is the computation of integrals of the form

$$I[f] := \int_a^b f(x)\omega(x)\, \mathrm{d}x \tag{3.39}$$

with *weight function* $\omega : [a,b] \to \mathbb{R}$, which is assumed to be integrable over $[a,b]$. In many situations, we are faced with the problem to recover $I[f]$ from a finite set of measurements $\{f(x_n)\}_{n=1}^N \subset \mathbb{R}$ at distinct points $\{x_n\}_{n=1}^N \subset [a,b]$. This problem is referred to as *numerical integration*. Henceforth, we roughly follow the presentation in [Gla19a, Gla19d] and Chapter 3.2 of [Gau97].

3.4.1 The basic idea: Quadrature rules

A universal approach is to make use of the data $\{(x_n, f(x_n))\}_{n=1}^N$ and to approximate the integral (3.39) by a finite sum

$$Q_N[f] := \sum_{n=1}^{N} w_n f(x_n) \tag{3.40}$$

which is called an *N-point QR*. The points $\{x_n\}_{n=1}^N$ are referred to as the *quadrature points* and the $\{w_n\}_{n=1}^N$ are referred to as the *quadrature weights*. A QR is uniquely determined by its quadrature points and quadrature weights. For the moment, we restrict ourselves to nonnegative weight functions ω. General weight functions will be addressed in Chapter 5.

3.4.2 What do we want? Exactness and stability

In many situations, QRs are required to be *exact*, at least for a certain finite dimensional subspace $V \subset C^0([a,b])$. A typical choice is $V = \mathbb{P}_d(\mathbb{R})$, i. e. the space of polynomials of degree at most d. This yields the following definition.

Definition 3.28 (Degree of exactness)
We say that a QR Q_N has *(polynomial) degree of exactness* d when the *exactness condition*

$$Q_N[p] = I[p] \quad \forall p \in \mathbb{P}_d \tag{3.41}$$

is satisfied; that is, the QR is exact whenever f is a polynomial of degree at most d.

It is possible — and sometimes desired — to consider exactness for other finite dimensional subspaces $V \subset C^0([a,b])$. Yet, for sake of simplicity, we will focus on exactness for polynomials in this work. Besides exactness, *stability* is a crucial property of QRs. Speaking heuristically, stability ensures that small changes in the input argument only yield small changes in the output value. See [TBI97, Lecture 14] for a more detailed discussion on stability. Let us consider a function f and a second function $\tilde{f}$, which might correspond to a perturbed version of f. If $|f(x) - \tilde{f}(x)| \leq \varepsilon$ for $x \in [a,b]$, we have

$$\left|I[f] - I[\tilde{f}]\right| \leq \int_a^b \left|\left(f(x) - \tilde{f}(x)\right)\omega(x)\right| \,\mathrm{d}x \leq \varepsilon K_\omega \tag{3.42}$$

for the weighted integral I, where we define

$$K_\omega := \int_a^b |\omega(x)| \,\mathrm{d}x. \tag{3.43}$$

Thus, the error growth of input errors is bounded by the factor K_ω. A similar behavior is desired for QRs approximating I. Denoting the vector of quadrature weights $(w_1, \dots, w_N)^T$ by $\mathbf{w}_N$, we have

$$\left|Q_N[f] - Q_N[\tilde{f}]\right| \leq \sum_{n=1}^{N} \left|w_n \left(f(x_n) - \tilde{f}(x_n)\right)\right| \leq \varepsilon\kappa(\mathbf{w}_N) \tag{3.44}$$

for the N-point QR Q_N, where we define

$$\kappa(\mathbf{w}_N) := \sum_{n=1}^{N} |w_n|. \tag{3.45}$$

$\kappa(\mathbf{w}_N)$ is a common stability measure for QRs. Unfortunately, the factor $\kappa(\mathbf{w}_N)$, which bounds the propagation of errors, might grow for increasing N. Hence, the QR might become less accurate even though more and more data is used. This problem is encountered by the concept of stability.

Definition 3.29 (Stability)
We call (a sequence of) N-point QRs $(Q_N)_{N\in\mathbb{N}}$ with quadrature weights $(\mathbf{w}_N)_{N\in\mathbb{N}}$ *stable* if $\kappa(\mathbf{w}_N)$ is uniformly bounded with respect to N, i. e., if

$$\sup_{N\in\mathbb{N}} \kappa(\mathbf{w}_N) < \infty \tag{3.46}$$

holds.

Remember that we are assuming a nonnegative weight function ω. Thus, if all quadrature weights are nonnegative and have degree of exactness 0, we get

$$\kappa(\mathbf{w}_N) = \sum_{n=1}^{N} w_n = Q_N[1] = I[1] = K_\omega. \tag{3.47}$$

This observation results in

Lemma 3.30
Let $(Q_N)_{N\in\mathbb{N}}$ *be a sequence of* N*-point QRs with nonnegative quadrature weights approximating the integral* (3.39) *with nonnegative weight function* ω. *If every* Q_N *has degree of exactness* 0, *the relation*

$$\kappa(\mathbf{w}_N) = K_\omega \tag{3.48}$$

holds. In particular, the sequence of QRs $(Q_N)_{N\in\mathbb{N}}$ *is stable.*

Proof. The assertion follows from (3.47) and from noting

$$\sup_{N\in\mathbb{N}} \kappa(\mathbf{w}_N) = K_\omega < \infty. \tag{3.49}$$

□

Otherwise, if one of the quadrature weights is negative, we have $\kappa(\mathbf{w}_N) > K_\omega$. Thus, $\kappa(\mathbf{w}_N) = K_\omega$ is the best possible stability measure we can achieve for a reasonable QRs (treating at least constants exactly).

3.4.3 Interpolatory quadrature rules

A class of QRs of particular interest are so-called interpolatory QRs.

Definition 3.31 (Interpolatory QRs)
We call an N-point QR Q_N *interpolatory* if Q_N has degree of exactness $d = N - 1$.

Interpolatory QRs are exactly the QRs which are obtained by polynomial interpolation. That is, f is replaced by its interpolation polynomial $f_{N-1} \in \mathbb{P}_{N-1}(\mathbb{R})$ with respect to the quadrature points $\{x_n\}_{n=1}^N$ and integration is performed over f_{N-1} instead of f then. When the interpolation polynomial f_{N-1} is given with respect to the Lagrange basis polynomials (3.7) as

$$f_{N-1}(x) = \sum_{n=1}^{N} f(x_n)\ell_n(x),$$

we have

$$I[f] \approx I[f_{N-1}] = \sum_{n=1}^{N} \underbrace{\left(\int_a^b \ell_n(x)\omega(x)\,\mathrm{d}x \right)}_{=w_n} f(x_n) = Q_N[f]. \tag{3.50}$$

While the Lagrange basis polynomials have the advantage of providing an explicit formula for the quadrature weights, also other bases are possible and often preferred. Especially from a computational point of view, these offer more robust alternatives for the computation of the quadrature weights, which can be derived from the exactness conditions (3.41). Let $\{\varphi_k\}_{k=1}^N$ be a basis of $\mathbb{P}_{N-1}(\mathbb{R})$. Then, if we write down the exactness condition for the basis elements φ_k, we obtain a system of linear equations,

$$A\mathbf{w}_N = \mathbf{m}, \tag{3.51}$$

for the quadrature weights $\mathbf{w}_N \in \mathbb{R}^N$. Here, the matrix

$$A = (\varphi_k(x_n))_{k,n=1}^N \in \mathbb{R}^{N \times N} \tag{3.52}$$

contains the function values of the basis elements at the quadrature points and the vector

$$\mathbf{m} = (I[\varphi_k])_{k=1}^N \in \mathbb{R}^N \tag{3.53}$$

contains the *moments* of the basis elements.

Remark 3.32. Given any N distinct quadrature points $x_1, \dots, x_N$, it is always possible to find a corresponding interpolatory QR, i. e. an N-point QR with degree of exactness of $d = N - 1$.

Example 3.33 (Newton–Cotes QRs)
For $\omega \equiv 1$ on $[-1, 1]$ and equally spaced quadrature points, interpolatory QRs have first been proposed by Newton in 1687 and implemented in detail by Cotes around 1712. Hence, these QRs are today known as *Newton–Cotes QRs*. Sometimes, we distinguish between *closed* Newton–Cotes QRs, for which the the endpoints a and b are included in the quadrature points, and *open* Newton–Cotes QRs, for which the endpoints are not included.

The question arises if we can do better. Can we find N-point QRs which achieve degree of exactness $d > N - 1$, for instance, by a clever choice of the quadrature points? The following theorem provides a simple and direct answer to this question. To formulate it, let us introduce the function

$$\nu_N(x) := \prod_{n=1}^{N} (x - x_n), \tag{3.54}$$

which is referred to as the *node polynomial*.

Theorem 3.34 (A necessary and sufficient criterion for higher degrees of exactness)
Let k be an integer with $0 \leq k \leq N$ and let Q_N be an N-point QR Q_N has degree of exactness $d = N - 1 + k$ if and only if the two following conditions are satisfied.

(a) Q_N is interpolatory.

(b) The node polynomial ν_N given by (3.54) fulfills

$$\int_a^b \nu_N(x)p(x)\omega(x)\,\mathrm{d}x = 0 \tag{3.55}$$

for all $p \in \mathbb{P}_{k-1}(\mathbb{R})$.

Proof. $\Longrightarrow$: Let Q_N have degree of exactness $d = N - 1 + k$. We have to show that Q_N satisfies the conditions (a) and (b). Note that condition (a) is trivial. For condition (b), we observe that the product $\nu_N p$ is a polynomial of degree at most $N + k - 1$. Hence, we have $I[\nu_N p] = Q_N[\nu_N p]$ and therefore

$$\int_a^b \nu_N(x)p(x)\omega(x)\,\mathrm{d}x = \sum_{n=1}^N w_n \nu_N(x_n)p(x_n) = 0,$$

since $\nu_N(x_n) = 0$ for $n = 1, \dots, N$. This proves condition (b).

$\Longleftarrow$: Now, let Q_N be an interpolatory QR for which condition (b) holds. We have to show that Q_N has degree of exactness $d = N - 1 + k$. Let p be a polynomial of degree at most $d = N - 1 + k$. We divide p by ν_N, so that

$$p = \nu_N q + r, \quad q \in \mathbb{P}_{k-1}(\mathbb{R}), \quad r \in \mathbb{P}_{N-1}(\mathbb{R}),$$

where q is the quotient and r is the remainder. Thus, we can rewrite the exact integral $I[p]$ as

$$\int_a^b p(x)\omega(x)\,\mathrm{d}x = \int_a^b \nu_N(x)q(x)\omega(x)\,\mathrm{d}x + \int_a^b r(x)\omega(x)\,\mathrm{d}x.$$

Note that the first integral vanishes due to condition (b). Furthermore, the second integral is handled exactly by the QR and therefore equals to

$$\sum_{n=1}^N w_n r(x_n) = \sum_{n=1}^N w_n \left[p(x_n) - \nu_N(x_n)q(x_n)\right] = \sum_{n=1}^N w_n p(x_n),$$

which is precisely $Q_N[p]$. The last equality again follows from $\nu_N(x_n) = 0$ for all $n = 1, \dots, N$. This proves that Q_N has degree of exactness $d = N - 1 + k$.

□

Remark 3.35 (Maximum degree of exactness). Note that condition (b) in Theorem 3.34 imposes k conditions on the quadrature points $x_1, \dots, x_N$. These must be chosen such that the node polynomial ν_N is orthogonal to $\mathbb{P}_{k-1}(\mathbb{R})$ with respect to the weight function ω. Assuming a nonnegative weight function ω, obviously k is bounded from above by N, i. e. $k \leq N$. Otherwise ν_N would need to be orthogonal to $\mathbb{P}_N(\mathbb{R})$. In particular, ν_N would need to be orthogonal to itself, which is impossible. Hence, $k = N$ is optimal and corresponds to a QRs of *maximum degree of exactness* $d_{\max} = 2N - 1$ for an N-point QR.

Example 3.36 (Gaussian QRs)
For the maximum degree of exactness $d_{\max} = 2N - 1$, i. e. choosing $k = N$ in Theorem 3.34, condition (b) requires the node polynomial ν_N to be orthogonal to $\mathbb{P}_{N-1}(\mathbb{R})$ with respect to the weight function ω. Note that this demand is met by choosing ν_N as the Nth-degree orthogonal polynomial $\pi_N(\cdot;\omega)$ belonging to the weight function ω. The resulting QR of maximum degree of exactness is called the *Gaussian QR* associated with the weight function ω. Its quadrature nodes are given as the roots of $\pi_N(\cdot;\omega)$ and its quadrature weights can be derived from (3.50), yielding

$$w_n = \int_a^b \ell_n(x)\omega(x)\,\mathrm{d}x = \int_a^b \frac{\pi_N(x;\omega)}{(x - x_n)\pi_N'(x_n;\omega)}\omega(x)\,\mathrm{d}x \tag{3.56}$$

for $n = 1, \dots, N$. This class of QRs was first developed by Gauss in 1814 [Gau14] for the special case of $\omega(x) = 1$ on $[-1, 1]$ and was extended to more general weight functions by Christoffel in 1877 [Chr77]; also see [Gau81]. As a results, Gaussian QRs are sometimes referred to as

Gauss–Christoffel QRs. Moreover, Gaussian QRs for the special case of $\omega(x) = 1$ on $[-1,1]$ are typically called *GLe-QRs*, since the Nth-degree orthogonal polynomial $\pi_N(\cdot;1)$ is given by the Nth-degree Legendre polynomial.

Example 3.37 (GLo-QRs)
In many situation, such as some of the later presented numerical methods for CLs, it is favorable to let both endpoints a and b — if these are finite — to serve as quadrature points, i.e. $x_1 = a$ and $x_N = b$. Then, following Theorem 3.34, we can only impose $k-2$ conditions on the remaining quadrature points $x_2,\dots,x_{N-1}$. These are taken to be roots of the $(N-2)$th-degree orthogonal polynomial $\pi_{N-2}(\cdot;\omega_{a,b})$ belonging to the weight function $\omega_{a,b}(x) = (x-a)(b-x)\omega(x)$. The resulting QR is called *GLo-QR*[1] and achieves a degree of exactness $d = 2N-3$.

In addition to providing a maximum degree of exactness, Gaussian QRs have some useful properties. Most notably, all their quadrature weights are positive.

Lemma 3.38
Gaussian QRs (see Example 3.36) have only positive quadrature weights.

Proof. Unfortunately, equation (3.56) allows no insight into the sign of the quadrature weights. Yet, the positivity of the quadrature weights can be noted immediately by an ingenious observation of Stieltjes: Let k be an integer with $1 \leq k \leq N$ and let ℓ_k be the kth Lagrange basis polynomial with respect to the set of quadrature points $\{x_n\}_{n=1}^N$. Remember that the Lagrange basis polynomials satisfy $\ell_k(x_n) = \delta_{kn}$. Thus, we have

$$w_k = \sum_{n=1}^{N} w_n \ell_k^2(x_n) = \int_a^b \ell_k^2(x)\,\mathrm{d}x > 0.$$

The last equality holds since $\ell_k^2 \in \mathbb{P}_{2N-2}(\mathbb{R})$ and since the Gaussian QRs has degree of exactness $d = 2N-1$. □

In particular, we can note from Lemma 3.38 that Gaussian QRs are stable (see Definition 3.29) and provide the best possible stability measure of $\kappa(\mathbf{w}_N) = K_\omega$. Of course, there are many more classes of QRs with different favorable properties. We refer to large body of literature [Bra77, GS83, KS06, DR07, BP11, Tre13].

3.5 Time integration

Often, we are faced with integrating a system of ordinary differential equations $\frac{\mathrm{d}}{\mathrm{d}t}u = L(u)$ in time. By now many different methods have been proposed for this purpose. For a general numerical treatment of systems of ordinary differential equations, we refer to the fundamental books [HNW91, WH96, HLW06]. For the problems considered in this thesis, additional properties are usually required from the numerical solution. In Chapter 2 we have already gathered some properties of CLs and their solutions. Most notable, the TV or other convex functional of the solution are often desired to not increase over time; see Chapter 2.8. This yields the concept of SSP methods, which will be addressed for explicit RK methods later in this section. For SSP time discretization, we refer to [Shu88, SO88, GS98, GST01, Ket08, GKS11] and references therein.

[1] Rehuel Lobatto (1797–1866) was a Dutch mathematician who stopped short of attaining an academic degree at the University of Amsterdam. He had to be satisfied with a low-level government position until he was appointed a teacher at the Polytechnical School of Delft in 1842. His work is relatively unknown, but he wrote several books, one of which [Lob52] contains the QRs now named after him.

3.5.1 The method of lines

In this work we consider only spatial discretization of the hyperbolic CL (2.2), leaving the problem continuous in time. This yields a system of ordinary differential equations,

$$\frac{\mathrm{d}}{\mathrm{d}t}u = L(u), \tag{3.57}$$

referred to as the *semidiscrete equation*. Here, $L(u)$ is a discretization of the spatial operator. This approach, i. e. where time dependent PDEs are reduced to a system of ordinary differential equations, is typically referred to as the *method of lines*; see [LeV02, Chapter 10.4].

3.5.2 Preferred method for time integration

Once we have discretized a hyperbolic CL in space, we return to solving the semidiscrete equation (3.57) in time. Popular choices of time integration methods include explicit TVD-RK methods [Shu88, GS98], by now also known as SSP-RK methods [GST01, Ket08]. These are addressed in more detail in the next subsection. In this work, if not stated otherwise, we will use the explicit TVD/SSP-RK method of third order using three stages (SSPRK(3,3)); see [GS98]:

Definition 3.39 (SSPRK(3,3))
Let u^n be the solution at time t^n. The solution u^{n+1} at time t^{n+1} is computed as

$$\begin{aligned} u^{(1)} &= u^n + \Delta t L(u^n), \\ u^{(2)} &= \frac{3}{4}u^n + \frac{1}{4}u^{(1)} + \frac{1}{4}\Delta t L(u^{(1)}), \\ u^{n+1} &= \frac{1}{3}u^n + \frac{2}{3}u^{(2)} + \frac{2}{3}\Delta t L(u^{(2)}). \end{aligned} \tag{3.58}$$

We note that L^2 stability for (3.58) is guaranteed for all time if it holds for the standard first order explicit Euler method, [GS98]. In [LT98] it was shown in the case of linear CLs that the L^2 stability is preserved in time for some choices of SSP-RK methods, including SSPRK(3,3). Also see [RGÖS16, RÖ18]. This is unfortunately generally not true in the nonlinear case, as the L^2 norm might increase after one iteration of the explicit Euler method if no dissipation is added to the numerical solution [ÖGR19]. Yet, strong stability of the explicit Euler method can be ensured by including a sufficient amount of dissipation to the semidiscretization of the hyperbolic CL.

3.5.3 Strong stability of explicit Runge–Kutta methods

We now show that the above SSPRK(3,3) method, in fact, is SSP. We will do so by presenting a sufficient criterion for strong stability of general explicit RK methods. The basic idea is to assume that the time discretization of the semidiscrete equation (3.57) by the first-order *forward Euler method*

$$u^{n+1} = u^n + \Delta t L(u^n) \tag{3.59}$$

is *strongly stable*, i. e. $\|u^{n+1}\| \leq \|u^n\|$, under a certain norm and a suitably restricted timestep Δt and then try to extend this property to higher-order methods for time discretizations. To such methods, we refer to as *SSP*. Explicit SSP methods were first developed in [Shu88, SO88] to ensure that the TV would not increase. As a consequence, the resulting time discretizations were termed TVD methods in [Shu88, SO88]. Yet, in fact, the same class of methods will preserve any convex functional of the solution and is therefore called SSP by now. Let us consider a general explicit m-stage RK method for (3.57) written in the form

$$\begin{aligned} u^{(0)} &= u^n, \\ u^{(i)} &= \sum_{k=0}^{i-1} \left(\alpha_{i,k} u^{(k)} + \Delta t \beta_{i,k} L(u^{(k)}) \right), \quad i = 1, \dots, m, \\ u^{n+1} &= u^{(m)} \end{aligned} \tag{3.60}$$

with $\alpha_{i,k} \geq 0$ and $\sum_{k=0}^{i-1} \alpha_{i,k} = 1$ for consistency. Then, if all the $\beta_{i,k}$'s are nonnegative as well, the intermediate stages $u^{(i)}$ in (3.60) amount to convex combinations of the forward Euler method with Δt replaced by $\frac{\beta_{i,k}}{\alpha_{i,k}} \Delta t$. This yields the following result.

Theorem 3.40 (A sufficient criterion for strong stability)
Let $\|\cdot\|$ *be any convex functional. If the forward Euler method* (3.59) *is strongly stable under the timestep restriction* $\Delta t \leq \Delta t_{\text{FE}}$, *i. e.*

$$\|u^n + \Delta t L(u^n)\| \leq \|u^n\| \tag{3.61}$$

for $\Delta t \leq \Delta t_{\text{FE}}$, *and if* $\alpha_{i,k}, \beta_{i,k} \geq 0$, *then the solution obtained by the* m*-stage RK method* (3.60) *satisfies the strong stability bound*

$$\left\|u^{n+1}\right\| \leq \|u^n\| \tag{3.62}$$

under the timestep restriction $\Delta t \leq c(\alpha, \beta) \Delta t_{\text{FE}}$ *with* $c(\alpha, \beta) = \min_{i,k} \frac{\alpha_{i,k}}{\beta_{i,k}}$. *Here, if* $\beta_{i,k} = 0$, *the corresponding expression is considered to be infinite.*

Proof. Considering the ith stage of the RK method (3.60), we observe

$$\left\|u^{(i)}\right\| = \left\| \sum_{k=0}^{i-1} \alpha_{i,k} \left(u^{(k)} + \Delta t \frac{\beta_{i,k}}{\alpha_{i,k}} L(u^{(k)}) \right) \right\|.$$

Thus, since $\|\cdot\|$ is convex and $\sum_{k=0}^{i-1} \alpha_{i,k} = 1$ by consistency, we get

$$\left\|u^{(i)}\right\| \leq \sum_{k=0}^{i-1} \alpha_{i,k} \left\| u^{(k)} + \Delta t \frac{\beta_{i,k}}{\alpha_{i,k}} L(u^{(k)}) \right\|.$$

Finally, note that by (3.61),

$$\left\| u^{(k)} + \Delta t \frac{\beta_{i,k}}{\alpha_{i,k}} L(u^{(k)}) \right\| \leq \left\|u^{(k)}\right\|$$

holds for $\Delta t \leq \frac{\beta_{i,k}}{\alpha_{i,k}} \Delta t_{\text{FE}}$. This yields

$$\left\|u^{n+1}\right\| \leq \|u^n\|$$

for $\Delta t \leq \frac{\beta_{i,k}}{\alpha_{i,k}} \Delta t_{\text{FE}}$ and therefore the assertion. □

In particular, we can note from Theorem 3.40 that the SSPRK(3,3) method (see Definition 3.39) is SSP. For more details on SSP methods for time discretization, we refer to a rich body of literature [Shu88, SO88, GS98, GST01, Ket08, GKS11]. However, it should be stressed that in general — for nonlinear problems — the whole theory of SSP methods is based on the assumption that the forward Euler method is strongly stable (i. e. (3.61) holds). Yet, in many situations this can not be guaranteed for the spatial discretization and additional dissipation might become necessary, especially in the presence of (shock) discontinuities. A large part of this thesis and the latter chapters is dedicated to the development and analysis of suitable techniques to do so.

CHAPTER 4

STABLE HIGH ORDER QUADRATURE RULES FOR EXPERIMENTAL DATA I: NONNEGATIVE WEIGHT FUNCTIONS

We present the first novel contribution of this thesis, which is the development and analysis of stable high order QRs for experimental data. Here, we focus on so-called LS-QRs. The material presented in this chapter resulted in the publication [Gla19a]. This work provides a more detailed description of the topic. For sake of simplicity, we only summarize the most important concepts and results here. Another recommendation are the original works of Wilson [Wil70b, Wil70a] on LS-QRs and the more recent work of Huybrechs [Huy09]. An application of LS-QRs to construct stable high order discretizations of DG methods can be found in [GÖ20]. During this chapter, the weight function ω is assumed to be nonnegative again. General weight functions are addressed in Chapter 5.

Outline

This chapter is organized as follows: We start in Chapter 4.1 with a brief motivation, especially addressing the shortcomings of interpolatory QRs regarding stability. In Chapter 4.2, we then introduce the class of LS-QRs. Moreover, by incorporating the concept of DOPs, we are able to provide a simple and explicit formula for the quadrature weights corresponding to LS-QRs. This explicit formula is utilized in the subsequent Chapter 4.3 to investigate and prove stability for LS-QRs. Finally, in Chapter 4.4, we present several numerical results. We close this chapter with some concluding thoughts in Chapter 4.5.

4.1 Motivation

In Chapter 3.4, we have seen that interpolatory QRs can be stable while providing a maximal degree of exactness. This is possible by choosing the quadrature points as roots of specific orthogonal polynomials, yielding the class of Gaussian QRs. Yet, in many applications, it is impractical — if not even impossible — to obtain data to fit these roots. For instance, experimental measurements are often performed at equidistant or even scattered points. Unfortunately, it is well-known that interpolatory QRs on equidistant and scattered points become increasingly unstable as the number of quadrature points is increased. This is demonstrated in Figure 4.1 for the Newton–Cotes QRs.

Hence, general interpolatory QRs can not be considered as a reasonable option in this situation. Composite QRs, on the other hand, are stable but only of limited degree of exactness. To achieve both, high degrees of exactness as well as stability, we have to take a new path. This

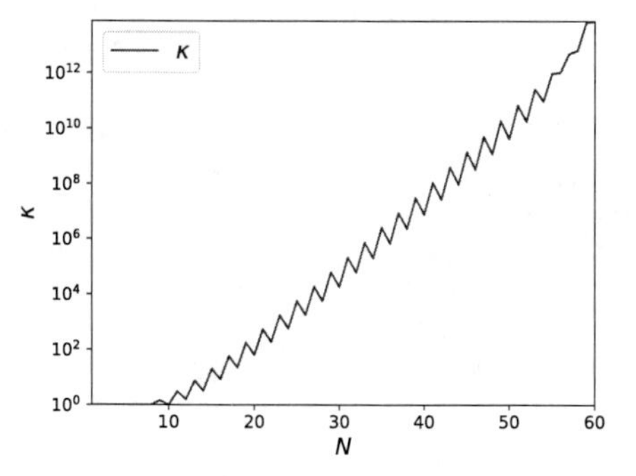

Figure 4.1: Stability values $\kappa(\mathbf{w}_N)$ for the Newton–Cotes QRs.

path leads us to the class of *LS-QRs*, which have been proposed by Wilson [Wil70b, Wil70a] in 1970 and further developed in [Huy09, Gla19a, Gla19d].

4.2 Least squares quadrature rules

The basic idea behind N-point LS-QRs is to aim for a degree of exactness d with $d \leq N-1$. Hence, LS-QRs generalize interpolatory QRs.

4.2.1 Formulation as a least squares problem

Let $\{\varphi_k\}_{k=0}^{d}$ be a basis of $\mathbb{P}_d(\mathbb{R})$ and let us choose $N > d+1$, i. e. a greater number of quadrature points than technically needed to construct an (interpolatory) QR with degree of exactness d. Then, the exactness condition (3.41) yields an underdetermined system of linear equations

$$A\mathbf{w}_N = \mathbf{m} \tag{4.1}$$

for the quadrature weights $\mathbf{w}_N \in \mathbb{R}^N$. The matrix $A \in \mathbb{R}^{(d+1)\times N}$ is given by

$$A = \left(\varphi_k(x_n)\right)_{k=0,n=1}^{d,N}, \tag{4.2}$$

while $\mathbf{m} \in \mathbb{R}^{d+1}$ denotes the vector of moments, that is

$$\mathbf{m} = \left(I[\varphi_k]\right)_{k=0}^{d}. \tag{4.3}$$

Here, I denotes the integral operator associated with the weight function ω, given by

$$I[f] = \int_a^b f(x)\omega(x)\,\mathrm{d}x.$$

The underdetermined system of linear equations (4.1) induces an $(N-d-1)$-dimensional affine linear subspace of solutions [GVL12], which we denote by

$$W := \left\{\mathbf{w}_N \in \mathbb{R}^N \mid A\mathbf{w}_N = \mathbf{m}\right\}.$$

Note that for every $\mathbf{w}_N \in W$, the resulting N-point QR has degree of exactness d. From W we want to determine the *LS solution* $\mathbf{w}_N^{\mathrm{LS}}$ which minimizes the Euclidean norm

$$\|\mathbf{w}_N\|_2 := \left(\sum_{n=1}^{N} |w_n|^2 \right)^{\frac{1}{2}}.$$

Thus, we want to determine $\mathbf{w}_N^{\mathrm{LS}}$ such that

$$\mathbf{w}_N^{\mathrm{LS}} = \arg\min_{\mathbf{w}_N \in W} \|\mathbf{w}_N\|_2 \tag{4.4}$$

is satisfied. The LS solution $\mathbf{w}_N^{\mathrm{LS}}$ exists and is unique [GVL12]. To the resulting QRs, we refer to as a LS-QR.

Definition 4.1 (LS-QRs)
We call an N-point QR Q_N a *LS-QR* of degree d if $d \leq N-1$ and the quadrature weights are given by (4.4).

At least formally, $\mathbf{w}_N^{\mathrm{LS}}$ can be obtained by

$$\mathbf{w}_N^{\mathrm{LS}} = A^T \mathbf{v}, \tag{4.5}$$

where $\mathbf{v}$ is the unique solution of the normal equation

$$AA^T \mathbf{v} = \mathbf{m}. \tag{4.6}$$

This yields

$$\mathbf{w}_N^{\mathrm{LS}} = A^T \left(AA^T \right)^{-1} \mathbf{m} \tag{4.7}$$

for the LS solution; see [CP76]. Here, $A^T(AA^T)^{-1}$ is the Moore–Penrose pseudoinverse A^+ of A; see [BIG03]. Yet, the above representation should not be solved numerically, since the normal equation (4.6) tends to be ill-conditioned.

4.2.2 Characterization by discrete orthonormal polynomials

The real advantage of this approach reveals when incorporating the concept of DOPs; see Chapter 3.2. Note that the matrix product AA^T in (4.7) can be identified as a Gram matrix,

$$AA^T = \begin{pmatrix} [\varphi_0, \varphi_0]_N & \dots & [\varphi_0, \varphi_d]_N \\ \vdots & & \vdots \\ [\varphi_d, \varphi_0]_N & \dots & [\varphi_d, \varphi_d]_N \end{pmatrix},$$

where $[\cdot, \cdot]_N$ denotes the discrete inner product

$$[u, v]_N := \sum_{n=1}^{N} u(x_n) v(x_n).$$

Moreover, let us denote the basis of DOPs corresponding to $[\cdot, \cdot]_N$ by $\{\pi_k(\cdot; N)\}_{k=0}^{d}$. When formulating the underdetermined system of linear equations (4.1) with respect to this basis of DOPs, we get

$$AA^T = I$$

and (4.7) reduces to

$$\mathbf{w}_N^{\text{LS}} = A^T \mathbf{m}.$$

Thus, we have

$$w_n^{\text{LS}} = \sum_{k=0}^{d} \pi_k(x_n; N) I\left[\pi_k(\cdot; N)\right], \quad n = 1, \dots, N. \tag{4.8}$$

Note that this formula is valid for any set of quadrature points as long as $N \geq d+1$ quadrature points are used.

Remark 4.2. Defining the quadrature weights of the LS-QR by equation (4.8), we are in need of computing the moments

$$I\left[\pi_k(\cdot; N)\right] = \int_a^b \pi_k(x; N) \omega(x) \, \mathrm{d}x \tag{4.9}$$

for general bases of DOPs $\{\pi_k(\cdot; N)\}_{k=0}^d$. Depending on the weight function ω and the basis $\{\pi_k(\cdot; N)\}_{k=0}^d$, the exact evaluation of (4.9) might be impractical. In our implementation, we have calculated the moments $I[\pi_k(\cdot; N)]$ numerically by using a GLe-QR

$$I[\pi_k(\cdot; N)] \approx \sum_{j=1}^{J} w_j^{\text{GL}} \pi_k(x_j^{\text{GL}}; N) \omega(x_j^{\text{GL}})$$

on a large set of GLe points $\{x_j^{\text{GL}}\}_{j=1}^J$. Note that for a nonnegative weight function ω we could also use a Gaussian QR associated with this weight function. Yet, the above approach also applies to general weight functions ω, for which no Gaussian QR is available.

4.2.3 Weighted least squares quadrature rules

To ensure stability of LS-QRs in the case of nonnegative weight functions it is convenient to consider LS solutions in a more general manner. More precise, we are interested in determining *weighted LS solutions* $\mathbf{w}_N^{\text{LS}}$ from W which minimize a weighted Euclidean norm

$$\|\mathbf{w}_N\|_{\mathbf{s}} := \left(\sum_{n=1}^{N} s_n |w_n|^2 \right)^{\frac{1}{2}}$$

with weights $s_n > 0$, $n = 1, \dots, N$. Note that

$$\|\mathbf{w}_N\|_{\mathbf{s}}^2 = \sum_{n=1}^{N} s_n |w_n|^2 = \sum_{n=1}^{N} |\sqrt{s_n} w_n|^2 = \left\| S^{1/2} \mathbf{w}_N \right\|_2^2,$$

where $S = \operatorname{diag}(s_1, \dots, s_N)$. Hence, the weighted LS solution satisfies

$$\mathbf{w}_N^{\text{LS}} = \underset{\mathbf{w}_N \in W}{\arg\min} \left\| S^{1/2} \mathbf{w}_N \right\|_2.$$

This yields

$$\mathbf{w}_N^{\text{LS}} = S^{-1} A^T \left(A S^{-1} A^T \right)^{-1} \mathbf{m} \tag{4.10}$$

for the weighted LS solution; see [CP76]. Yet, once more, we can characterize the solution by using DOPs. This time, let us consider the (weighted) discrete inner product

$$[u, v]_{N,\mathbf{r}} := \sum_{n=1}^{N} r_n u(x_n) v(x_n) \tag{4.11}$$

with $r_n := 1/s_n > 0$, $n = 1,\dots,N$, and let us denote the basis of DOPs corresponding to $[\cdot,\cdot]_{N,\mathbf{r}}$ by $\{\pi_k(\cdot;N,\mathbf{r})\}_{k=0}^d$. Again, the matrix product $AS^{-1}A^T$ in (4.10) can be identified as a Gram matrix,

$$AS^{-1}A^T = ARA^T = \begin{pmatrix} [\varphi_0,\varphi_0]_{N,\mathbf{r}} & \cdots & [\varphi_0,\varphi_d]_{N,\mathbf{r}} \\ \vdots & & \vdots \\ [\varphi_d,\varphi_0]_{N,\mathbf{r}} & \cdots & [\varphi_d,\varphi_d]_{N,\mathbf{r}} \end{pmatrix}.$$

Further, formulating the underdetermined system of linear equations (4.1) with respect to this basis of DOPs, we get

$$AS^{-1}A^T = I$$

and therefore

$$\mathbf{w}_N^{\mathrm{LS}} = RA^T\mathbf{m}.$$

Thus, we have

$$w_N^{\mathrm{LS}} = r_n \sum_{k=0}^{d} \pi_k(x_n;N,\mathbf{r}) I\left[\pi_k(\cdot;N,\mathbf{r})\right], \quad n = 1,\dots,N, \tag{4.12}$$

for the weighted LS solution. To the resulting N-point QR with quadrature weights $\mathbf{w}_N^{\mathrm{LS}}$, we refer to as a *weighted LS-QR* if we want to exhibit that a weighted discrete inner product has been used. Often, however, we will call these QRs simply LS-QRs as well.

4.3 Stability of least squares quadrature rules for nonnegative weight functions

Still, let ω be a nonnegative weight function which is assumed to be integrable over $[a,b]$. We show that for each degree of exactness $d \in \mathbb{N}$, we find a LS-QR which has nonnegative-only quadrature weights if a sufficiently large number $N \geq d+1$ of quadrature points is used. In particular, this implies stability of these LS-QRs. To the best of the author's knowledge, the proves in this chapter have been given only in [Gla19a] so far.

4.3.1 Main result and consequences

In more detail, our main result is the following Theorem from [Gla19a].

Theorem 4.3
Let $d \in \mathbb{N}_0$ and let $[\cdot,\cdot]_{N,\mathbf{r}}$ be a discrete inner product on $\mathbb{P}_d(\mathbb{R})$ such that

$$\lim_{N\to\infty} [u,v]_{N,\mathbf{r}} = \langle u, v\rangle \quad \forall u,v \in \mathbb{P}_d(\mathbb{R}). \tag{4.13}$$

Then, there exists an $N_1 \in \mathbb{N}$ such that

$$w_n^{\mathrm{LS}} := r_n \sum_{k=0}^{d} \pi_k(x_n;N,\mathbf{r}) I\left[\pi_k(\cdot;N,\mathbf{r})\right] \geq 0$$

for all $N \geq N_1$ and $n = 1,\dots,N$.

Thus, for a suitable sequence of discrete inner products, we are able to construct QRs of arbitrarily high degrees of exactness and nonnegative quadrature weights. An immediate consequence of this is stability of the resulting LS-QRs.

Corollary 4.4
Let $d \in \mathbb{N}_0$ and let $[\cdot,\cdot]_{N,\mathbf{r}}$ be a discrete inner product on $\mathbb{P}_d(\mathbb{R})$ such that

$$\lim_{N\to\infty} [u,v]_{N,\mathbf{r}} = \langle u,\, v\rangle \quad \forall u,v \in \mathbb{P}_d(\mathbb{R}).$$

Then, the (weighted) LS-QR with degree of exactness $d \in \mathbb{N}$ which results from minimizing $\left\| R^{-1/2}\mathbf{w}_N \right\|_2$ is stable. Further, there exists an $N_1 \in \mathbb{N}$ such that $\kappa(\mathbf{w}_N^{\mathrm{LS}}) = I[1]$ for all $N \geq N_1$.

It only remains to address how the weights r_n in the discrete inner product should be chosen so that (4.13) is satisfied. There are many possible choices and, in fact, these can be used to enhance the convergence. Yet, the most simple choice is $r_n = (b-a)/N$, which corresponds to a simple composite mid point rule. In the later numerical tests, this discrete inner product will sometimes be referred to as the *standard inner product* and the resulting LS-QRs is said to be the *standard LS-QR*. It is obvious that also every other convergent QRs with positive quadrature weights does the job. A comparison of different QRs, corresponding inner products, and the resulting LS-QRs has been given in [Gla19a]. Chapter 4.4.1 provides a short snapshot of these investigations to, at least, demonstrate the potential gain from utilizing these additional degrees of freedom.

4.3.2 Proof of the main result

Here, the continuous inner product of interest is given by

$$\langle u,\, v\rangle = \int_a^b u(x)v(x)\omega(x)\,\mathrm{d}x. \tag{4.14}$$

The induced basis of continuous orthonormal polynomials is denoted by $\{\pi_k(\cdot;\omega)\}_{k=0}^d$ and assumed to be constructed by Gram–Schmidt orthogonalization, i. e.

$$\begin{aligned} \tilde{\pi}_k(x;\omega) &= e_k(x) - \sum_{l=0}^{k-1} \langle e_k,\, \pi_l(\cdot;\omega)\rangle\, \pi_l(x;\omega), \\ \pi_k(x;\omega) &= \frac{\tilde{\pi}_k(x;\omega)}{\|\tilde{\pi}_k(\cdot;\omega)\|} \end{aligned} \tag{4.15}$$

with $e_k(x) := x^k$. The same assumption holds for the bases of DOPs, $\{\pi_k(\cdot;N,\mathbf{r})\}_{k=0}^d$, which are induced by the discrete inner product (4.11). In particular; all bases consist of normed monic polynomials with $\deg \pi_k = k$, $k = 0,\dots,d$. The above assumptions are made to ensure uniform convergence of the DOPs to the continuous orthonormal polynomials. Hence, let us start with two preliminary results on the convergence of discrete inner products and the induced bases of DOPs.

Lemma 4.5
Let $\langle\cdot,\,\cdot\rangle$ be the continuous inner product given by (4.14) and let $d \in \mathbb{N}$. Further, for $N \in \mathbb{N}$, let $[\cdot,\cdot]_{N,\mathbf{r}}$ be a discrete inner product given by (4.11) such that

$$\lim_{N\to\infty} [u,v]_{N,\mathbf{r}} = \langle u,\, v\rangle \quad \forall u,v \in \mathbb{P}_d(\mathbb{R}). \tag{4.16}$$

and let $(u_N)_{N\in\mathbb{N}}$ and $(v_N)_{N\in\mathbb{N}}$ be two sequences in $\mathbb{P}_d(\mathbb{R})$ with

$$u_N \to u, \quad v_N \to v \quad \text{in } L^\infty([a,b]) \tag{4.17}$$

for $N \to \infty$, *where* $u, v \in \mathbb{P}_d(\mathbb{R})$. *Then,*

$$\lim_{N\to\infty} [u_N, v_N]_{N,\mathbf{r}} = \langle u,\, v\rangle$$

holds.

Proof. Let us note that

$$\begin{aligned}|\langle u,\, v\rangle - [u_N, v_N]_{N,\mathbf{r}}| \leq & |\langle u,\, v\rangle - [u, v]_{N,\mathbf{r}}| + |[u, v]_{N,\mathbf{r}} - [u_N, v]_{N,\mathbf{r}}| \\ & + |[u_N, v]_{N,\mathbf{r}} - [u_N, v_N]_{N,\mathbf{r}}| \,.\end{aligned}$$

The first term, $|\langle u,\, v\rangle - [u, v]_{N,\mathbf{r}}|$, converges to zero because of (4.16). For the second term, the Cauchy–Schwarz inequality yields

$$|[u, v]_{N,\mathbf{r}} - [u_N, v]_{N,\mathbf{r}}|^2 = |[u - u_N, v]_{N,\mathbf{r}}|^2 \leq \|u - u_N\|_{N,\mathbf{r}}^2 \, \|v\|_{N,\mathbf{r}}^2 \,,$$

where $\|v\|_{N,\mathbf{r}}^2 \to \|v\|^2$ for $N \to \infty$, due to (4.16), and

$$\begin{aligned}\|u - u_N\|_{N,\mathbf{r}}^2 &= \sum_{n=1}^{N} r_n |u(x_n) - u_N(x_n)|^2 \\ &\leq \|1\|_{N,\mathbf{r}}^2 \, \|u - u_N\|_{L^\infty([a,b])}^2 \to 0\end{aligned}$$

for $N \to \infty$, due to (4.17). Thus, the second term converges to zero as well. A similar argument can be used to show that the third term converges to zero, which yields the assertion. □

Next, we show that, when the discrete inner product (4.11) converges to the continuous inner product (4.14) for all polynomials of degree at most d, the DOPs $\pi_k(\cdot; N, \mathbf{r})$ converge uniformly to the continuous orthonormal polynomials $\pi_k(\cdot; \omega)$.

Lemma 4.6

Let $\langle \cdot,\, \cdot\rangle$ *be the continuous inner product given by* (4.14) *and let* $d \in \mathbb{N}$. *Moreover, for* $N \in \mathbb{N}$, *let* $[\cdot, \cdot]_{N,\mathbf{r}}$ *be a discrete inner product given by* (4.11) *such that*

$$\lim_{N\to\infty} [u, v]_{N,\mathbf{r}} = \langle u,\, v\rangle \quad \forall u, v \in \mathbb{P}_d(\mathbb{R}).$$

By the Gram–Schmidt orthogonalization, for $k = 0, \dots, d$, *the discrete inner products induce a sequence of* k*th DOPs* $(\pi_k(\cdot; N, \mathbf{r}))_{N\in\mathbb{N}}$ *and the continuous inner product induces a single* k*th orthonormal polynomial* $\pi_k(\cdot; \omega)$. *Then, we have*

$$\pi_k(\cdot; N, \mathbf{r}) \to \pi_k(\cdot; \omega) \quad \textit{in } L^\infty([a, b])$$

for $N \to \infty$ *and* $k = 0, \dots, d$.

Proof. We proof the assertion by induction. First, we show that the assertion holds for $k = 0$. Consulting the Gram–Schmidt orthogonalization (4.15), we have

$$\tilde{\pi}_0(\cdot; \omega) = \tilde{\pi}_0(\cdot; N, \mathbf{r}) = e_0 \equiv 1.$$

Hence, the very first orthonormal polynomials are respectively given by

$$\pi_0(\cdot; \omega) \equiv \frac{1}{\|1\|}, \quad \pi_0(\cdot; N, \mathbf{r}) \equiv \frac{1}{\|1\|_{N,\mathbf{r}}}.$$

This yields

$$\|\pi_0(\cdot;\omega) - \pi_0(\cdot;N,\mathbf{r})\|_{L^\infty([a,b])} = \frac{\|1\|_{N,\mathbf{r}} - \|1\|}{\|1\| \, \|1\|_{N,\mathbf{r}}} \to 0$$

for $N \to \infty$ and therefore the assertion for $k = 0$. Next, assuming the assertion holds for the first $k-1$ orthonormal polynomials, we show that it also holds for the kth orthonormal polynomial. In combination with the first part of this proof ($k = 0$), this yields the assertion for all $k = 0, \dots, d$. Thus, let

$$\pi_l(\cdot;N,\mathbf{r}) \to \pi_l(\cdot;\omega) \text{ in } L^\infty([a,b]), \quad N \to \infty,$$

hold for $l = 0, \dots, k-1$. By the Gram–Schmidt orthogonalization, the kth orthonormal polynomials are given by (4.15). Lemma 4.5 gives us

$$[e_k, \pi_l(\cdot;N,\mathbf{r})]_{N,\mathbf{r}} \to \langle e_k, \pi_l(\cdot;\omega)\rangle, \quad N \to \infty,$$

for $l = 0, \dots, k-1$ and therefore

$$\tilde{\pi}_k(\cdot;N,\mathbf{r}) \to \tilde{\pi}_k(\cdot;\omega) \text{ in } L^\infty([a,b])$$

for $N \to \infty$. Further, Lemma 4.5 yields $\|\tilde{\pi}_k(\cdot;N,\mathbf{r})\|_N \to \|\tilde{\pi}_k(\cdot;\omega)\|$ for $N \to \infty$. This implies

$$\pi_k(\cdot;N,\mathbf{r}) \to \pi_k(\cdot;\omega) \text{ in } L^\infty([a,b])$$

for $N \to \infty$, which completes the proof. □

Theorem 4.7 (Theorem 4.3)
Let $d \in \mathbb{N}_0$ and let $[\cdot,\cdot]_{N,\mathbf{r}}$ be a discrete inner product on $\mathbb{P}_d(\mathbb{R})$ such that

$$\lim_{N\to\infty} [u,v]_{N,\mathbf{r}} = \langle u, v\rangle \quad \forall u,v \in \mathbb{P}_d(\mathbb{R}). \tag{4.18}$$

Then, there exists an $N_1 \in \mathbb{N}$ such that

$$w_n^{\text{LS}} := r_n \sum_{k=0}^{d} \pi_k(x_n;N,\mathbf{r}) I\left[\pi_k(\cdot;N,\mathbf{r})\right] \geq 0$$

for all $N \geq N_1$ and $n = 1, \dots, N$.

Proof. Let us define

$$\varepsilon_k := [\pi_k(\cdot;N,\mathbf{r}), 1]_{N,\mathbf{r}} - \langle \pi_k(\cdot;N,\mathbf{r}), 1\rangle. \tag{4.19}$$

Then, we have

$$\varepsilon_k = \delta_{k,0}[\pi_0(\cdot;N,\mathbf{r}),1]_{N,\mathbf{r}} - I[\pi_k(\cdot;N,\mathbf{r})]$$

and therefore

$$\begin{aligned} w_n^{\text{LS}} &= r_n \sum_{k=0}^{d} \pi_k(x_n;N,\mathbf{r}) \left(\delta_{k,0}[\pi_0(\cdot;N,\mathbf{r}),1]_{N,\mathbf{r}} - \varepsilon_k\right) \\ &= r_n \left(\pi_0(x_n;N,\mathbf{r})[\pi_0(\cdot;N,\mathbf{r}),1]_{N,\mathbf{r}} - \sum_{k=0}^{d} \varepsilon_k \pi_k(x_n;N,\mathbf{r}) \right). \end{aligned}$$

Assuming the basis of DOPs $\{\pi_k(\cdot;N,\mathbf{r})\}_{k=0}^d$ is ordered by degree, the first polynomial is given by $\pi_0(\cdot;N,\mathbf{r}) \equiv 1/\|1\|_{N,\mathbf{r}}$. This yields

$$\pi_0(x_n;N,\mathbf{r})[\pi_0(\cdot;N,\mathbf{r}),1]_{N,\mathbf{r}} = [\pi_0(\cdot;N,\mathbf{r}), \pi_0(\cdot;N,\mathbf{r})]_{N,\mathbf{r}} = 1.$$

Hence, the assertion $w_n^{\mathrm{LS}} \geq 0$ is equivalent to

$$\sum_{k=0}^{d} \varepsilon_k \pi_k(x_n; N, \mathbf{r}) \leq 1.$$

By (4.18) the discrete inner products $[\cdot,\cdot]_{N,\mathbf{r}}$ converge to the continuous inner product $\langle \cdot, \cdot \rangle$ for all polynomials of degree at most d and Lemma 4.6 implies that

$$\pi_k(\cdot; N, \mathbf{r}) \to \pi_k(\cdot; \omega) \quad \text{in } L^\infty([a,b]) \tag{4.20}$$

for $N \to \infty$ and $k = 0, \dots, d$. In particular, the polynomials are uniformly bounded; that is there exists a constant $C > 0$ such that

$$|\pi_k(x; N, \mathbf{r})| \leq C$$

for all $x \in [a,b]$ and $k = 0, \dots, d$. Thus, we get

$$\sum_{k=0}^{d} \varepsilon_k \pi_k(x_n; N, \mathbf{r}) \leq C \sum_{k=0}^{d} |\varepsilon_k| .$$

Finally, since uniform convergence (4.20) holds, Lemma 4.5 yields $\varepsilon_k \to 0$, $N \to \infty$, for $k = 0, \dots, d$. Hence, there exists an $N_1 \in \mathbb{N}_0$ such that

$$|\varepsilon_k| \leq \frac{1}{C(d+1)}, \quad k = 0, \dots, d,$$

for $N \geq N_1$. This implies

$$\sum_{k=0}^{d} \varepsilon_k \pi_k(x_n; N, \mathbf{r}) \leq 1$$

and therefore the assertion. □

4.4 Numerical tests

In what follows, a comparative study for different choices in the construction of LS-QRs is provided. Several discrete inner products, yielding different LS-QRs, are investigated in a way that allows us to give clear recommendation what kind of LS-QR should be used in a given situation. Moreover, we compare the accuracy of different LS-QRs and demonstrate that stability for LS-QRs essentially is a $N \approx Cd^2$ process for commonly used integrators. Note that in the prior main result (Theorem 4.3) the ratio between N and d has not been quantified.

4.4.1 Comparison of different inner products

We start by providing a demonstration on how large N has to be chosen for a LS-QR to be stable with $\kappa(\mathbf{w}_N^{\mathrm{LS}}) = I[1]$. In the following, we call such QRs *perfectly stable.* In Figure 4.2(a), the stability value κ is illustrated for a fixed degree of exactness of $d = 50$ and an increasing number of quadrature points N on $[a,b] = [0,1]$. The continuous integral operator I which is approximated by the QR has the weight function $\omega \equiv 1$. Here, the straight (blue) line corresponds to $\kappa(\mathbf{w}_N^{\mathrm{LS}})$ for the LS-QR resulting from the standard discrete inner product with weights $r_n = \frac{b-a}{N}$ on equidistant quadrature points. The dashed (red) line, on the other hand, corresponds to $\kappa(\mathbf{w}_N^{\text{LS}})$ for the LS-QR resulting from the standard discrete inner product with weights $r_n = \frac{b-a}{N}$ on a set of scattered points. In our implementation, this set of scattered points is constructed by iteratively adding a new uniformly distributed point to the set of already existing points. Note that the resulting discrete inner product therefore corresponds to Monte–Carlo integration. Thus, convergence to the continuous inner product is ensured by the law of large numbers. Moreover, Figure 4.2(b) illustrates the minimal number N of quadrature points which is needed for perfect stability of the above LS-QRs with degree of exactness d. It

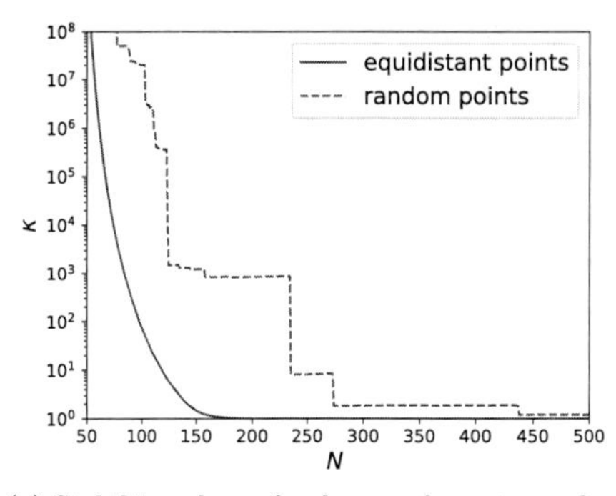

(a) Stability value κ for degree of exactness $d = 50$ and an increasing number of equidistant as well as uniformly distributed points N..

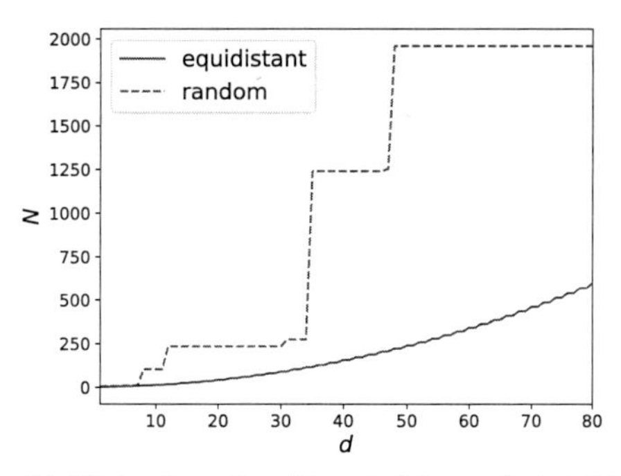

(b) Minimal number N needed for perfect stability of the LS-QR corresponding to the standard discrete inner product.

Figure 4.2: Demonstration of the ratio between the degree of exactness d and the number N of (equidistant as well as scattered) quadrature points for perfect stability to hold for the resulting LS-QR with respect to the standard discrete inner product with $r_n = \frac{b-a}{N}$.

is demonstrated by Figure 4.2(b) that the minimal number N of quadrature points needed for a perfectly stable QR might be quite large, especially for scattered points. Next, we therefore aim to investigate the possibility to reduce N by going over to other discrete inner products. This is achieved by comparing LS-QRs resulting from different discrete inner products (weights $\mathbf{r}$). We first do so in the case of equidistant points. Taking a look back at the proof of Theorem 4.3, we note that the aim essentially has to be to reduce the sum over the absolute value of the errors ε_k given by (4.19). There are essentially two ways to do so:

1. The first $\tilde{d}+1$ errors $\varepsilon_0, \ldots, \varepsilon_{\tilde{d}}$ vanish completely if the discrete inner product induces a QR with degree of exactness $\tilde{d}$.

2. The order of convergence for $\varepsilon_k \to 0$ can be increased by utilizing discrete inner products with faster convergence to the continuous inner product.

Both mechanisms are met by composite interpolatory QRs [DR07], at least up to a certain degree. A comparison of the composite rules for the first eight Newton–Cotes rules on equidistant points is provided by Figure 4.3. Note that composite Newton–Cotes of higher degrees are not reasonable, since negative weights arise then. Here, the resulting LS-QRs are compared with respect to the minimal number of quadrature points that is required for the rule to be perfectly stable. In Figure 4.3(a), the degree of the composite Newton–Cotes rule which provides the smallest number N of quadrature points in order to be perfectly stable is plotted against an increasing degree of exactness d. In our implementation, we set p to be the smallest degree, if multiple composite Newton–Cotes rules provide the same number N. When having the choice between different discrete inner products providing the same advantage, we think it is preferable to decide for the simpler one.

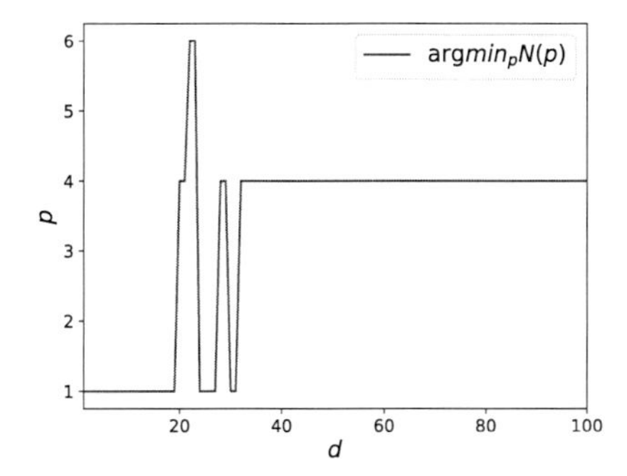

(a) Number of points p used in the composite Newton–Cotes rule which provides the smallest number N of quadrature points.

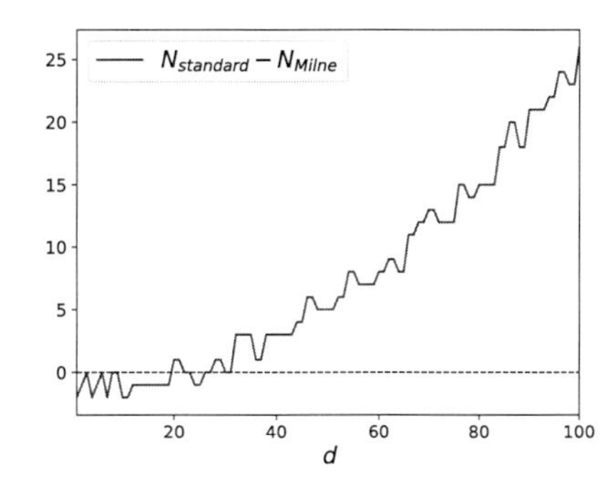

(b) Difference in the minimal number N of quadrature points between the standard discrete inner product and the composite 4-point Newton–Cotes rule.

Figure 4.3: A comparative study of LS-QRs resulting from different composite Newton–Cotes rules as discrete inner products.

Figure 4.3(a) illustrates that the midpoint rule (i. e. the standard discrete inner product) and the composite 4-point Newton–Cotes rule (also known as the *Simpson's* 3/8 *rule*) provide the best results for almost all degrees of exactness d. The minimal numbers N of quadrature points required for these rules to be perfectly stable are further compared in Figure 4.3(b). Here, the difference in N between the two QRs is plotted against an increasing degree of exactness d. A positive value means that the composite 4-point Newton–Cotes rule provides a smaller number N, while a negative value means that the midpoint rule provides a smaller number N. Lumping things together, we recommend to

- utilize the midpoint rule for the underlying discrete inner product for degrees of exactness $d \leq 20$,
- utilize the composite 4-point Newton–Cotes rule for the underlying discrete inner product for all degrees of exactness $d > 20$.

Finally, we note that for scattered quadrature points only the standard inner product (midpoint rule) and the composite trapezoidal rule (composite 2-point Newton-Cotes rule) seem reasonable. Note that both rules just differ slightly at the end points. In accordance to this, Figure 4.4 illustrates only a slight difference between the minimal numbers N of quadrature points needed for the LS-QRs to be perfectly stable.

Yet, a small advantage can be observed for the composite trapezoidal rule. The advantage of using the composite trapezoidal rule will become more wide-ranging when comparing the accuracy in Chapter 4.4.4.

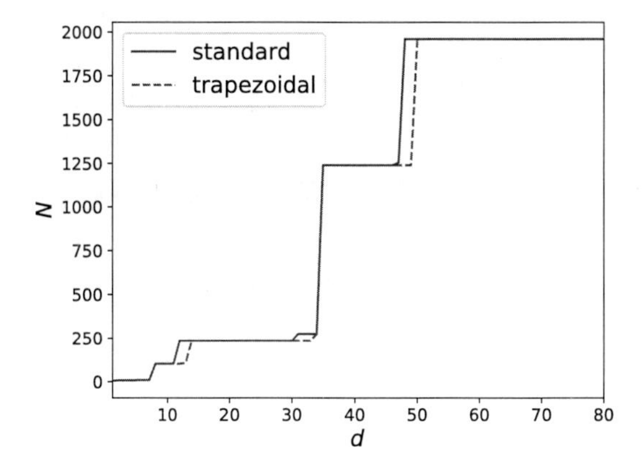

Figure 4.4: Minimal number N of scattered quadrature points needed for the LS-QRs corresponding to the standard discrete inner product and the composite trapezoidal rule to be perfectly stable.

4.4.2 Minimal number of quadrature points for different weight functions

Wilson [Wil70b] proved that for the standard discrete inner product corresponding to the weight function $\omega(x) = 1$ perfect stability essentially holds for $N \approx Cd^2$. We demonstrate that the same behavior also holds for other nonnegative weight functions and numerically determine the respective constants C. In the subsequent numerical tests, we can observe perfect stability to hold for $N \approx Cd^2$ with fairly small constants $C < 1$ for all tested weight functions. Figure 4.5 shows the number N of equidistant quadrature points needed for perfect stability to hold for the standard LS-QR and different weight functions. Here, the (blue) straight line corresponds

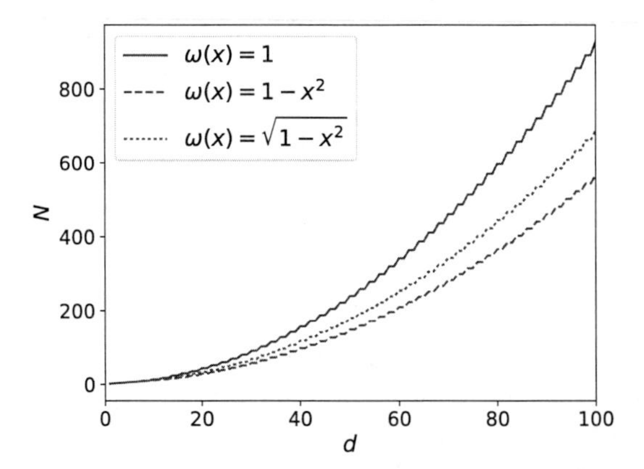

Figure 4.5: Minimal number N of equidistant quadrature points for the standard LS-QR to be perfectly stable.

to the Legendre weight function $\omega(x) = 1$ on $[-1, 1]$, the (red) dashed line corresponds to the Jacobi ($\alpha = \beta = 1$) weight function $\omega(x) = 1 - x^2$ on $[-1, 1]$, and the (green) dotted line corresponds to the Chebyshev weight function of the second kind $\omega(x) = \sqrt{1-x^2}$ on $[-1,1]$. All three functions indicate that perfect stability of the corresponding LS-QR is a d^2 process, i. e. $N \approx Cd^2$ is sufficient for perfect stability to hold. This can be examined by fitting the parameters C and s in the model $N = Cd^s$ in the sense of LS. Table 4.1 lists these parameters for three different weight functions.

$\omega(x)$	C	s
1	0.1	1.95
$1-x^2$	0.07	1.93
$\sqrt{1-x^2}$	0.8	1.94

Table 4.1: LS fit of the parameters C and s in the model $N = Cd^s$ for the different weight functions ω.

Note that the results presented in Table 4.1 are in accordance with Wilson's work [Wil70b], where he has shown that stability for the standard LS-QR for weight function $\omega(x) = 1$ essentially is a d^2 process. Here, we have gathered evidence that this ratio between d and N holds for other weight functions as well.

4.4.3 Accuracy on equidistant points

Next, let us investigate accuracy of LS-QRs on equidistant points for the weight function $\omega(x) = 1$ on $[-1,1]$. The errors of the standard LS-QR ($r_n = (b-a)/N$) and the LS-QR for Simpson's 3/8 rule are compared to the errors of the (interpolatory) Newton–Cotes QR as well as the composite trapezoidal rule. Figure 4.6 shows the corresponding errors as a function in d for four different functions f. The errors for the (interpolatory) Newton–Cotes rule are illustrated by the (blue) straight line, while the (red) dashed line corresponds to the composite trapezoidal rule. The errors of the standard LS-QR are illustrated by the (green) dotted line and the errors of the LS-QR for Simpson's 3/8 rule are illustrated by the (black) straight line. These are respectively labeled as "DLS (standard)" and "DLS (composite 4-point NC rule)", where DLS abbreviates (discrete) LS. For the number of equidistant quadrature points, we have chosen $N = d$ for the (interpolatory) Newton–Cotes rule and N equal to the minimal number of quadrature points needed for perfect stability to hold for the LS-QRs. To allow a fair comparison with the LS-QRs, we have also chosen the number of quadrature points for the composite trapezoidal rule to be equal to the number of points for the standard LS-QR. Note that the (interpolatory) Newton–Cotes rule diverges in all tests. The composite trapezoidal rule is demonstrated to converge in all four cases, even though the convergence is fairly slow. At the same time, both LS-QRs show formidably fast convergence in all test cases. Furthermore, the LS-QRs provide the smallest errors from the start in Figure 4.6(a) and 4.6(d), and at least the smallest errors for degrees $d \geq 25$ in 4.6(b) and 4.6(c). Note that in Figure 4.6(a) both LS-QRs reach machine precision, i. e. $2^{-53} \approx 1.1 \cdot 10^{-16}$ for double precision, already for $d = 40$. In our tests we have not been able to observe noticeable differences in the accuracy of the LS-QRs based on the standard inner product and Simpson's 3/8 rule. Finally, it should be stressed that both discrete LS-QRs do not just provide a remarkably faster convergence in all test cases, but are also able — in contrast to the composite trapezoidal rule — to exactly integrate polynomials up to degree d.

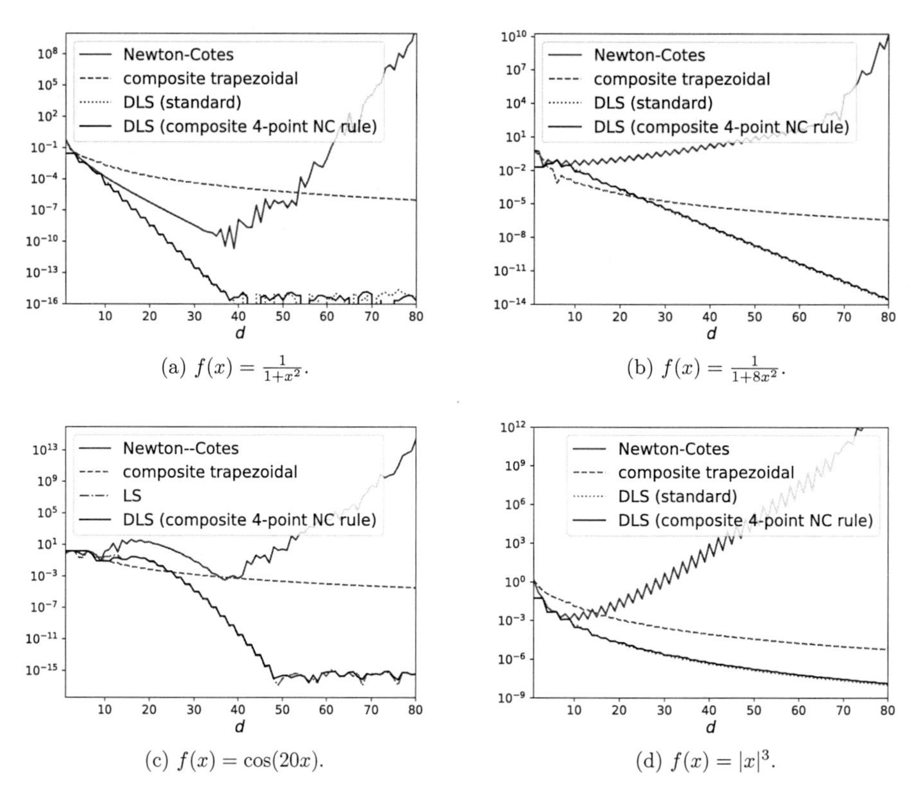

Figure 4.6: Errors for the Newton–Cotes and composite trapezoidal rule as well as for the standard LS-QR and the LS-QR based on Simpson's 3/8 rule. Equidistant quadrature points on $[-1, 1]$ have been used.

4.4.4 Accuracy on scattered points

Finally, we perform the same analysis as in Chapter 4.4.3 for sets of scattered quadrature points. Figure 4.7 illustrates the errors in the same manner as Figure 4.6. This time, the LS-QR based on the composite trapezoidal rule on scattered points is included instead of the LS-QR based Simpson's 3/8 rule.

For a fair comparison, all rules are applied to the same number of $N = d^2$ scattered points in $[-1, 1]$. As demonstrated in Figure 4.7, the interpolatory QR (still labeled as "Newton–Cotes") again diverges in all test cases. The composite trapezoidal rule still converges, but now even slower than in the case of equidistant quadrature points. We note that both LS-QRs still show a formidable rate of convergence, even though they are now applied to a set of scattered quadrature points. Using a more challenging set of scattered points, in fact, we are now able to note an advantage in accuracy for the LS-QR based on the composite trapezoidal rule compared to the standard LS-QR (which is based on the midpoint rule). In most tests for scattered points, the LS-QR based on the composite trapezoidal rule provides more accurate results.

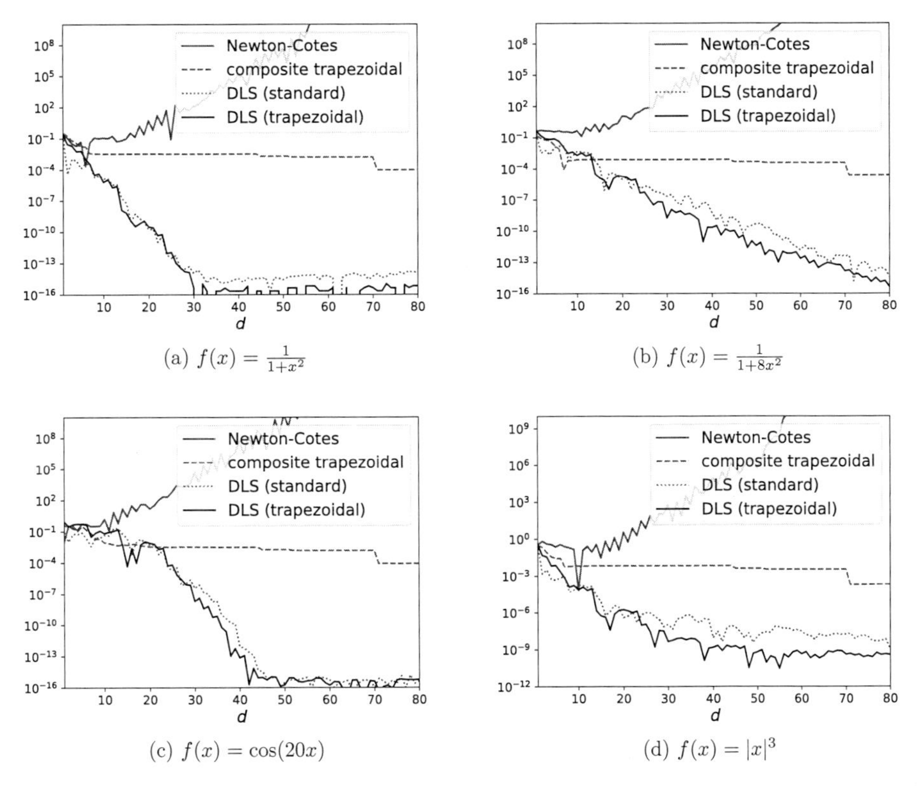

(a) $f(x) = \frac{1}{1+x^2}$

(b) $f(x) = \frac{1}{1+8x^2}$

(c) $f(x) = \cos(20x)$

(d) $f(x) = |x|^3$

Figure 4.7: Errors for the interpolatory QR and the composite trapezoidal rule as well as for the standard LS-QR and the LS-QR based on the composite trapezoidal rules. Scattered quadrature points on $[-1, 1]$ have been used.

4.5 Concluding thoughts and outlook

In this Chapter, we have developed and investigated the class of LS-QRs, which is especially suited for experimental data obtained at equidistant or even scattered locations. Further, we have compared LS-QRs resulting from different discrete inner products and observed that these can be used to reduce the minimal number of quadrature points needed for perfect stability. This was shown to be especially favorable for scattered quadrature points. An additional advantage is that known errors or uncertainties of the data can be incorporated by adapting the weights in the underlying discrete inner product. In particular, future work will focus on the development of (perfectly) stable LS-QRs for multiple measurement vectors (MMVs) of the unknown function f; see [EM09, GS19] and references therein. Another promising field of application, which has not been fully explored yet, is the construction of stable numerical methods for partial differential equations. Especially in FD and some recent DG methods, high degrees of exactness of QRs are vital to mimic integration by parts on a discrete level, a property also known as SBP. The LS-QRs constructed in this chapter can be seen as a method to construct so called SBP operators — a hot topic in computational fluid dynamics

[FHZ14, SN14, Gas13, CS17, GÖ20] — on many (wild) sets of collocation points.

CHAPTER 5

STABLE HIGH ORDER QUADRATURE RULES FOR EXPERIMENTAL DATA II: GENERAL WEIGHT FUNCTIONS

So far, we have always assumed the weight function ω to be nonnegative and all QRs discussed so far have build up on this fact. In this chapter, however, we extend our study of LS-QRs to integrals of the form

$$I[f] = \int_a^b f(x)\omega(x)\,\mathrm{d}x \tag{5.1}$$

with general weight function $\omega : [a,b] \to \mathbb{R}$, only assumed to be integrable over $[a,b]$. To the best of the author's knowledge, stable high order QRs for experimental data and general weight functions have not been proposed before. The material presented in this chapter resulted in the publication [Gla19d].

Outline

This chapter is organized as follows: After a short motivation of general weight functions in Chapter 5.1, we revisit the concept of stability for QRs in Chapter 5.2, this time addressing general weight functions. In particular, we introduce the concept of sign-consistency. In Chapter 5.3, we show that the LS-QRs introduced in Chapter 4 are also stable in the case of general weight functions. This time, however, we can not utilize the connection between stability and nonnegativity of the quadrature weights. So-called sign-consistent QRs with high degrees of exactness, on the other hand, are constructed in Chapter 5.4 by a nonnegative least squares (NNLS) formulation of the exactness condition (3.41). Finally, stable and sign-consistent (sequences of) QRs are compared by numerical tests in Chapter 5.5. We end this chapter with some concluding thoughts and an outlook on future work in Chapter 5.6.

5.1 Motivation

General weight functions with mixed signs can arise in many situations; For instance in the weak form of the Schrödinger equation, integrals of the form

$$I[f] = \int_a^b f(x)V(x)\,\mathrm{d}x$$

arise, where V is a potential with possibly mixed signs. Another field of application are (highly) oscillatory integrals of the form

$$I[f] = \int_a^b f(x) e^{i\omega g(x)} \,\mathrm{d}x,$$

where g is a smooth oscillator and ω is a frequency parameter ranging from 0 to ∞. See [IN04, INO06, HO09] and references therein. In what follows, we investigate stability concepts for QRs approximating integrals (5.1) with general weight functions ω. For nonnegative weight functions stability of (a sequence of) QRs essentially follows from the QRs to have nonnegative-only quadrature weights. For general weight functions, however, this connection breaks down and we have to consider stable and sign-consistent (the quadrature weights have the same sign as the weight function at the quadrature points) QRs as separated classes. Following these observations, we propose two different procedures to construct stable QRs as well as sign-consistent QRs with high degrees of exactness on equidistant and scattered quadrature points for integrals with general weight functions. The first procedure has already been introduced in the prior Chapter 4 and yields the class of LS-QRs. These QRs build up on a LS formulation of the exactness condition (3.41). The procedure to construct sign-consistent QRs with high degrees of exactness builds up on a NNLS formulation of the exactness conditions (3.41), resulting in so-called NNLS-QRs. These have already been investigated by Huybrechs [Huy09] for positive weight functions. Here, we modify the procedure to handle general weight functions and present a discussion of their stability.

5.2 Stability concepts for general weight functions

We have already addressed stability of QRs in Chapter 3.4, though only for nonnegative weight functions. For these, we have noted that if all quadrature weights are nonnegative and the QRs have degree of exactness 0, we have $\kappa(\mathbf{w}_N) = K_\omega$. The constant K_ω is defined by

$$K_\omega = \int_a^b |\omega(x)| \,\mathrm{d}x \tag{5.2}$$

and, for a nonnegative weight function ω, we therefore have $K_\omega = I[1]$. If the quadrature weights have mixed signs, however, we get $\kappa(\mathbf{w}_N) > I[1]$. This connection between perfect stability of QRs and their quadrature weights to be nonnegative has been utilized in the last chapter to construct stable LS-QRs with high degrees of exactness on equidistant and scattered quadrature points. Yet, nonnegative weights are only reasonable if also the weight function ω in (5.1) is nonnegative. Enforcing nonnegative quadrature weights for general weight functions can not be considered as reasonable. Instead, it seems more convenient to fit the sign of a quadrature weight w_n to the sign of the weight function ω at the corresponding quadrature point x_n. In this work, we refer to QRs with such quadrature weights as sign-consistent QRs.

Definition 5.1
Let Q_N be an N-point QR with quadrature points $\{x_n\}_{n=1}^N$ and quadrature weights $\{w_n\}_{n=1}^N$, which approximates the integral (5.1) with weight function $\omega : [a,b] \to \mathbb{R}$. We call the QR Q_N *sign-consistent* if

$$\operatorname{sign}(w_n) = \operatorname{sign}(\omega(x_n)) \tag{5.3}$$

holds for all $n = 1, \ldots, N$.

In Definition 5.1, $\operatorname{sign}(\cdot)$ denotes the usual *sign function* defined as

$$\operatorname{sign}: \mathbb{R} \to \mathbb{R}, \quad \operatorname{sign}(x) := \begin{cases} -1 & \text{if } x < 0, \\ 1 & \text{if } x \geq 0. \end{cases} \tag{5.4}$$

The following lemma could be considered as a reformulation of Lemma 3.30.

Lemma 5.2
Let $(Q_N)_{N \in \mathbb{N}}$ be a sequence of N-point QRs approximating the integral (5.1) *with nonnegative weight function ω. If every Q_N has degree of exactness* 0 *and is sign-consistent, the sequence of QRs $(Q_N)_{N \in \mathbb{N}}$ is stable.*

Remark 5.3. Sign-consistency can be observed to be crucial for ensuring certain degrees of exactness for general weight functions. For instance, assume an N-point QR with nonnegative-only quadrature weights $w_n \geq 0$, $n = 1, \dots, N$. If such a QR is used to approximate an integral operator I with weight function ω such that

$$I[1] = \int_a^b \omega(x) \, \mathrm{d}x < 0, \tag{5.5}$$

the QR Q_N can not even have degree of exactness $d = 0$. Note that in this case, the QR Q_N is *not* sign-consistent and we have $Q_N[1] \geq 0$, yielding $Q_N[1] \neq I[1]$.

Of course, there are stable (sequences of) QRs which are not sign-consistent, for instance some composite Newton–Cotes rules. For nonnegative weight functions, sign-consistency can therefore be regarded as a stronger property than stability. For general weight functions, however, this connection breaks down. Still, a weaker connection can be shown and is stated in the following Theorem.

Theorem 5.4
Let $(Q_N)_{N \in \mathbb{N}}$ be a sequence of N-point QRs approximating the integral (3.39) *with weight function ω. Moreover, let every Q_N be sign-consistent and assume that*

$$\lim_{N \to \infty} Q_N[\operatorname{sign} \circ \omega] = I[\operatorname{sign} \circ \omega] \tag{5.6}$$

holds. Then, we have

$$\lim_{N \to \infty} \kappa(\mathbf{w}_N) = K_\omega. \tag{5.7}$$

Proof. Since every Q_N is sign-consistent and since $|r| = \operatorname{sign}(r) r$ holds for $r \in \mathbb{R}$, we have

$$\kappa(\mathbf{w}_N) = \sum_{n=1}^N \operatorname{sign}(w_n) w_n = \sum_{n=1}^N \operatorname{sign}(\omega(x_n)) w_n = Q_N[\operatorname{sign} \circ \omega]. \tag{5.8}$$

Next, assumption (5.6) provides us with

$$\lim_{N \to \infty} \kappa(\mathbf{w}_N) = I[\operatorname{sign} \circ \omega]. \tag{5.9}$$

Finally, this yields

$$\lim_{N \to \infty} \kappa(\mathbf{w}_N) = \int_a^b \operatorname{sign}(\omega(x)) \omega(x) \, \mathrm{d}x = K_\omega, \tag{5.10}$$

and therefore the assertion. □

Note that for a nonnegative weight function $\omega \geq 0$, (5.6) can be replaced with

$$\lim_{N \to \infty} Q_N[1] = I[1], \tag{5.11}$$

which is a fairly weak assumption. Yet, if assumption (5.6) is satisfied for a more general weight function ω depends on the weight function itself (e. g. if only finitely many jump discontinuities occur in sign $\circ$ ω) and how well the QRs Q_N can handle these discontinuities. Note that assumption (5.6) could be ensured by adding $Q_N[\text{sign} \circ \omega] = I[\text{sign} \circ \omega]$ to the exactness conditions — assuming the resulting system of linear equations remains solvable. This will be investigated in future works. Here, we treat stable (sequences of) QRs and sign-consistent QRs as two separate classes.

5.3 Stability of least squares quadrature rules for general weight functions

Let us consider the (unweighted) LS solution of the underdetermined system of linear equations (4.1), given by

$$\mathbf{w}_N^{\text{LS}} = A^T \left(A A^T \right)^{-1} \mathbf{m}$$

for a general basis of $\mathbb{P}_d(\mathbb{R})$ and simpler by

$$w_n^{\text{LS}} = \sum_{k=0}^{d} \pi_k(x_n; N) I \left[\pi_k(\cdot; N) \right], \quad n = 1, \dots, N, \tag{5.12}$$

when a basis of DOPs is used. See Chapter 4.2 for more details. For the discrete inner product $[u, v]_N = \sum_{n=1}^{N} u(x_n) v(x_n)$ with equidistant quadrature points $\{x_n\}_{n=1}^{N}$ a basis of DOPs can be derived explicitly from the discrete Chebyshev polynomials [Gau04, Chapter 1.5.2]. We will make use of this by consulting certain results for the discrete Chebyshev polynomials for the latter proof of stability of LS-QRs for general weight functions. Using scattered quadrature points, however, often no explicit formula is known for bases of DOPs. Again, we recommend to construct bases of DOPs using the modified Gram–Schmidt process.

5.3.1 Main results

Our main result of this chapter is the following theorem, which ensures stability for LS-QRs on equidistant quadrature points with any fixed degree of exactness.

Theorem 5.5
Let $d \in \mathbb{N}$ and let Q_N be an N-point LS-QR with equidistant quadrature points in $[a, b]$, degree of exactness d, and quadrature weights given by (5.12). Then, the sequence of LS-QRs $(Q_N)_{N>d}$ is stable; that is

$$\sup_{N>d} \kappa(\mathbf{w}_N^{\text{LS}}) < \infty$$

holds.

For scattered quadrature points, we note the following theorem.

Theorem 5.6
Let $d \in \mathbb{N}$ and let Q_N be an N-point LS-QR with distinct quadrature points in $[a,b]$, degree of exactness d, and quadrature weights given by (5.12). Moreover, let $\{\pi_k(\cdot;N)\}_{k=0}^{d}$ be the basis of DOPs corresponding to the discrete inner product $[\cdot,\cdot]_N$ and assume that there are constants $C(d) > 0$ and $N(d) > d$ such that

$$\|\pi_k(\cdot;N)\|_\infty^2 \leq \frac{C(d)}{N} \quad \forall k = 0,\dots,d \;\; \forall N \geq N(d). \tag{5.13}$$

Then, the sequence of LS-QRs $(Q_N)_{N>d}$ is stable; that is

$$\sup_{N>d} \kappa(\mathbf{w}_N^{\mathrm{LS}}) < \infty$$

holds.

As we will see in Chapter 5.3.3, the proof of Theorem 5.5 relies on the bound

$$\|\pi_k(\cdot;N)\|_\infty^2 \leq \frac{2k+1}{N}$$

to hold for the DOPs on equidistant points and for sufficiently large N. This results is derived from a similar bound to hold for the discrete Chebyshev polynomials. Unfortunately, we have not been able to find similar results for DOPs on nonequidistant points in the literature. Thus, we have added assumption (5.13) to Theorem 5.6 in order to guarantee this bound somewhat artificially.

5.3.2 Preliminaries on discrete Chebyshev polynomials

Using equidistant quadrature points in $[a,b]$ with $x_1 = a$ and $x_N = b$ essentially results in the normalized discrete Chebyshev polynomials [Gau04, Ch. 1.5.2] as basis of DOPs. Here, we recall some of their properties which come in useful to prove Theorem 5.5. The discrete Chebyshev polynomials arise as a special case of the Hahn polynomials [Hah49, Sze39]. For $\alpha, \beta > -1$ and $k = 0,\dots,N-1$, the *Hahn polynomials* may be defined in terms of a generalized hypergeometric series as

$$\begin{aligned} Q_k(x;\alpha,\beta,N-1) &= {}_3F_2(-k,-x,k+\alpha+\beta+1;\alpha+1,-N+1;1) \\ &= \sum_{j=0}^{k} \frac{(-k)_j(k+\alpha+\beta+1)_j(-x)_j}{(\alpha+1)_j(-N+1)_j j!} \end{aligned}$$

on $[0,N-1]$, where we have used the *Pochhammer symbol*

$$(a)_0 = 1, \quad (a)_j = a(a+1)\dots((a+j-1)).$$

The Hahn polynomials are orthogonal on $[0,N-1]$ with respect to the (discrete) inner product

$$\langle f, g\rangle_\rho := \sum_{n=0}^{N-1} f(n)g(n)\rho(n)$$

with weight function

$$\rho(x) := \binom{x+\alpha}{x}\binom{N-1-x+\beta}{N-1-x}.$$

They are normalized by

$$\|Q_k(\cdot, \alpha, \beta, N-1)\|_\rho^2 = h_k := \frac{(-1)^k (k+\alpha+\beta+1)_N (\beta+1)_k k!}{(2k+\alpha+\beta+1)(\alpha+1)_k (-N+1)_k (N-1)!}.$$

Further, Dette [Det95] proved that for $\alpha + \beta > -1$ and

$$k \leq k(\alpha, \beta, N-1) := -\frac{1}{2}\left(\alpha+\beta-1-\sqrt{(\alpha+\beta+1)(\alpha+\beta+2N-1)}\right)$$

the Hahn polynomials are bounded by

$$\max_{x \in [0,N-1]} |Q_k(x, \alpha, \beta, N-1)| \leq \max\left\{1, \frac{(\beta+1)_k}{(\alpha+1)_k}\right\}.$$

Here, we choose $\alpha = \beta = 0$, which results in the *discrete Chebyshev polynomials* on $[0, N-1]$. We normalize and transform them to the interval $[a, b]$, resulting in the polynomials

$$\pi_k(\cdot; N) : [a,b] \to \mathbb{R}, \quad \pi_k(x; N) = \frac{1}{\sqrt{h_k}} Q_k\left(\frac{N-1}{b-a}(x-a), 0, 0, N-1\right). \tag{5.14}$$

These polynomials form a basis of DOPs with respect to the discrete inner product $[u, v]_N = \sum_{n=1}^N u(x_n)v(x_n)$, assuming equidistant quadrature points $\{x_n\}_{n=1}^N$ in $[a, b]$. Further, for

$$k \leq k(N) := \frac{1}{2}\left(1 + \sqrt{2N-1}\right) \tag{5.15}$$

they are bounded by

$$\max_{x \in [a,b]} |\varphi_k(x)| \leq \frac{1}{\sqrt{h_k}}. \tag{5.16}$$

Finally, we note that

$$\begin{aligned}
(k+1)_N &= (k+1)(k+1)\dots(k+N) \\
&= \frac{(N+k)!}{k!}, \\
(-N+1)_k &= (-N+1)(-N+2)\dots(-N+k) \\
&= (-1)^k (N-1)(N-2)\dots(N-k) \\
&= (-1)^k \frac{(N-1)!}{(N-k-1)!}.
\end{aligned}$$

Thus, we have

$$h_k = \frac{(-1)^k (k+1)_N k!}{(2k+1)(-N+1)_k (N-1)!} = \frac{(N+k)!(N-k-1)!}{(2k+1)(N-1)!(N-1)!} \tag{5.17}$$

for $\alpha = \beta = 0$.

5.3.3 Proofs of the main results

Now, let us prove our main results, i. e. Theorems 5.5 and 5.6, and therefore stability of LS-QRs for general weight functions.

Theorem 5.7 (Theorem 5.5)
Let $d \in \mathbb{N}$ and let Q_N be an N-point LS-QR with equidistant quadrature points in $[a,b]$, degree of exactness d, and quadrature weights given by (5.12). Then, the sequence of LS-QRs $(Q_N)_{N>d}$ is stable; that is

$$\sup_{N>d} \kappa(\mathbf{w}_N^{\mathrm{LS}}) < \infty$$

holds.

Proof. Let $d \in \mathbb{N}$ and $N > d$. First, we note that

$$\begin{aligned}
\kappa(\mathbf{w}_N^{\mathrm{LS}}) &= \sum_{n=1}^{N} |w_n^{\mathrm{LS}}| \\
&\overset{(5.12)}{=} \sum_{n=1}^{N} \left| \sum_{k=0}^{d} \pi_k(x_n;N) I[\pi_k(\cdot;N)] \right| \\
&\leq \sum_{n=1}^{N} \sum_{k=0}^{d} |\pi_k(x_n;N)| \, |I[\pi_k(\cdot;N)]|
\end{aligned}$$

and that the DOPs $\pi_k(\cdot;N)$ are given by the transformed and normalized discrete Chebyshev polynomials; see (5.14). Hence, a 'quick and dirty' estimate provides us with

$$\begin{aligned}
|\pi_k(x_n;N)| &\leq \max_{x\in[a,b]} |\pi_k(x;N)| =: \|\pi_k(\cdot;N)\|_\infty, \\
|I[\pi_k(\cdot;N)]| &\leq \int_a^b |\pi_k(x;N)| \, |\omega(x)| \, \mathrm{d}x \leq \|\pi_k(\cdot;N)\|_\infty K_\omega.
\end{aligned}$$

Thus, we have

$$\kappa(\mathbf{w}_N^{\mathrm{LS}}) \leq \sum_{n=1}^{N} \sum_{k=0}^{d} \|\pi_k(\cdot;N)\|_\infty^2 K_\omega = N K_\omega \sum_{k=0}^{d} \|\pi_k(\cdot;N)\|_\infty^2 .$$

Choosing $N \geq N(d) := \frac{1}{2}[(2d-1)^2+1]$, condition (5.15) is satisfied for all $k = 0,\dots,d$ and the DOPs $\pi_k(\cdot;N)$ are bounded by

$$\|\pi_k(\cdot;N)\|_\infty \leq \frac{1}{\sqrt{h_k}}, \tag{5.18}$$

see (5.16). Further, this yields

$$\begin{aligned}
\|\pi_k(\cdot;N)\|_\infty^2 &\leq \frac{1}{h_k} \\
&\overset{(5.17)}{=} \frac{(2k+1)(N-1)!(N-1)!}{(N+k)!(N-k-1)!} \\
&= \frac{(2k+1)(N-1)!k!}{(N+k)!} \binom{N-1}{k}.
\end{aligned} \tag{5.19}$$

Next, we note the simple inequalities

$$\binom{N-1}{k} \leq \frac{(N-1)^k}{k!}$$

and

$$\begin{aligned}\frac{(N-1)!}{(N+k)!} &= \frac{(N-1)!}{(N-1)!N(N+1)\dots(N+k)} \\ &= \frac{1}{N(N+1)\dots(N+k)} \\ &\leq \frac{1}{N^{k+1}}.\end{aligned}$$

Incorporating these inequalities into (5.19), we get

$$\begin{aligned}\|\pi_k(\cdot;N)\|_\infty^2 &\leq \frac{(2k+1)(N-1)!k!}{(N+k)!}\binom{N-1}{k} \\ &\leq \frac{(2k+1)k!(N-1)^k}{N^{k+1}k!} \\ &< \frac{(2k+1)}{N},\end{aligned}$$

yielding

$$\begin{aligned}\kappa(\mathbf{w}_N^{\text{LS}}) &\leq NK_\omega \sum_{k=0}^{d} \|\pi_k(x_n;N)\|_\infty^2 \\ &\leq NK_\omega \sum_{k=0}^{d} \frac{(2k+1)}{N} \\ &= K_\omega \sum_{k=0}^{d}(2k+1) \\ &= K_\omega(d+1)^2\end{aligned}$$

for $N \geq N(d)$. Finally, the claim follows from

$$\kappa(\mathbf{w}_N^{\text{LS}}) \leq \max\left\{\kappa(\mathbf{w}_{d+1}^{\text{LS}}),\dots,\kappa(\mathbf{w}_{N(d)-1}^{\text{LS}}), K_\omega(d+1)^2\right\}.$$

□

Remark 5.8. We note that in the above proof we have essentially bounded the stability value $\kappa(\mathbf{w}_N^{\text{LS}})$ by $K_\omega(d+1)^2$. For increasing degree of exactness d, this constant becomes fairly large and one might say that the LS-QRs are only *pseudostable* for general weights. In this case, round-off errors are still bounded, but might be magnified by a large factor. In our numerical tests, however, we have observed the stability value $\kappa(\mathbf{w}_N^{\text{LS}})$ to converge to the continuous bound K_ω given by (5.2) for all degrees of exactness d. This indicates that the LS-QRs have even more desirable stability properties than we have proven in Theorem 5.5.

As already stressed above, the proof of Theorem 5.5 relies on the bound

$$\|\pi_k(\cdot;N)\|_\infty^2 \leq \frac{2k+1}{N}$$

to hold for the DOPs on equidistant points and for sufficiently large N. Unfortunately, we have not been able to find similar results for DOPs on nonequidistant points in the literature. Yet, by assuming a similar bound, we can also prove

Theorem 5.9 (Theorem 5.6)
Let $d \in \mathbb{N}$ and let Q_N be an N-point LS-QR with distinct quadrature points in $[a,b]$, degree of exactness d, and quadrature weights given by (5.12). Moreover, let $\{\pi_k(\cdot;N)\}_{k=0}^d$ be the basis

of DOPs corresponding to the discrete inner product $[\cdot,\cdot]_N$ *and assume that there are constants* $C(d) > 0$ *and* $N(d) > d$ *such that*

$$\|\pi_k(\cdot;N)\|_\infty^2 \leq \frac{C(d)}{N} \quad \forall k = 0,\dots,d \;\; \forall N \geq N(d). \tag{5.20}$$

Then, the sequence of LS-QRs $(Q_N)_{N>d}$ *is stable; that is*

$$\sup_{N>d} \kappa(\mathbf{w}_N^{\mathrm{LS}}) < \infty$$

holds.

Proof. The proof is analogous to the one of Theorem 5.5. Let $d \in \mathbb{N}$ and $N > d$. Again, we have

$$\kappa(\mathbf{w}_N^{\mathrm{LS}}) \leq N K_\omega \sum_{k=0}^{d} \|\pi_k(\cdot;N)\|_\infty^2 .$$

Assuming there are constants $C(d) > 0$ and $N(d) > d$ such that (5.20) holds yields

$$\|\pi_k(\cdot;N)\|_\infty^2 \leq \frac{C(d)}{N}$$

for all $k = 0,\dots,d$ and $N \geq N(d)$. Thus, we get

$$\begin{aligned}
\kappa(\mathbf{w}_N^{\mathrm{LS}}) &\leq N K_\omega \sum_{k=0}^{d} \|\pi_k(\cdot;N)\|_\infty^2 \\
&\leq N K_\omega \sum_{k=0}^{d} \frac{C(d)}{N} \\
&\leq K_\omega (d+1) C(d)
\end{aligned}$$

for $N \geq N(d)$. Once more, the claim follows from

$$\kappa(\mathbf{w}_N^{\mathrm{LS}}) \leq \max\left\{\kappa(\mathbf{w}_{d+1}^{\mathrm{LS}}),\dots,\kappa(\mathbf{w}_{N(d)-1}^{\mathrm{LS}}), K_\omega(d+1)C(d)\right\}.$$

□

5.4 Nonnegative least squares quadrature rules

Next, let us construct sign-consistent QRs with high degrees of exactness for scattered data and general weight functions. The idea for the construction of stable LS-QRs in Chapter 5.3 was to ensure a fixed degree of exactness d and to optimize the stability value $\kappa(\mathbf{w}_N)$ then. The idea behind our construction of sign-consistent QRs is somewhat reversed. First, we ensure that the QR is sign-consistent and only afterwards we 'optimize' the degree of exactness. This approach has already been proposed by Huybrechs [Huy09] for nonnegative weight functions. Here, we extend it to general weight functions.

5.4.1 The nonnegative least squares problem

The so-called *NNLS* problem is a constrained LS problem, where the solution entries are not allowed to become negative. Following [LH95], the problem is to

$$\text{minimize } \|B\mathbf{x} - \mathbf{c}\|_2 \quad \text{subject to} \quad x_n \geq 0, \; n = 1, \dots, N, \tag{5.21}$$

with $B \in \mathbb{R}^{M \times N}$, $\mathbf{x} \in \mathbb{R}^N$, and $\mathbf{c} \in \mathbb{R}^M$. Moreover, $\|\cdot\|_2$ again denotes the Euclidean norm, this time in $\mathbb{R}^M$. The NNLS problem (5.21) always has a solution [LH95]. Note that in our setting, the matrix B is underdetermined, which results in infinitely many solutions. In this case, we sought a solution with maximal sparsity, i. e. the solution vector has as many zero entries as possible.

5.4.2 Formulation as a nonnegative least squares problem

When concerned with sign-consistent QRs with degree of exactness d for integrals with general weight functions ω, we want to

$$\text{minimize } \|A\mathbf{w} - \mathbf{m}\|_2 \tag{5.22}$$

subject to mixed inequality constraints

$$\begin{cases} w_n \geq 0 & \text{if} \quad \omega(x_n) \geq 0 \\ w_n < 0 & \text{if} \quad \omega(x_n) < 0 \end{cases}, \quad n = 1, \dots, N. \tag{5.23}$$

Again, the matrix A and the vector $\mathbf{m}$ are respectively given by (4.2) and (4.3) and contain the function values of the basis elements at the quadrature points and the corresponding moments. Nevertheless, we are able to formulate this problem as a usual NNLS problem (5.21) by introducing an auxiliary *sign-matrix*

$$S = \operatorname{diag}\left(\operatorname{sgn}\left(\omega(x_1)\right), \dots, \operatorname{sgn}\left(\omega(x_N)\right)\right).$$

Then, the LS problem (5.22) with mixed inequality constraints (5.23) can be compactly rewritten as

$$\text{minimize } \|A\mathbf{w} - \mathbf{m}\|_2 \quad \text{subject to} \quad S\mathbf{w} \geq 0, \tag{5.24}$$

where $S\mathbf{w} \geq 0$ is understood in an element-wise sense. Finally, by going over to an auxiliary vector

$$\mathbf{u} := S\mathbf{w},$$

problem (5.24) can be reformulated as a usual NNLS problem,

$$\text{minimize } \|AS\mathbf{u} - \boldsymbol{m}\|_2 \quad \text{subject to} \quad u_n \geq 0, \; n = 1, \dots, N, \tag{5.25}$$

since $\mathbf{w} = S^{-1}\mathbf{u} = S\mathbf{u}$. Let us denote the vector of quadrature weights which result from the NNLS problem (5.25) by $\mathbf{w}_N^{\text{NNLS}}$. To the corresponding QR

$$Q_N[f] := \sum_{n=1}^{N} w_n^{\text{NNLS}} f(x_n)$$

we refer to as the N-points *NNLS-QR* with quadrature points $\{x_n\}_{n=1}^N$ and with *approximate degree of exactness d*. Note that if

$$0 < ||A\mathbf{w} - \mathbf{m}||_2 < \varepsilon,$$

the NNLS-QR is not exact for all polynomials up to degree d. Thus, we say that Q_N has "approximate degree of exactness d". Yet, for $\varepsilon \approx 0$, we can expect the NNLS-QR to only have small errors. This is further discussed in Chapter 5.5.

5.5 Numerical results

Finally, we numerically investigate the proposed LS-QR and NNLS-QR. We do so for two different weight functions

$$\omega(x) = x\sqrt{1-x^3} \quad \text{and} \quad \omega(x) = \cos(20\pi x) \tag{5.26}$$

with mixed signs on $[-1, 1]$ as well as for equidistant and scattered quadrature points. The scattered quadrature points are obtained by adding white Gaussian noise to the set of equidistant quadrature points $\{x_n\}_{n=1}^N$. Thus, the scattered quadrature points $\{\tilde{x}_n\}_{n=1}^N$ are given by

$$\tilde{x}_1 = -1, \quad \tilde{x}_N = 1, \quad \tilde{x}_n = x_n + Z_n \quad \text{with} \quad Z_n \in \mathcal{N}\left(0; (4N)^{-2}\right), \tag{5.27}$$

for $n = 2, \dots, N-1$, where the Z_n are independent, identically distributed, and assumed to not be correlated with the x_n.

5.5.1 Implementation details

The subsequent numerical tests have been computed in MATLAB with double precision. This yields a machine precision of 2^{-52} or approximately $2.22 \cdot 10^{-16}$. Further, the LS quadrature weights given by (4.8) are computed as follows:

1. Matrix A and therefore the nodal values $\pi_k(x_n; N)$ are computed by applying the Gram–Schmidt process to an initial basis of Legendre polynomials.
2. The moments $I[\pi_k(\cdot; N)]$ are computed by determining the nodal values of the DOPs $\pi_k(\cdot; N)$ (again by the Gram–Schmidt process) and of the weight function ω at a set of J GLe points and utilizing the GLe-QR applied to the integrand $\pi_k(\cdot; N)\omega(\cdot)$ then.
3. Finally, the LS quadrature weights are obtained by summing up the products $\pi_k(x_n; N)I[\pi_k(\cdot; N)]$ over k; see (4.8).

The NNLS quadrature weights, on the other hand, are computed by the available MATLAB implementation (*lsqnonneg*) of the following NNLS algorithm [LH95, Chapter 23]. Also see [Huy09] or the MATLAB documentation of *lsqnonneg*.

5.5.2 Stability

We start by investigating stability of the proposed QRs. Figure 5.1 illustrates the stability measure $\kappa(\mathbf{w}_N)$ of the LS-QR and the NNLS-QR with degree of exactness $d = 10, 20$ and an increasing number N of equidistant as well as scattered quadrature points.

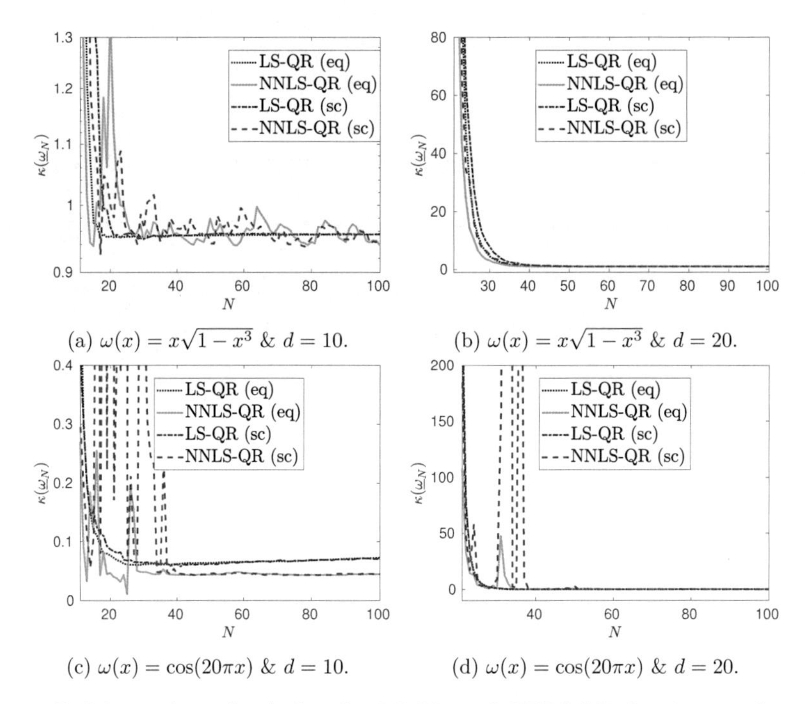

(a) $\omega(x) = x\sqrt{1-x^3}$ & $d = 10$.

(b) $\omega(x) = x\sqrt{1-x^3}$ & $d = 20$.

(c) $\omega(x) = \cos(20\pi x)$ & $d = 10$.

(d) $\omega(x) = \cos(20\pi x)$ & $d = 20$.

Figure 5.1: Stability value $\kappa(\mathbf{w}_N)$ for the LS-QR and NNLS-QR for degree of exactness $d = 10, 20$, equidistant (eq) as well as scattered (sc) quadrature points, and two different weight functions.

In accordance to Theorem 5.5 (and Theorem 5.6), we note that the stability value for the LS-QR is bounded for increasing N in all cases. Further, we can observe that the stability value seems to converge to the bound K_ω given by the continuous integral in (5.2). This indicates that the LS-QR might have even more desirable stability properties than we have proven in Theorem 5.5. A sharper bound which shows the convergence of $\kappa(\mathbf{w}_N^{\text{LS}})$ to K_ω would be highly desirable. We look forward to more refined results in this direction. A similar behavior can be observed for the NNLS-QR on equidistant as well as scattered quadrature points. Yet, the NNLS-QR shows a slightly oscillatory profile in some tests. Finally, we note that in Figure 5.1(c) the NNLS-QR provides even better values for the stability value than the LS-QR.

5.5.3 Sign-consistency

Next, we investigate sign-consistency of the proposed LS-QR and NNLS-QR. We measure sign-consistency by the following *sign-consistency measure*

$$S_\omega(\mathbf{w}_N) = \frac{1}{N} \sum_{n=1}^{N} |\text{sgn}(w_n) - \text{sgn}(\omega(x_n))| \,. \tag{5.28}$$

It is evident that an N-point QR Q_N with quadrature weights $\mathbf{w}_N$ is sign-consistent if and only if $S_\omega(\mathbf{w}_N) = 0$ holds. Figure 5.2 illustrates the sign-consistency measure for the LS-QR and the NNLS-QR for the two weight functions (5.26), degrees of exactness $d = 10, 20$, and equidistant as well as scattered quadrature points.

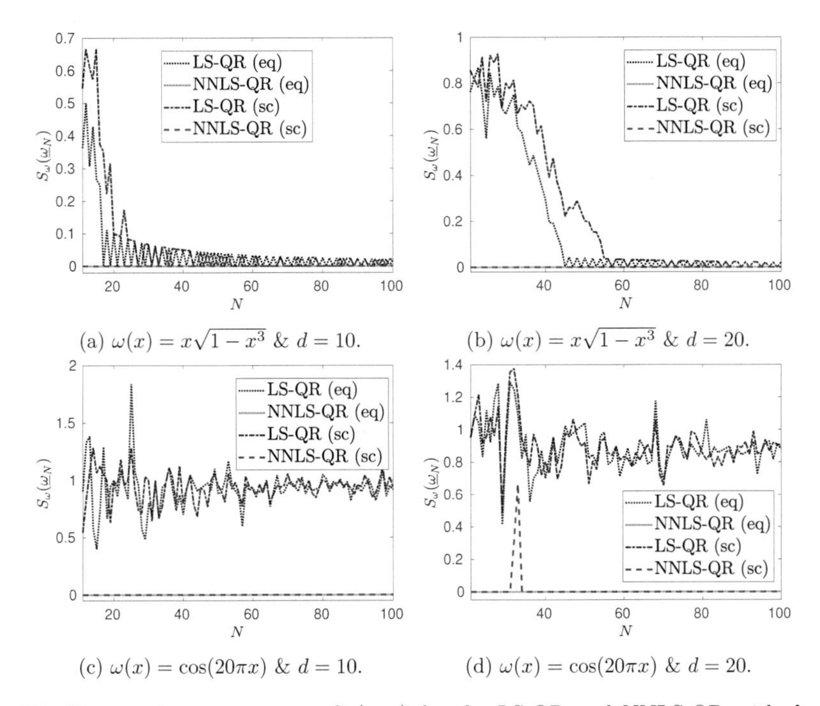

(a) $\omega(x) = x\sqrt{1-x^3}$ & $d = 10$.

(b) $\omega(x) = x\sqrt{1-x^3}$ & $d = 20$.

(c) $\omega(x) = \cos(20\pi x)$ & $d = 10$.

(d) $\omega(x) = \cos(20\pi x)$ & $d = 20$.

Figure 5.2: Sign-consistency measure $S_\omega(\mathbf{w}_N)$ for the LS-QR and NNLS-QR with degree of exactness $d = 10, 20$, equidistant (eq) as well as scattered (sc) quadrature points, and two different weight functions.

We can observe that, in fact, the NNLS-QR always provides a sign-consistency measure of $S_\omega(\mathbf{w}_N) = 0$ (except for a short outlier in Figure 5.2(d)). The LS-QR, on the other hand, is demonstrated to violate sign-consistency in many cases. Together with our observations from Chapter 5.5.2, this indicates that the sign-consistent NNLS-QR might provide overall superior stability properties than the LS-QR.

5.5.4 Exactness

Next, we investigate exactness of the LS-QR and the NNLS-QR. Figure 5.3 displays the *exactness measure*

$$||A\mathbf{w}_N - \mathbf{m}||_2$$

of both QRs for the same set of parameters as before. Note that for an N-point QR Q_N with quadrature weights $\mathbf{w}_N$ the minimal value for the exactness measure, $||A\mathbf{w}_N - \mathbf{m}||_2 = 0$, holds if and only if Q_N has degree of exactness d.

In accordance to their construction as LS solutions of the linear system of exactness conditions (3.51), the LS-QR provides a minimal value of $||A\mathbf{w}_N - \mathbf{m}||_2 = 0$ in all cases. The NNLS-QRs, on the other hand, violate the degree of exactness and are demonstrated to have a nonzero exactness measure if small numbers N of quadrature points are used. Yet, for fixed degree of exactness d and increasing N, also the NNLS-QR provides degree of exactness d.

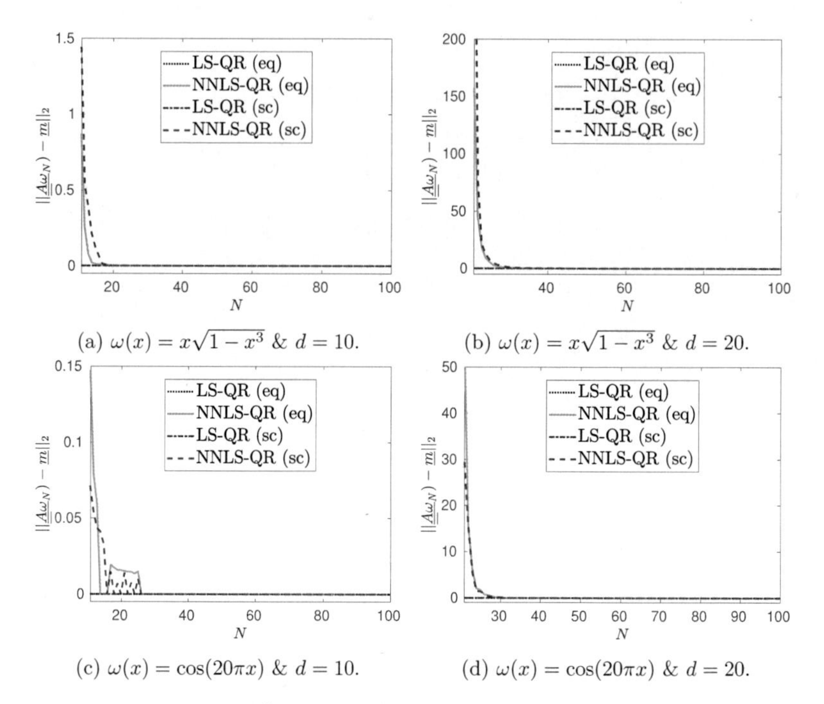

(a) $\omega(x) = x\sqrt{1-x^3}$ & $d = 10$.

(b) $\omega(x) = x\sqrt{1-x^3}$ & $d = 20$.

(c) $\omega(x) = \cos(20\pi x)$ & $d = 10$.

(d) $\omega(x) = \cos(20\pi x)$ & $d = 20$.

Figure 5.3: Exactness measure $||A\mathbf{w}_N - \mathbf{m}||_2$ for the LS-QR and NNLS-QR with degree of exactness $d = 10, 20$, equidistant (eq) as well as scattered (sc) quadrature points, and two different weight functions.

5.5.5 Accuracy for increasing N

We now come to investigate accuracy of the LS-QR and the NNLS-QR for fixed degrees of exactness d and an increasing number N of quadrature points. We compare both QRs with a generalized composite trapezoidal rule applied to the unweighted Riemann integral with integrator $f\omega$. Further, we consider the two test functions

$$f(x) = |x|^3 \quad \text{and} \quad f(x) = e^x. \tag{5.29}$$

In the following, the errors

$$|Q_N[f] - I[f]|$$

for fixed degrees of exactness d and an increasing number N of quadrature points are illustrated. Figure 5.4 shows the errors for the first test function $f(x) = |x|^3$ and equidistant quadrature points while Figure 5.5 provides the errors for the second test function $f(x) = e^x$ and equidistant quadrature points. Figure 5.6 and Figure 5.7 illustrate the same errors for scattered quadrature points. These are once more constructed by adding white Gaussian noise to the set of equidistant quadrature points; see (5.27).

In Figure 5.4, where the first test function $f(x) = |x|^3$ is investigated for equidistant quadrature points, the LS-QR and NNLS-QR are demonstrated to provide more accurate results than the generalized composite trapezoidal rule in nearly all cases. Yet, the NNLS-QR is observed to oscillate in their accuracy and to be less accurate for small numbers N of quadrature points.

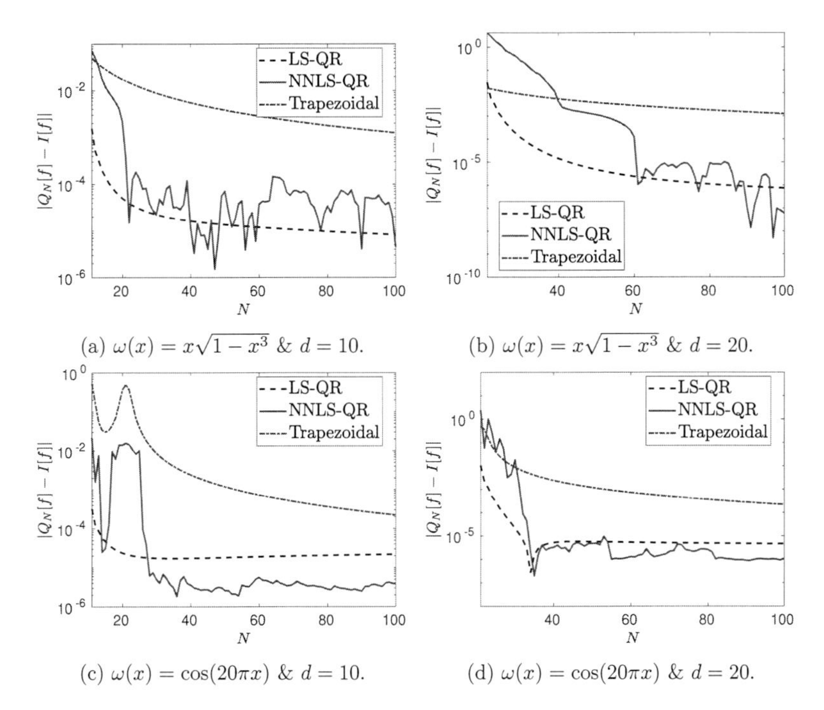

(a) $\omega(x) = x\sqrt{1-x^3}$ & $d = 10$.

(b) $\omega(x) = x\sqrt{1-x^3}$ & $d = 20$.

(c) $\omega(x) = \cos(20\pi x)$ & $d = 10$.

(d) $\omega(x) = \cos(20\pi x)$ & $d = 20$.

Figure 5.4: Errors for the generalized composite trapezoidal rule, the LS-QR, and NNLS-QR with degree of exactness $d = 10, 20$ on equidistant quadrature points. Test function $f(x) = |x|^3$.

Note that for an insufficiently small number N of quadrature points, the exactness condition (3.41) is not satisfied by the NNLS-QR and therefore not even constants might be handled accurately. For a sufficiently large number of quadrature points, however, the NNLS-QR provides similar accurate (sometimes even more accurate) results as the LS-QR.

Figure 5.5 demonstrates that the difference in accuracy between the generalized composite trapezoidal rule and the two proposed QRs becomes even more noticeable when the second test function $f(x) = e^x$ is investigated on equidistant quadrature points. The LS-QR and the NNLS-QR yield highly accurate results which are up to 10^{12} times more accurate than the ones obtained by the generalized composite trapezoidal rule at the same set of quadrature points. Again, the NNLS-QR shows oscillations in the accuracy and only becomes highly accurate when a sufficiently large number of quadrature points is used.

Next, in Figure 5.6, we consider the test function $f(x) = |x|^3$ on scattered quadrature points. For scattered quadrature points the results and differences between the different QRs become less clear. Yet, the LS-QR and the NNLS-QR are observed to provide more accurate results than the generalized composite trapezoidal rule in most cases. Figure 5.7 illustrates the results for the second test function $f(x) = e^x$ on scattered quadrature points. In this test, once more, the LS-QR and NNLS-QR (for sufficiently great N) are observed to provide significantly more accurate results than the generalized composite trapezoidal rule. Even on scattered quadrature points, the LS-QR and the NNLS-QR are up to 10^{12} times more accurate. Yet, the oscillations in the accuracy of the NNLS-QR get heavier on scattered quadrature points. Again, the NNLS-QR is demonstrated to provide highly accurate results only when a sufficiently great

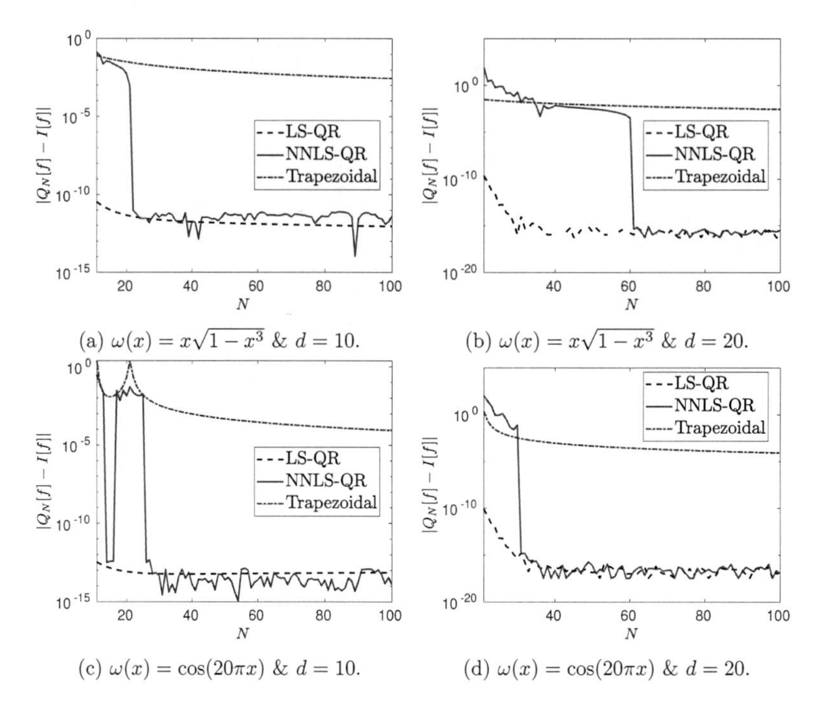

(a) $\omega(x) = x\sqrt{1-x^3}$ & $d = 10$. (b) $\omega(x) = x\sqrt{1-x^3}$ & $d = 20$.

(c) $\omega(x) = \cos(20\pi x)$ & $d = 10$. (d) $\omega(x) = \cos(20\pi x)$ & $d = 20$.

Figure 5.5: Errors for the generalized composite trapezoidal rule, the LS-QR, and NNLS-QR with degree of exactness $d = 10, 20$ on equidistant quadrature points. Test function $f(x) = e^x$.

number N of quadrature points is used.

5.5.6 Accuracy for increasing d

Above, we have considered accuracy for fixed degree of exactness d and an increasing number N of quadrature points. Here, we investigate accuracy for an increasing degree of exactness d and adaptively choose $N = N(d) = \frac{1}{2}\left[(2d-1)^2 + 1\right]$. This choice is motivated by Theorem 5.5. The most crucial step in the proof of Theorem 5.5 has been inequality (5.18), which holds for all $k = 0, \dots, d$ if $N \geq N(d) = \frac{1}{2}\left[(2d-1)^2 + 1\right]$ is chosen.

Figure 5.8 illustrates the results of the LS-QR, the NNLS-QR, and the generalized composite trapezoidal rule for both test functions and weight functions as before on equidistant quadrature points. We observe that the LS-QR and the NNLS-QR provide more accurate results than the generalized composite trapezoidal rule in all cases. These are up to 10^{12} times more accurate.

Similar observations can be made for scattered quadrature points and are displayed in Figure 5.9. Using the same set of equidistant or scattered quadrature points, the LS-QR and the NNLS-QR are able to provide highly accurate results even for general weight functions.

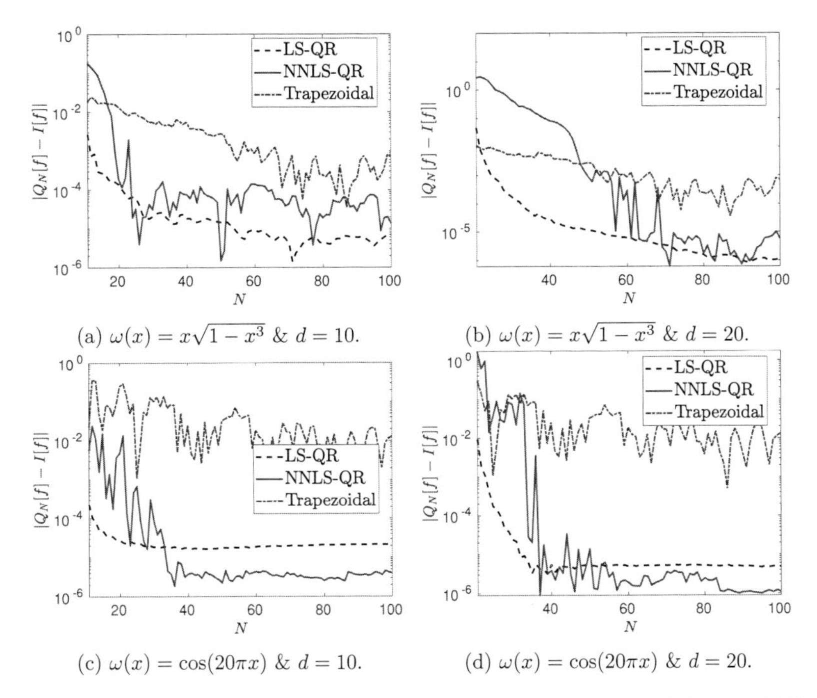

(a) $\omega(x) = x\sqrt{1-x^3}$ & $d = 10$.

(b) $\omega(x) = x\sqrt{1-x^3}$ & $d = 20$.

(c) $\omega(x) = \cos(20\pi x)$ & $d = 10$.

(d) $\omega(x) = \cos(20\pi x)$ & $d = 20$.

Figure 5.6: Errors for the generalized composite trapezoidal rule, the LS-QR, and NNLS-QR with degree of exactness $d = 10, 20$ on scattered quadrature points. Test function $f(x) = |x|^3$.

5.5.7 Ratio between d and N

Finally, we address the ratio between the degree of exactness d and the number of equidistant quadrature points N that is needed for stability (and exactness) for the LS-QRs (and NNLS-QRs). In [Wil70a], Wilson showed that stability of LS-QRs for the weight function $\omega \equiv 1$ essentially is an d^2 process; that is $N \approx Cd^2$ equidistant quadrature points are needed for $\kappa(\mathbf{w}_N^{\text{LS}}) = K_\omega$ to hold. In [Huy09] and [Gla19a], the same ratio between d and N has been observed for other positive weight functions, including $\omega(x) = 1 - x^2$ and $\omega(x) = \sqrt{1-x^2}$. Here, we demonstrate that a similar ratio also holds for more general weight functions.

$\omega(x)$	1	$1-x^2$	$\sqrt{1-x^2}$	$x\sqrt{1-x^3}$	$\cos(20\pi x)$
s	1.65	1.45	1.56	1.63	1.94
C	0.22	0.32	0.25	0.26	0.08

Table 5.1: LS fit of the parameters C and s in the model $N = Cd^s$ for the LS-QR and different weight functions ω. N refers to the minimal number of equidistant quadrature points such that the quadrature weights of the LS-QR with degree of exactness d satisfy $\kappa(\mathbf{w}_N^{\text{LS}}) \leq 2K_\omega$.

Table 5.1 lists the results of a LS fit of the parameters C and s in the model $N = Cd^s$ for the LS-QR and different weight functions ω. Here, N refers to the minimal number of

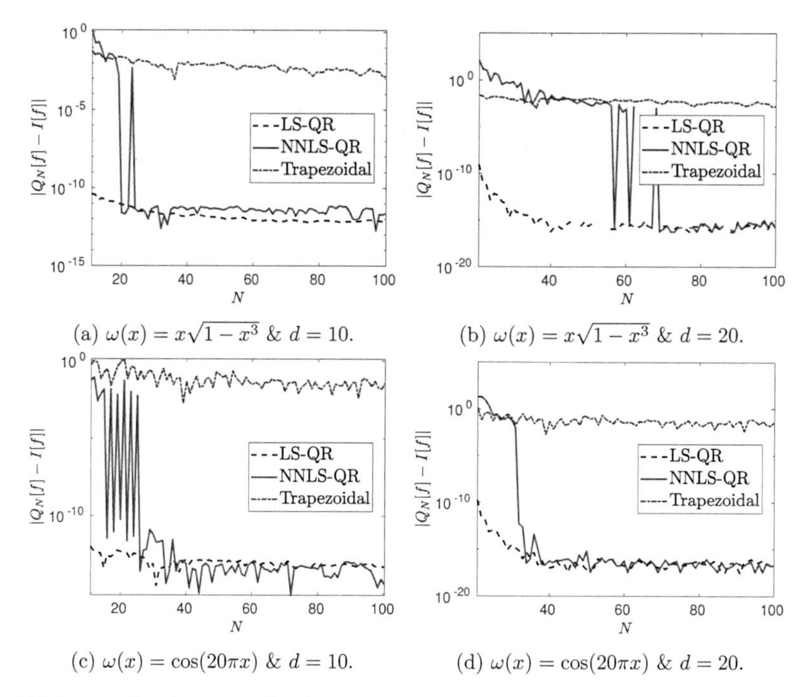

(a) $\omega(x) = x\sqrt{1-x^3}$ & $d = 10$.

(b) $\omega(x) = x\sqrt{1-x^3}$ & $d = 20$.

(c) $\omega(x) = \cos(20\pi x)$ & $d = 10$.

(d) $\omega(x) = \cos(20\pi x)$ & $d = 20$.

Figure 5.7: Errors for the generalized composite trapezoidal rule, the LS-QR, and NNLS-QR with degree of exactness $d = 10, 20$ on scattered quadrature points. Test function $f(x) = e^x$.

equidistant quadrature points such that the quadrature weights of the LS-QR with degree of exactness d satisfy $\kappa(\mathbf{w}_N^{\text{LS}}) \leq 2K_\omega$, where d ranges from 0 to 40. Note that for general weight functions we have only been able to prove that $\kappa(\mathbf{w}_N^{\text{LS}})$ is uniformly bounded with respect to N. Yet, in contrast to positive weight functions [Gla19a], it is not ensured that $\kappa(\mathbf{w}_N^{\text{LS}}) \leq K_\omega$ holds for a sufficiently large N. In all numerical test, we observed $\kappa(\mathbf{w}_N^{\text{LS}}) \leq 2K_\omega$ to hold for a sufficiently large N, however. In particular, this yields stability of the LS-QR. As a result of using $\kappa(\mathbf{w}_N^{\text{LS}}) \leq 2K_\omega$ instead of $\kappa(\mathbf{w}_N^{\text{LS}}) \leq K_\omega$ as a criterion to determine N, we observe slightly smaller exponent parameters $s < 2$ in Table 5.1. Yet — and more important — we can note from Table 5.1 that the ratios between d and N are similar for positive and general weight functions. Similar ratios also hold for the NNLS-QR and are reported in Table 5.2. Here, N refers to the minimal number of equidistant quadrature points such that the NNLS-QR with

$\omega(x)$	1	$1-x^2$	$\sqrt{1-x^2}$	$x\sqrt{1-x^3}$	$\cos(20\pi x)$
s	1.76	1.66	1.70	1.66	1.68
C	0.19	0.30	0.35	0.41	0.27

Table 5.2: LS fit of the parameters C and s in the model $N = Cd^s$ for the NNLS-QR and different weight functions ω. N refers to the minimal number of equidistant quadrature points such that the quadrature weights of the NNLS-QR with approximate degree of exactness d satisfy $\kappa(\mathbf{w}_N^{\text{NNLS}}) \leq 2K_\omega$ as well as $||A\mathbf{w}_N^{\text{NNLS}} - \mathbf{m}||_2 \leq 10^{-14}$.

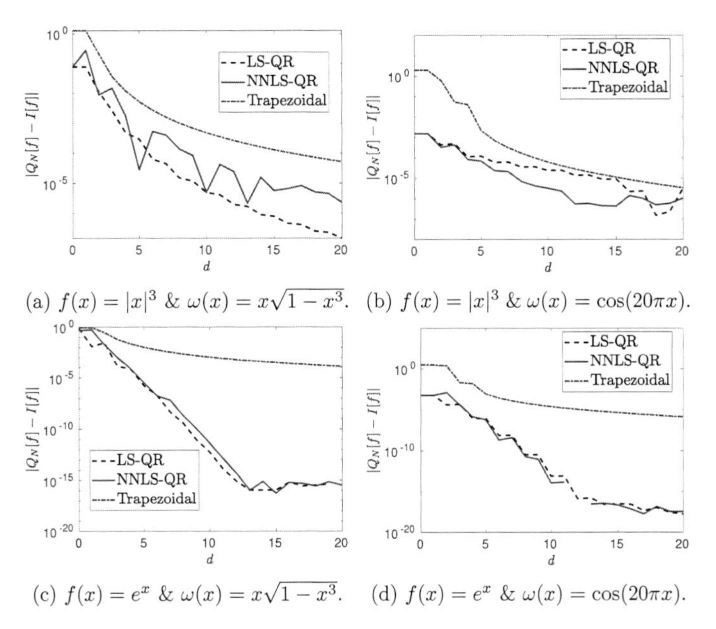

(a) $f(x) = |x|^3$ & $\omega(x) = x\sqrt{1 - x^3}$. (b) $f(x) = |x|^3$ & $\omega(x) = \cos(20\pi x)$.

(c) $f(x) = e^x$ & $\omega(x) = x\sqrt{1 - x^3}$. (d) $f(x) = e^x$ & $\omega(x) = \cos(20\pi x)$.

Figure 5.8: Errors for the generalized composite trapezoidal rule, the LS-QR, and NNLS-QR on equidistant quadrature points.

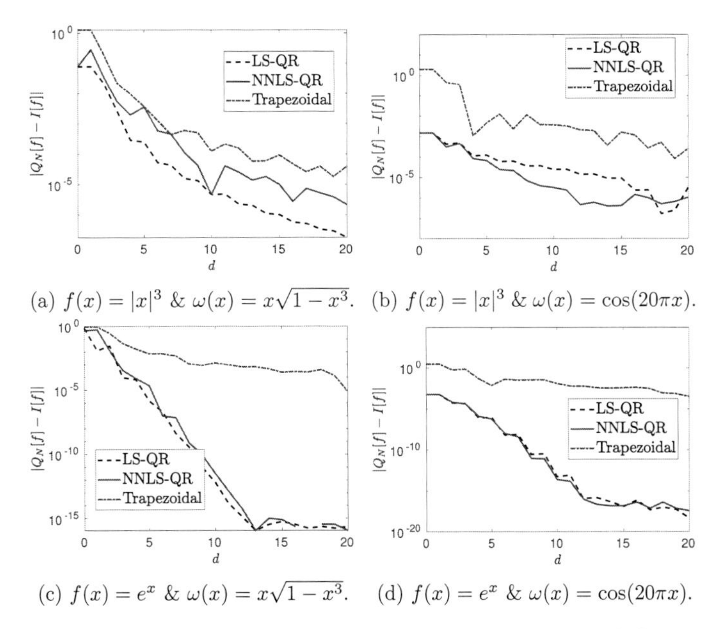

(a) $f(x) = |x|^3$ & $\omega(x) = x\sqrt{1 - x^3}$. (b) $f(x) = |x|^3$ & $\omega(x) = \cos(20\pi x)$.

(c) $f(x) = e^x$ & $\omega(x) = x\sqrt{1 - x^3}$. (d) $f(x) = e^x$ & $\omega(x) = \cos(20\pi x)$.

Figure 5.9: Errors for the generalized composite trapezoidal rule, the LS-QR, and NNLS-QR on scattered quadrature points.

approximate degree of exactness d is stable as well as exact. Again, stability is checked by the condition $\kappa(\mathbf{w}_N^{\text{NNLS}}) \leq 2K_\omega$. Exactness, on the other hand, is checked by the condition $||A\mathbf{w}_N^{\text{NNLS}} - \mathbf{m}||_2 \leq 10^{-14}$. We use the bound 10^{-14} instead of 0 to encounter possible round-off errors in our implementation; see Chapter 5.5.1.

5.6 Concluding thoughts and outlook

In this chapter, we have investigated stability of QRs for integrals with general weight functions, possibly having mixed signs. Such weight functions arise, for instance, in the weak form of the Schrödinger equation and for (highly) oscillatory integrals. In contrast to nonnegative weight functions, stability of QRs for general weight functions is not guaranteed by nonnegative-only quadrature weights anymore (as discussed in Chapter 3.4). Therefore, we have proposed to treat stable QRs (for which round-off errors due to inexact arithmetics are uniformly bounded with respect to the number N of quadrature points) and sign-consistent QRs (for which the signs of the quadrature weights match the signs of the weight function at the corresponding quadrature points) as two separated classes. Moreover, we have proposed two different procedures to construct such QRs for general weight functions. In particular, our procedures allow us to construct stable high order QRs on equidistant and even scattered quadrature points. This is especially beneficial since in many applications it can be impractical, if not even impossible, to obtain data to fit known QRs. Numerical tests demonstrate that both QRs, are able to provide highly accurate results. Future work will focus on the extension of the proposed QRs to higher dimensions.

CHAPTER 6

High order numerical methods for conservation laws

In the last decades, great efforts have been made to develop accurate and stable numerical methods for time dependent PDEs, especially for hyperbolic CLs (2.2). Traditionally, low order numerical schemes, for instance classical finite volume (FV) methods, have been used to solve hyperbolic CLs, particularly in industrial applications. But since they become quite costly for high accuracy or long time simulations, there is a rising demand of high order (third and above) methods. These have the potential of providing accurate solutions at reasonable costs. In this chapter, we briefly review two modern high order methods for the numerical treatment of hyperbolic CLs. Up to now, many different methods have been proposed and investigated. We mention FD methods, FV methods, finite element (FE) methods, (weighted) essentially nonoscillatory methods (WENO), spectral methods, RBF methods, continuous Galerkin (CG) methods, spectral difference (SD) methods, active flux (AF) methods, residual distribution (RD) methods, and countless more. Of course, addressing all of these methods would heavily exceed the scope of this work. Thus, we restrict ourself to DG and FR methods. Both methods can be considered as two state-of-the-art high order schemes. It will also be these two methods for which most of the shock capturing strategies are developed and investigated in the later chapters. Both methods utilize a partition of the computational domain into smaller elements and approximate the solution by piecewise polynomials, which are allowed to be discontinuous at element interfaces. Their most interesting difference might be that the DG method solves a CL in its weak form, whereas the FR method discretizes a CL in its differential form. While the weak form allows a somewhat natural coupling of neighboring elements by numerical fluxes, discretizing the differential form of a CL is often considered to be simpler and therefore to be favored by practitioners. However, for most of the latter proposed shock capturing strategies it is only crucial that both methods use a piecewise polynomial representation of the numerical solution.

Outline

This chapter is organized as follows: In Chapter 6.1 we present a short description of DG methods, first of the so-called analytical DG method and afterwards for one of its most commonly used discretizations. The FR method, on the other hand, is discussed in Chapter 6.2. Besides a description of the method, we also provide a brief outlook on some recent stability results for FR methods. Even though we do not address these results in detail, they indirectly emerged from the present thesis and have been published in [ÖGR18].

6.1 Discontinuous Galerkin spectral element methods

Let us consider the particularly popular class of DGSEMs for hyperbolic CLs

$$\partial_t \boldsymbol{u} + \partial_t \boldsymbol{f}(\boldsymbol{u}) = 0 \tag{6.1}$$

on a computational domain Ω. Henceforth, we assume suitable IC and BCs. For more information about DG methods, we refer to the original series of works [CS91, CS89, CLS89, CHS90, CS98] by Cockburn, Shu, and co-authors and the monograph [HW07] of Hesthaven and Warburton. For the presentation here, we roughly follow the works [GÖ20, GG19a, GNJA$^+$19].

6.1.1 Introduction

DG methods were first introduced 1973 by Reed and Hill [RH73] to solve the hyperbolic neutron transport equation in a nuclear reactor and were put on mathematically solid ground by Cockburn, Shu, and co-authors in a series of papers [CS91, CS89, CLS89, CHS90, CS98] around 1990. In [CS89] and [CHS90] it was proven that the resulting class of DG methods is (formally) high order accurate in smooth regions, TV bounded (TVB) in one space dimension, and maximum-norm bounded in any number of space dimensions. Further, Jiang and Shu proved in [JS94] a cell entropy inequality for the square entropy

$$\eta(u) = \frac{u^2}{2}$$

of linear as well as nonlinear scalar CLs for the DG method. It should be noted that this result does not need any nonlinear limiting as introduced in [CS89] and [CHS90]. The cell entropy inequality makes the DG method consistent with the entropy condition (2.26) and implies L^2 stability of the scheme. Yet, Jiang and Shu's proof relies on exact evaluation of integrals and therefore only applies to the *analytical* DG method, where all arising integrals are evaluated exactly; see Chapter 6.1.2. Unfortunately, exact evaluation of integrals is often computationally impractical or — depending on the nonlinearity of the flux function f — even impossible and is usually replaced by numerical QRs. Then, exactness (at least up to machine precision) can be guaranteed by a sufficiently great number of quadrature points. This was investigated, for instance, by Kirby and Karniadakis in [KK03]. Another alternative are so-called DGSEMs [HW07]. In these methods, the solution u as well as the flux f are approximated by interpolation polynomials in each element and the corresponding interpolation points are further matched with the quadrature points; see Chapter 6.1.3. This results in highly efficient operators [Kop09]. However, a typical problem of such discretizations of the DG method is their instability, especially for nonlinear CLs, due to the reduced accuracy of the integrals. There are several possible stabilization methods in the literature, such as minmod-type limiting [CS89, CHS90], AV methods [PP06, KWH11, RGÖS18, GNJA$^+$19], modal filtering [Van91, HK08, GÖS18, RGÖS18], FV sub-cells [HCP12, DZLD14, SM14, MO16], and many more. Yet, in [Gas13, KG14, GWK16b] Gassner, Kopriva, and co-authors have been able to construct a DGSEM which is L^2 stable for certain linear (variable coefficient) as well as nonlinear CLs by utilizing skew-symmetric formulations of the CL (6.1) together with SBP operators, which were first used and investigated in FD methods [KS74, Str94, Ols95a, Ols95b, NC99]. It should be stressed that the theoretical stability (in the sense of a provable L^2 norm inequality) as well as the numerical stability (in the sense of stable interpolation polynomials and QRs) of the DGSEM heavily rely on the usage of GLo points and quadrature weights, which include the boundary nodes and are more dense there; see [Gas13, KG14]. In [RÖS16] these results have been extended to GLe points and quadrature weights, which do not include the boundary nodes but are still more dense there. A more general approach, ensuring entropy stability by using decoupled SBP operators was recently proposed in [CDRFC19] by Chen and coauthors. Their approach is quite general and will be addressed in greater detail in Chapter 7.1.

6.1.2 The analytical discontinuous Galerkin method

Decoupling space and time by the method of lines (see Chapter 3.5), the DG method provides a semidiscretization of the CL (6.1) on a computational domain Ω. For sake of simplicity, we only explain the idea behind DG methods for scalar CLs. Yet, the extension to systems can be easily achieved by applying the subsequently described procedure to every component of a system. For a scalar CL (6.1), the DG method is obtained in the following way: First, the computational domain Ω is subdivided into smaller elements Ω_i, $i = 1, \dots, I$, with $\Omega = \bigcup_{i=1}^{I} \overline{\Omega_i}$. In one dimension, the elements are given by simple subintervals

$$\Omega_i = \left(x_{i-\frac{1}{2}}, x_{i+\frac{1}{2}} \right),$$

which are typically mapped to a single reference element $\Omega_{\text{ref}} = (-1, 1)$, where all computations are performed. In two or more space dimensions, the elements might be given by triangles, tetrahedrons or other simple geometric objects. The CL (6.1) is solved in a weak form for every element then. On an element Ω_i, equation (6.1) is multiplied by a test function v and integrated in space. Integration by parts yields the weak form

$$\int_{\Omega_i} (\partial_t u) v \,\mathrm{d}x - \int_{\Omega_i} f(u)(\partial_x v) \,\mathrm{d}x + f\left(u(x_{i+\frac{1}{2}})\right) v(x_{i+\frac{1}{2}}) - f\left(u(x_{i-\frac{1}{2}})\right) v(x_{i-\frac{1}{2}}) = 0.$$

Now assume that both the solution u as well as the test function v come from a finite dimensional approximation space V_h, which is usually chosen to be the space of piecewise polynomials of degree at most K, i. e.

$$V_h = \left\{ v : \Omega \to \mathbb{R} \mid v^i := v|_{\Omega_i} \in \mathbb{P}_K(\Omega_i) \right\}.$$

It should be stressed that for this choice $u, v \in V_h$ might be discontinuous at element interfaces. Thus, the boundary terms $f\left(u(x_{i\pm\frac{1}{2}})\right)$ and $v(x_{i\pm\frac{1}{2}})$ are not well defined. We distinguish between values from inside Ω_i, e. g.

$$u_{i-\frac{1}{2}}^{+} = u^i(x_{i-\frac{1}{2}}), \quad u_{i+\frac{1}{2}}^{-} = u^i(x_{i+\frac{1}{2}}), \quad v_{i-\frac{1}{2}}^{+} = v^i(x_{i-\frac{1}{2}}), \quad v_{i+\frac{1}{2}}^{-} = v^i(x_{i+\frac{1}{2}})$$

and values from the neighboring elements $\Omega_{i-i}, \Omega_{i+i}$, e. g.

$$u_{i-\frac{1}{2}}^{-} = u^{i-1}(x_{i-\frac{1}{2}}), \quad u_{i+\frac{1}{2}}^{+} = u^{i+1}(x_{i+\frac{1}{2}}), \quad v_{i-\frac{1}{2}}^{-} = v^{i-1}(x_{i-\frac{1}{2}}), \quad v_{i+\frac{1}{2}}^{+} = v^{i+1}(x_{i+\frac{1}{2}}).$$

From conservation and stability (upwinding) considerations, we take a single valued *numerical flux*

$$f_{i+\frac{1}{2}}^{\text{num}} = f^{\text{num}}\left(u_{i+\frac{1}{2}}^{-}, u_{i+\frac{1}{2}}^{+} \right)$$

to replace $f\left(u(x_{i+\frac{1}{2}})\right)$. The numerical flux is consistent ($f^{\text{num}}(u, u) = f(u)$), Lipschitz continuous, and monotone (f^{num} is nondecreasing in the first argument and nonincreasing in the second argument). Examples of commonly used numerical fluxes can be found in [CS89] and [Tor13]. Hence, the DG method is: Find $u \in V_h$ such that

$$\int_{\Omega_i} (\partial_t u) v \, \mathrm{d}x - \int_{\Omega_i} f(u)(\partial_x v) \, \mathrm{d}x + f^{\mathrm{num}}_{i+\frac{1}{2}} v^-_{i+\frac{1}{2}} - f^{\mathrm{num}}_{i-\frac{1}{2}} v^+_{i-\frac{1}{2}} = 0 \tag{6.2}$$

for all $v \in V_h$ and $i = 1, \dots, I$. Note that in (6.2) all integrals are assumed to be evaluated exactly. We therefore refer to (6.2) as the *analytical DG method*. Finally, the analytical DG method (6.2) can be rewritten as a system of ODEs

$$\frac{\mathrm{d}}{\mathrm{d}t} u = L(u)$$

and the approximation $u \in V_h$ is evolved over time by some explicit SSP-RK method, as discussed in Chapter 3.5.

6.1.3 The discontinuous Galerkin collocation spectral element method

Next, we revisit the DGSEM. The DGSEM can be considered as a special realization of the (analytic) DG method. In the DGSEM u and f are both approximated by interpolation polynomials in each element and the corresponding interpolation points are matched with the quadrature points. Thus, we assume $u, v, f(u) \in V_h$; that is all functions are approximated by piecewise polynomials of degree up to K. Further, it is convenient to transform equation (6.1) for each element $\Omega_i = \left(x_{i-\frac{1}{2}}, x_{i+\frac{1}{2}}\right)$ to the reference element $\Omega_{\mathrm{ref}} = (-1, 1)$ via the linear map

$$x_i(\xi) = \overline{x}_i + \frac{\Delta x_i}{2} \xi, \tag{6.3}$$

where $\overline{x}_i$ is the center of the element Ω_i and Δx_i is its length, respectively given by

$$\overline{x}_i := \frac{1}{2}(x_{i+\frac{1}{2}} + x_{i-\frac{1}{2}}), \quad \Delta x_i := x_{i+\frac{1}{2}} - x_{i-\frac{1}{2}}.$$

This results in the transformed equation

$$\frac{\Delta x_i}{2} \partial_t u + \partial_\xi f(u) = 0 \tag{6.4}$$

on $\Omega_{\mathrm{ref}} = (-1, 1)$. By defining a set of $K + 1$ distinct interpolation points $\{\xi_k\}_{k=0}^K$ in the reference element, $u^i = u|_{\Omega_i}$ and $f^i = f(u)|_{\Omega_i}$ are computed by polynomial interpolation as

$$u^i(\xi, t) = \sum_{k=0}^{K} u^i_k(t) \ell_k(\xi), \quad f^i(\xi, t) = \sum_{k=0}^{K} f^i_k \ell_k(\xi)$$

with time depended nodal values $u^i_k(t) = u^i(t, \xi_k)$ and $f^i_k = f(u^i_k)$ at the interpolation points and corresponding Lagrange basis functions ℓ_k, which are defined by

$$\ell_k(\xi) = \prod_{j=0, j \neq k}^{K} \frac{\xi - \xi_j}{\xi_k - \xi_j} \tag{6.5}$$

and satisfy the cardinal property $\ell_k(x_j) = \delta_{kj}$. Also see Chapter 3.1 for more details on polynomial interpolation and related techniques. Here, for sake of simplicity, we will only focus on GLo points (see Example 3.37 in Chapter 3.4.3) for the interpolation as well as quadrature points $\{\xi_j\}_{j=0}^K$. We denote the associated quadrature weights by $\{w_j\}_{j=0}^K$. Analytic integration is replaced by the resulting QR

$$\int_{-1}^{1} g(\xi)\,\mathrm{d}\xi \approx \sum_{j=0}^{K} w_j g(\xi_j) \tag{6.6}$$

then. Besides the GLo points, the GLe points are another typical choice for the collocation approach. Inserting these approximations into the DG formulation (6.2) and choosing $v^i = \ell_k$, we get the *DGSEM*: Solve

$$w_k \frac{\Delta x_i}{2} \frac{\mathrm{d}}{\mathrm{d}t} u_k^i - \sum_{j=0}^{K} w_j f_j^i \ell_k'(\xi_j) + f_{i+\frac{1}{2}}^{\mathrm{num}} \ell_k(1) - f_{i-\frac{1}{2}}^{\mathrm{num}} \ell_k(-1) = 0 \tag{6.7}$$

for all $k = 0, \ldots, K$ and $i = 1, \ldots, I$. Here, ℓ_k' denotes the spatial derivative $\partial_\xi \ell_k$. Note that (6.7) is directly solved for the nodal degrees of freedom $\{u_k^i\}_{k=0,i=1}^{K,I}$ of the numerical solution $u \in V_h$.

6.1.4 Matrix vector notation

We note two alternative formulations for the DGSEM using discrete inner products and a matrix vector representation, respectively. Note that the QR (6.6) induces a discrete inner product

$$[u, v]_M = \sum_{k=0}^{K} w_k u(\xi_k) v(\xi_k) = \mathbf{u}^T M \mathbf{v}$$

with *mass matrix*

$$M = \mathrm{diag}(w_0, \ldots, w_K) \tag{6.8}$$

and vectors of nodal degrees of freedom

$$\mathbf{u} = (u_0, \ldots, u_K)^T, \quad \mathbf{v} = (v_0, \ldots, v_K)^T.$$

Using the discrete inner product, the spacial approximation (6.7) becomes

$$\frac{\Delta x_i}{2} \left[\frac{\mathrm{d}}{\mathrm{d}t} u^i, \ell_k \right]_M = \left[f^i, \ell_k' \right]_M - \left(f^{\mathrm{num}} \ell_k \right) \Big|_{-1}^{1}$$

for $k = 0, \ldots, K$ and $i = 1, \ldots, I$. Finally going over to a matrix vector representation and utilizing the cardinal property of the Lagrange basis functions, the DG approximation can be compactly rewritten in its *weak form* as

$$\frac{\Delta x_i}{2} M \frac{\mathrm{d}}{\mathrm{d}t} \mathbf{u}^i = D^T M \mathbf{f}^i - R^T B \boldsymbol{f}_i^{\mathrm{num}}, \quad i = 1, \ldots, I, \tag{6.9}$$

where

$$D = \left(\ell_j'(\xi_k) \right)_{k,j=0}^{K}, \quad R = \begin{pmatrix} 0 & \ldots & 0 & 1 \\ 1 & 0 & \ldots & 0 \end{pmatrix}, \quad B = \begin{pmatrix} -1 & 0 \\ 0 & 1 \end{pmatrix} \tag{6.10}$$

are respectively the *differentiation matrix*, *restriction matrix*, and *boundary matrix*. Moreover, let $\boldsymbol{f}_i^{\mathrm{num}}$ denote the vector containing the values of the numerical flux at the element boundaries of Ω_i. Note that applying integration by parts a second time to (6.2) would result in the *strong form*

$$\frac{\Delta x_i}{2} M \frac{\mathrm{d}}{\mathrm{d}t} \mathbf{u}^i = -MD\mathbf{f}^i - R^T B \left(\boldsymbol{f}_i^{\mathrm{num}} - R\mathbf{f}^i \right), \quad i = 1, \ldots, I, \tag{6.11}$$

of the DGSEM. Both forms are equivalent when using SBP operators, which satisfy

$$MD + D^T M = R^T B R. \tag{6.12}$$

The strong form (6.11) can further be recovered as a special case of the subsequent FR schemes; see [AJ11, YW13, DGMM+14].

6.2 Flux reconstruction methods

FR methods have been established in [Huy07] as a framework for high order semidiscretizations which is able to recover some well known schemes. These include SD and certain DG methods with special choices of the parameters. See, for instance, [AJ11, YW13, DGMM+14] and references therein. Later, Huynh et al. reviewed FR methods and coined the common name *correction procedure via reconstruction* (CPR) [HWV14]. Today, both names (FR and CPR) are used in the scientific community, but since the name "flux reconstruction" is more common, it will be used here.

6.2.1 General idea

Similar to the DGSEM, FR methods are semidiscretizations using a polynomial approximation on elements. In order to describe the basic idea, again a scalar CL (6.1) in one space dimension equipped with appropriate IC and BCs is considered. In this case, again, the computational domain $\Omega \subset \mathbb{R}$ is subdivided into smaller elements Ω_i for $i = 1, \ldots, I$ such that $\Omega = \bigcup_{i=1}^I \overline{\Omega_i}$. The elements Ω_i are mapped to a single reference element $\Omega_{\mathrm{ref}} = (-1, 1)$ where all computations are performed. Next, on each element Ω_i, the solution u is approximated by a polynomial of degree up to K. Usually, polynomial interpolation is used. For the corresponding basis of Lagrange basis polynomials, the coefficients of $u^i = u|_{\Omega_i}$ are the nodal values $\mathbf{u}^i = (u_1^i, \ldots, u_K^i)^T$, where $u_k^i = u^i(\xi_k, t)$, $k = 0, \ldots, K$, and $\{\xi_k\}_{k=0}^K$ are the distinct interpolation points in $[-1, 1]$. Since FR methods are polynomial collocation schemes, the flux $f(u)$ is approximated by a polynomial f^i interpolating at the nodes ξ_k as well, that is

$$f_k^i := f^i(\xi_k) = f(u_k^i).$$

Using a discrete differentiation matrix D as in (6.10), the spatial derivative of f^i is $D\mathbf{f}^i$. Yet, since the numerical solution will probably be discontinuous across elements, the discrete flux will also have jump discontinuities. Hence, in order to include the effects of neighboring elements, a correction of the discontinuous flux is performed. Similar to the DG method, the numerical solution u^i is interpolated to the left and right boundary yielding the values

$$u_{i-\frac{1}{2}}^+ = u^i(x_{i-\frac{1}{2}}), \quad u_{i+\frac{1}{2}}^- = u^i(x_{i+\frac{1}{2}}).$$

Thus, at each boundary node, there are two values u^-, u^+ of the elements on the left and right hand side, respectively. A continuous corrected flux with the common value of a numerical flux $f^{\mathrm{num}}(u^-, u^+)$ (Riemann solver) at the boundary shall be used, analogously to the prior

DG methods. Moreover, the polynomial approximation of the flux f^i is interpolated to the boundaries yielding the values

$$f^+_{i-\frac{1}{2}} = f^i(x_{i-\frac{1}{2}}), \quad f^-_{i+\frac{1}{2}} = f^i(x_{i+\frac{1}{2}}).$$

Next, left and right correction functions g_L, g_R are introduced on the reference element. These functions are symmetric, i. e. $g_L(\xi) = g_R(-\xi)$, approximate zero in $(-1,1)$, and fulfill

$$g_L(-1) = g_R(1) = 1, \quad g_L(1) = g_R(-1) = 0$$

Then, let us denote the vectors of nodal values of $\partial_\xi g_L$ and $\partial_\xi g_R$ by $\mathbf{g}'_L$ and $\mathbf{g}'_R$. Finally, the semidiscrete equation of the FR method is given by

$$\frac{\Delta x_i}{2}\frac{\mathrm{d}}{\mathrm{d}t}\mathbf{u}^i = -D\mathbf{f}^i - (f^{\text{num}}_{i-\frac{1}{2}} - f^+_{i-\frac{1}{2}})\mathbf{g}'_L - (f^{\text{num}}_{i+\frac{1}{2}} - f^+_{i+\frac{1}{2}})\mathbf{g}'_R \tag{6.13}$$

for $i = 1, \dots, I$. If we use the restriction matrix R performing interpolation to the boundary as in (6.10), a correction matrix $C = (\mathbf{g}'_L, \mathbf{g}'_R)$, and writing $\boldsymbol{f}^{\text{num}}_i = (f^{\text{num}}_{i-\frac{1}{2}}, f^{\text{num}}_{i+\frac{1}{2}})^T$, this can be reformulated as

$$\frac{\Delta x_i}{2}\frac{\mathrm{d}}{\mathrm{d}t}\mathbf{u}^i = -D\mathbf{f}^i - C\left(\boldsymbol{f}^{\text{num}}_i - R\mathbf{f}^i\right), \quad i = 1, \dots, I, \tag{6.14}$$

by using matrix vector notation.

6.2.2 Flux reconstruction and summation by parts operators

Using a nodal basis associated with a suitable QR using quadrature weights $w_0, \dots, w_K$, the mass matrix $M = \operatorname{diag}(w_0, \dots, w_K)$ is diagonal and corresponds to an SBP operator; that is using the boundary matrix $B = \operatorname{diag}(-1, 1)$, integration by parts

$$\int_\Omega u\,\partial_x v + \int_\Omega \partial_x u\,v = u\,v\big|_{\partial\Omega} \tag{6.15}$$

is mimicked on a discrete level by SBP

$$\mathbf{u}^T M D \mathbf{v} + \mathbf{u}^T D^T M \mathbf{v} = \mathbf{u}^T R^T B R \mathbf{v}, \tag{6.16}$$

since the SBP property

$$MD + D^T M = R^T B R \tag{6.17}$$

is fulfilled. Here, either a GLe basis without boundary nodes or a GLo basis including both boundary nodes is used. The associated QRs are exact for polynomials of degree up to $\leq 2K+1$ and $\leq 2K-1$, respectively. The mass matrix M is traditionally called *norm matrix* in the FD framework whereas the DG community prefers the name mass matrix. Further, the correction term $C\left(\boldsymbol{f}^{\text{num}}_i - R\mathbf{f}^i\right)$ in the semidiscretization (6.14) corresponds to a simultaneous approximation term (SAT) in the framework of FD-SBP methods. In general, the canonical choice of the correction matrix C presented in [RÖS16] is

$$C = M^{-1} R^T B. \tag{6.18}$$

However, using different choices yields the full range of energy stable schemes derived in [VCJ11, VFWJ15]. For a detailed introduction to FR methods and SBP operators, we recommend the works [Huy07, HWV14, RÖS16, FHZ14, SN14] and references therein.

6.2.3 Outlook: New stability results for Burgers' equation using a polynomial chaos approach

Since the specific numerical method is not of importance for the later proposed shock capturing techniques, we do not provide much more details about the FR (and DG) methods and their properties, such as stability. Their decisive common feature is that they are (high order) discontinuous SE methods. Yet, we close this chapter by at least briefly pointing out a novel stability result for FR schemes. This result indirectly emerged from the present thesis and has been published in [ÖGR18]. In this work, the Burgers' equation with uncertain IC and BCs has been considered using a polynomial chaos approach. This approach yields a hyperbolic system of deterministic equations which has been discretized by a FR method using SBP operators. Due to the usage of split-forms, the mayor challenge is to construct entropy stable numerical fluxes then. In [ÖGR18], for the first time, such numerical fluxes have been constructed for all systems resulting from the polynomial chaos approach for Burgers' equation.

CHAPTER 7

Two novel high order methods

In what follows, we present two high order methods for hyperbolic CLs. To the best of our knowledge, both methods have not been proposed before and the material presented here resulted in the publications [GÖ20] and [GG19b]. The first method is addressed in Chapter 7.1 and consists of a new type of discretization of DG methods. This discretization builds up on DLS approximations and the novel LS-QRs investigated in Chapter 4 and will be proven to be stable even on equidistant and scattered collocation points. The second method, on the other hand, discusses stability of RBF methods. Here, we start by demonstrating that usual RBF methods, derived from the differential form (2.1) of a CL, are not stable. We then proceed by proposing novel RBFs methods based on the weak form of a CL and prove their stability.

Outline

This chapter is organized as follows: In the first instance, the present chapter can be divided into Chapter 7.1, proposing new stable discretization of DG methods on equidistant and scattered collocation points, and Chapter 7.2, investigating conservation and L^2 stability of RBF methods for hyperbolic CLs. More precise outlines for the individual chapters will be given at the end of their corresponding motivations, i. e. at the end of chapters 7.1.1 and 7.2.1.

7.1 Stable discretizations of discontinuous Galerkin methods on equidistant and scattered points

In this chapter, we propose and investigate a new stable discretization of DG methods on equidistant and scattered collocation points. We do so by incorporating the concept of DLS into the DG framework. DLS approximations allow us to construct stable and high order accurate approximations on arbitrary collocation points, while LS-QRs enable their stable and exact numerical integration; also see Chapter 4. Both methods are computed efficiently by using bases of DOPs; see Chapter 3.2 and 4.2.2. Thus, the proposed discretization generalizes known classes of discretizations of the DG method, such as the DGSEM; see 6.1.3. We are able to prove conservation and linear L^2 stability of the proposed discretization. Finally, numerical tests investigate their accuracy and demonstrate their extension to nonlinear CLs, hyperbolic systems, longtime simulations, and a variable coefficient problem in two space dimensions. The material presented in this chapter resulted in the publication [GÖ20] and, to the best of our knowledge, has not been proposed somewhere else.

7.1.1 Motivation

In Chapter 6.1, we have already introduced the DG method and the DGSEM methods as an especially efficient discretization. Following the idea of collocation, where the interpolation points for u and f are matched with the quadrature points, usually GLe or GLo collocation points are used. Here, we describe a first step towards L^2 stable DGSEM on equidistant and even scattered collocation points. When using equidistant points for the (interpolation) polynomials approximating u and f in the DGSEM, we run into the Runge phenomenon [Run01]. Adapting an idea of Gelb et al. [GPR08] from spectral collocation methods, we tackle this problem by making a somewhat maverick generalization. Instead of usual polynomial approximations by interpolation on $K+1$ points, we build the polynomial approximations by the method of DLS, where the data at a larger number of $N > K+1$ points is used. The method therefore utilizes more information of the underlying function. By going over to a larger number of nodal values than technically needed, polynomial DLS approximations are known to provide high accuracy while successfully suppressing Runge oscillations, even on equidistant points. Care also has to be taken when performing numerical integration on equidistant or even scattered collocation points. To prove conservation as well as L^2 stability, QRs which exactly evaluate polynomials of degree up to $2K$ are needed. Even though interpolatory QRs of arbitrary high degrees of exactness can be constructed on equidistant (and scattered) points, they quickly become unstable; see Chapter 4. Thus, for an exact as well as stable evaluation of integrals, we propose to use LS-QRs (see Chapter 4). To the best of our knowledge, LS-QRs have not been investigated in a PDE solver before. Utilizing DLS approximations together with stable high order LS-QRs, in fact, we are able to prove conservation and linear L^2 stability of the resulting discretization of the DG method on equidistant and scattered points.

Outline

The subsequent material is organized as follows. In Chapter 7.1.2, we briefly revisit the concept of DLS approximations. LS-QRs, on the other hand, have already been discussed in chapters 4 and 5. Here, we are interested in the simple case of a weight function $\omega \equiv 1$. Building up on concepts DLS approximations and LS-QRs, in Chapter 7.1.3, we propose a stable high order discretization of the DG method on equidistant and scattered points then. The resulting discretization of the DG method is referred to as the *discontinuous Galerkin discrete least squares (DGDLS) method* and generalizes the usual DGSEM. Conservation as well as linear L^2 stability of the DGDLS method are proven in Chapter 7.1.4. In Chapter 7.1.5, numerical tests are presented. In particular, we investigate accuracy of the method and address the extension to systems of CLs, longtime simulations, and a variable coefficient problem in two spatial dimensions. We close our study in Chapter 7.1.6 with concluding thoughts and outlook.

7.1.2 Discrete least squares approximations

Originally, DLS approximations were born from the wish to fit a (linear) mathematical model to given observations. In contrast to polynomial interpolation [PTK11], DLS approximations can be stable and highly accurate on equidistant and even scattered points. Here, we use DLS approximations to fit polynomials $f_{K,N}$ of degree at most K to a larger number of N observations, given by nodal values at the collocation points. Let f be a function on $[-1,1]$ which we only know at a set of $N+1$ distinct points $\{\xi_n\}_{n=0}^N$ in $[-1,1]$. The problem is to find a polynomial $f_{K,N} \in \mathbb{P}_K(\mathbb{R})$ such that $f_{K,N} - f$ is minimized; that is

$$\|f_{K,N} - f\| \leq \|v - f\| \quad \forall v \in \mathbb{P}_K(\mathbb{R})$$

holds for some norm $\|\cdot\|$. Here, we consider norms which are induced by (weighted) discrete inner products

$$[u, v]_{\mathbf{r}} := \sum_{n=0}^{N} r_n u(\xi_n) v(\xi_n) \tag{7.1}$$

with $r_n > 0$, $n = 0, \dots, N$. As discussed in Chapter 3.1.2, the best approximation $f_{K,N}$ exists and is unique. Since this best approximation corresponds to a discrete inner product of the form (7.1), we refer to it as a *DLS approximation*. Moreover, let us denote the basis of DOPs corresponding to (7.1) by $\{\pi_k(\cdot; \mathbf{r})\}_{k=0}^{K}$. Then, consulting Theorem 3.6, the DLS approximation is simply given by

$$f_{K,N}(\xi) = \sum_{k=0}^{K} [f, \pi_k(\cdot; \mathbf{r})]_{\mathbf{r}} \, \pi_k(\xi; \mathbf{r}). \tag{7.2}$$

When computed by using bases of DOPs, DLS approximations are sometimes referred to as *DOP-DLS approximations*; see [GPR08].

Next, let us address the weights r_n in (7.1) as well as the quadrature weights w_n which are used to approximate continuous integrals by QRs. In what follows, we aim for stable high order discretizations of the DG method on equidistant and even scattered collocation points. Thus, we propose to use the LS-QRs proposed in Chapter 4. If a sufficiently large number of collocation/quadrature points is used, these are able to provide perfectly stable QRs with arbitrary degree of exactness. Remember that the LS quadrature weights are explicitly given by

$$w_n^* = \sum_{k=0}^{K} \pi_k(x_n; \mathbf{1}) I\left[\pi_k(\cdot; \mathbf{1})\right], \quad n = 0, \dots, N,$$

where $\{\pi_k(\cdot; \mathbf{1})\}_{k=0}^{K}$ is the basis of DOP with respect to the discrete inner product

$$[u, v]_{\mathbf{1}} := \sum_{n=0}^{N} u(\xi_n) v(\xi_n).$$

We note that other discrete inner products are possible — and have been investigated in Chapter 4 — but for sake of simplicity we do not include them here. It should also be stressed that the above LS-QRs are constructed for the constant weight function $\omega \equiv 1$. Henceforth, we simply denote the DOPs $\pi_k(\cdot; \mathbf{r})$ by φ_k.

7.1.3 Proposed discretization of the discontinuous Galerkin method

Starting point of our discretization is the analytical DG method (6.2). Transformed to the reference element $\Omega_{\text{ref}} = (-1, 1)$, the task is to find $u^i \in \mathbb{P}_K(\mathbb{R})$ such that

$$\frac{\Delta x_i}{2} \int_{-1}^{1} \dot{u}^i v \, \mathrm{d}\xi - \int_{-1}^{1} f(u^i) v' \, \mathrm{d}\xi + f_{i+\frac{1}{2}}^{\text{num}} v_{i+\frac{1}{2}}^{-} - f_{i-\frac{1}{2}}^{\text{num}} v_{i-\frac{1}{2}}^{+} = 0 \tag{7.3}$$

holds for all $v \in \mathbb{P}_K(\mathbb{R})$. Here, $\dot{u}^i$ denotes the temporal derivative $\partial_t u^i$ and v' denotes the spatial derivative $\partial_\xi v$. We initiate our discretization by replacing $f(u^i)$ in (7.3) with a polynomial $f^i \in \mathbb{P}_K(\mathbb{R})$, given by the DLS approximation

$$f(u^i) \approx f^i = \sum_{k=0}^{K} \hat{f}_{k,N}^i \, \varphi_k(\xi) \quad \text{with} \quad \hat{f}_{k,N}^i = [f, \varphi_k(\cdot)]_{\mathbf{r}} \tag{7.4}$$

For $N = K$, this is the usual polynomial interpolation. Yet, in our discretization, N might be chosen larger than K. Next, the involved integrals are replaced by LS-QRs

$$Q_N[g] = \sum_{n=0}^{N} w_n^* g(\xi_n) \approx \int_{-1}^{1} g(\xi) \, \mathrm{d}\xi, \tag{7.5}$$

as discussed above in Chapter 7.1.2. Utilizing the discrete inner product $[u, v]_{\mathbf{w}^*}$ results in the discretization

$$\frac{\Delta x_i}{2} [\dot{u}^i, v]_{\mathbf{w}^*} = [f^i, v']_{\mathbf{w}^*} - \left(f_{i+\frac{1}{2}}^{\text{num}} v_{i+\frac{1}{2}}^- - f_{i-\frac{1}{2}}^{\text{num}} v_{i-\frac{1}{2}}^+ \right). \tag{7.6}$$

$N \in \mathbb{N}$ and the quadrature weights $\mathbf{w}^* \in \mathbb{R}^{N+1}$ are chosen such that the resulting LS-QR is perfectly stable ($\kappa(\mathbf{w}^*) = 1$) and provides degree of exactness $2K$. This will be crucial to prove conservation and linear stability of the resulting discretization. Further, we follow the idea of collocation and match the quadrature points with the points at which the nodal values of $f(u^i)$ are used to construct the DLS approximation $f^i \in \mathbb{P}_K(\mathbb{R})$. This results in a more efficient implementation of the proposed discretization. For sake of simplicity, we use the LS quadrature weights $\mathbf{w}^*$ also for the DLS approximation (7.4), i. e. $r_n = w_n^*$ in (7.1). Different choices are possible but will not be investigated here. Finally, we include bases of DOPs $\{\varphi_k\}_{k=0}^K$ with respect to the discrete inner product

$$[u, v]_{\mathbf{w}^*} = \sum_{n=0}^{N} w_n^* u(\xi_n) v(\xi_n).$$

This allows us to avoid computing a mass matrix on the left hand side of (7.6). When choosing $v = \varphi_l$, we have

$$[\dot{u}^i, v]_{\mathbf{w}^*} = \sum_{k=0}^{K} \frac{\mathrm{d}}{\mathrm{d}t} \hat{u}_k^i [\varphi_k, \varphi_l]_{\mathbf{w}^*} = \frac{\mathrm{d}}{\mathrm{d}t} \hat{u}_l^i \tag{7.7}$$

and (7.6) becomes a system of $K + 1$ ordinary differential equations

$$\frac{\Delta x_i}{2} \frac{\mathrm{d}}{\mathrm{d}t} \hat{u}_l^i = [f^i, \varphi_l']_{\mathbf{w}^*} - \left(f_{i+\frac{1}{2}}^{\text{num}} \varphi_l(1) - f_{i-\frac{1}{2}}^{\text{num}} \varphi_l(-1) \right), \quad l = 0, \dots, K. \tag{7.8}$$

Moreover, the discrete inner product on the right hand side of (7.8) is given by

$$[f^i, \varphi_l']_{\mathbf{w}^*} = \sum_{k=0}^{K} \hat{f}_{k,N}^i [\varphi_k, \varphi_l']_{\mathbf{w}^*}. \tag{7.9}$$

For the sake of brevity, we will refer to the DLS based discretization (7.8) of the DG method as the DGDLS method. A main part of the DGDLS method is to initially determine a suitable vector of LS quadrature weights $\mathbf{w}^*$ and a corresponding basis of DOPs. Figure 7.1 provides a flowchart which summarizes this procedure.

Remark 7.1. Computing the derivative $\frac{\mathrm{d}}{\mathrm{d}t} \hat{u}_l^i$ for a fixed $l \in \{0, \dots, K\}$ in the DGDLS method has the following complexity: First note that the inner products $[\varphi_k, \varphi_l']_{\mathbf{w}^*}$ in (7.9) are computed once-for-all a priori. Thus, the computation of a flux coefficients $\hat{f}_{k,N}^i$ is performed in $\mathcal{O}(N)$ operations and the whole set $\{\hat{f}_{k,N}^i\}_{k=0}^K$, corresponding to a fixed element Ω_i, is computed in $\mathcal{O}(NK)$ operations. This also yields $\frac{\mathrm{d}}{\mathrm{d}t} \hat{u}_l^i$ in the DGDLS method (7.8) to be computed in $\mathcal{O}(NK)$ operations. In the subsequent numerical tests, we found the choice $N = 2K$ to be sufficient for a stable computation on equidistant points. From point of complexity, this compares to a DG method using over-integration with $2K$ GLo or GLe points and is by a factor 2 less efficient than a collocation-type DGSEM method using $K + 1$ GLo or GLe points.

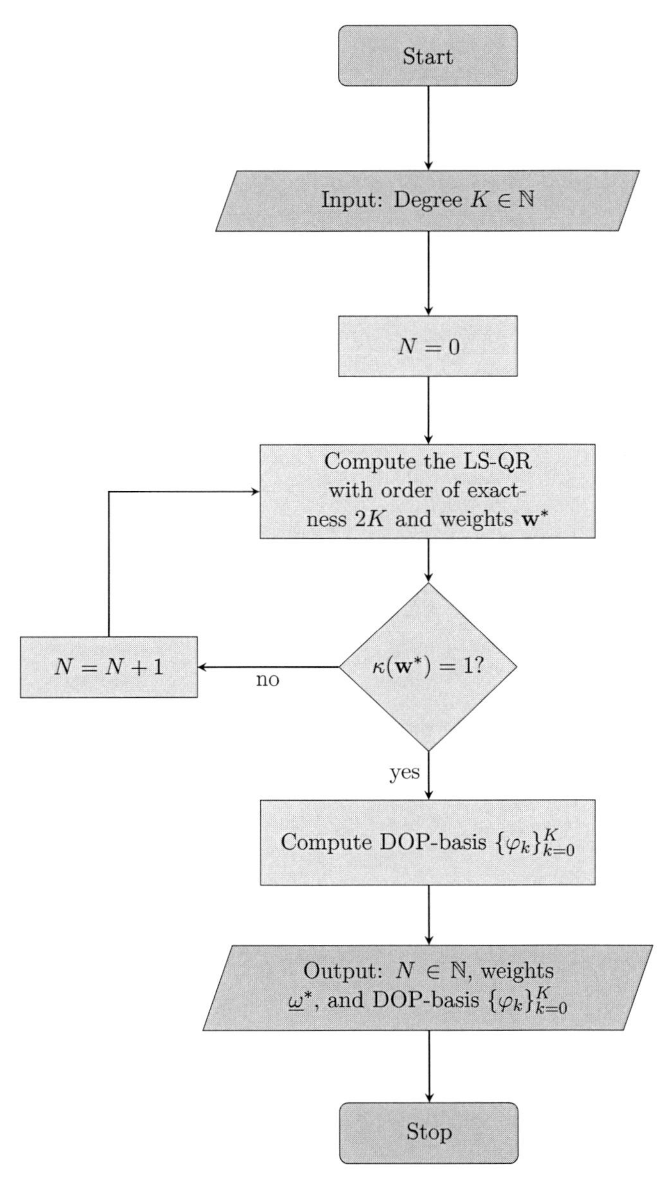

Figure 7.1: Flowchart describing the initial construction of suitable weights $\mathbf{w}^*$ and a DOP basis $\{\varphi_k\}_{k=0}^K$.

Remark 7.2. In many methods it is desirable to avoid mass matrices and their inversion; see [Abg17, ABT16]. In the proposed discretization of the DG method this is possible by using bases of DOPs. CG methods might benefit even more from the combination of DLS approximations and QRs with bases of DOPs. Of course, the restriction to a continuous approximation space has to be regarded. Nevertheless, we believe that the application of DOPs could have a positive impact on these schemes. Future work will investigate this possibility.

7.1.4 Conservation and stability

Above, we have proposed a new stable high order discretization of DG methods on equidistant and scattered collocation points. Here, we prove conservation and linear L^2 stability of this discretization, i. e. of the DGDLS method. Finally, we also give an outlook on entropy stability.

Conservation

Let $u \in V_h$ be a numerical solution of the scalar CL

$$\partial_t u + \partial_x f(u) = 0 \tag{7.10}$$

consisting of a piecewise polynomial of degree at most K on $\Omega = [a,b]$. Note that the total amount of the conserved variable u only changes due to the flux across the element boundaries, i. e.,

$$\frac{\mathrm{d}}{\mathrm{d}t} \int_\Omega u \,\mathrm{d}x = -\left[f\left(u(b)\right) - f\left(u(a)\right) \right] \tag{7.11}$$

holds for solutions of (7.10). We now show that the numerical solution $u \in V_h$ produced by our discretization (7.6) of the DG method also fulfills this property of *conservation.*

The contribution due to a single element Ω_i, transformed to the reference element, is

$$\frac{\mathrm{d}}{\mathrm{d}t} \int_{\Omega_i} u \,\mathrm{d}x = \frac{\Delta x_i}{2} \int_{-1}^{1} \dot{u}^i \,\mathrm{d}\xi.$$

Here, $u^i = u^i(t,\xi)$ is the transformation of the numerical solution $u = u(t,x)$ on Ω_i to $\Omega_{\text{ref}} = (-1,1)$. Note that by choosing $v = 1$ in our discretization (7.6), we get

$$\frac{\Delta x_i}{2} \left[\dot{u}^i, 1 \right]_{\mathbf{w}^*} = -\left(f^{\text{num}}_{i+\frac{1}{2}} - f^{\text{num}}_{i-\frac{1}{2}} \right)$$

and since the weights $\mathbf{w}^*$ are related to a QR on the collocation points $\{\xi_n\}_{n=0}^N$ with degree of exactness $2K$,

$$\left[\dot{u}^i, 1 \right]_{\mathbf{w}^*} = Q_N[\dot{u}^i] = \int_{-1}^{1} \dot{u}^i \,\mathrm{d}\xi$$

holds. Thus, we have

$$\frac{\mathrm{d}}{\mathrm{d}t} \int_{\Omega_i} u \,\mathrm{d}x = \frac{\Delta x_i}{2} \int_{-1}^{1} \dot{u}^i \,\mathrm{d}\xi = \frac{\Delta x_i}{2} \left[\dot{u}^i, 1 \right]_{\mathbf{w}^*} = -\left(f^{\text{num}}_{i+\frac{1}{2}} - f^{\text{num}}_{i-\frac{1}{2}} \right).$$

Finally, summing up over all elements, the rate of change of the total amount of the numerical solution $u \in V_h$ is given by

$$\frac{\mathrm{d}}{\mathrm{d}t} \int_\Omega u \,\mathrm{d}x = -\left(f^{\text{num}}_b - f^{\text{num}}_a \right), \tag{7.12}$$

where $f^{\text{num}}_b = f^{\text{num}}_{i+\frac{1}{2}}$ is the numerical flux at the right boundary of Ω and $f^{\text{num}}_a = f^{\text{num}}_{\frac{1}{2}}$ is the numerical flux at the left boundary of Ω. For a consistent single valued numerical flux, (7.12) is consistent with (7.11). Note that for periodic BCs we have

$$\frac{\mathrm{d}}{\mathrm{d}t} \int_\Omega u \,\mathrm{d}x = 0.$$

This means that for an isolated system, the numerical method is able to exactly mimic conservation of mass.

L^2 stability

Another fundamental design principle for numerical methods is stability. It is well-known that the squared L^2 norm of physically reasonable (entropy) solutions of (7.10) with periodic BCs does not increase over time, i. e.

$$\frac{\mathrm{d}}{\mathrm{d}t} \|u\|_{L^2}^2 \leq 0, \tag{7.13}$$

where a strict inequality reflects the presence of shock waves. For scalar CLs, the squared L^2 norm is also referred to as the *energy*. We now prove that for the linear advection equation

$$\partial_t u + \lambda \partial_x u = 0$$

with constant velocity $\lambda > 0$ on $\Omega = [a, b]$, *(strong) L^2 stability* (7.13) is also fulfilled by our discretization (7.6) of the DG method. Therefor, let us assume periodic BCs and let $u \in V_h$ be the numerical solution consisting of a piecewise polynomial of degree at most K. We start by noting that the rate of change of the squared L^2 norm can be expressed as

$$\frac{\mathrm{d}}{\mathrm{d}t} \|u\|_{L^2}^2 = 2 \int_\Omega (\partial_t u) u \,\mathrm{d}x.$$

In a single element Ω_i, transformed to the reference element $\Omega_{\text{ref}} = (-1, 1)$, we have

$$\int_{\Omega_i} (\partial_t u) u \,\mathrm{d}x = \frac{\Delta x_i}{2} \int_{-1}^{1} \dot{u}^i u^i \,\mathrm{d}\xi. \tag{7.14}$$

This time choosing $v = u^i$ in (7.6), results in

$$\begin{aligned} 2 \int_{\Omega_i} (\partial_t u) u \,\mathrm{d}x &= \Delta x_i \left[\dot{u}^i, u^i \right]_{\mathbf{w}^*} \\ &= 2 \left[\lambda u^i, (u^i)' \right]_{\mathbf{w}^*} - 2 \left(f_{i+\frac{1}{2}}^{\text{num}} \lambda u^i(1) - f_{i-\frac{1}{2}}^{\text{num}} \lambda u^i(-1) \right), \end{aligned}$$

since $\mathbf{w}^*$ provides a QR with degree of exactness $2K$. Moreover, we have

$$\begin{aligned} 2 \left[\lambda u^i, (u^i)' \right]_{\mathbf{w}^*} &= 2\lambda \int_{-1}^{1} u^i (\partial_\xi u^i) \,\mathrm{d}\xi \\ &= \lambda \left([u^i(1)]^2 - [u^i(-1)]^2 \right) \end{aligned}$$

and therefore

$$\begin{aligned} 2 \int_{\Omega_i} (\partial_t u) u \,\mathrm{d}x &= \lambda \left([u^i(1)]^2 - [u^i(-1)]^2 \right) - 2 \left(f_{i+\frac{1}{2}}^{\text{num}} \lambda u^i(1) - f_{i-\frac{1}{2}}^{\text{num}} \lambda u^i(-1) \right) \\ &= \lambda u^i(1) \left[\lambda u^i(1) - 2 f_{i+\frac{1}{2}}^{\text{num}} \right] - \lambda u^i(-1) \left[\lambda u^i(-1) - 2 f_{i-\frac{1}{2}}^{\text{num}} \right]. \end{aligned}$$

Finally, summing up over all elements, the global rate of change cuts down to a sum of local contributions

$$\lambda u_- \left(\lambda u_- - 2 f^{\text{num}} \right) - \lambda u_+ \left(u_+ - 2 f^{\text{num}} \right),$$

where the interface between two neighboring elements is considered. Here, u_- denotes the value $u^i(1)$ from the left element, u_+ denotes the value $u^{i+1}(-1)$ from the right element, and $f^{\text{num}} = f^{\text{num}}(u_-, u_+)$ denotes the single valued numerical flux at the interface. Using a usual full upwind numerical flux, i. e.

$$f^{\text{num}}(u_-, u_+) = \lambda u_- \tag{7.15}$$

(assuming $\lambda > 0$), we get

$$\lambda u_- \left(\lambda u_- - 2 f^{\text{num}} \right) - \lambda u_+ \left(\lambda u_+ - 2 f^{\text{num}} \right) = -\lambda^2 (u_- - u_+)^2 \leq 0.$$

Hence, the proposed discretization (7.6) of the DG method is L^2 stable for the linear advection equation.

An Outlook on entropy stability

Besides L^2 stability, often entropy stability is desired as well. For the DGSEM on GLo points (including the boundary nodes) a skew-symmetric formulation and SBP operators have been used to prove entropy stability [Gas13, GWK16b, GWK16a]. Recently, Chan, Fernandez, and Carpenter [CDRFC19] proved entropy stability also for a collocation based DG method on GLe nodes (not including the boundary nodes) using a decoupled SBP formulation. Moreover, they were able to extend their approach to general selections of quadrature points and various bases. Thus, once we can ensure the SBP property for the proposed DLS discretization, we could follow the results of Chan, Fernandez, and Carpenter to ensure entropy stability also for the proposed DLS based discretization of the DG method.

Another option to address entropy stability could be the introduction of entropy correction terms as proposed in [Abg18]. For degree of freedom $k \in \{0, \dots, K\}$ and element $i \in I$, the correction term is given by

$$r_k^i = \alpha(\hat{u}_k^i - \overline{u}^i), \tag{7.16}$$

where

$$\overline{u} := \frac{1}{K+1} \sum_{k=0}^{K} \hat{u}_k^i \quad \text{and} \quad \alpha := \frac{E}{\sum_{k=0}^{N} (\hat{u}_k^i - \overline{u}^i)^2}.$$

Here, E is the so-called *entropy error*, which can be calculated using an entropy numerical flux $\hat{g}^{\text{num}}$ and an entropy variable v^i, which is set equal to the solution u^i for sake of simplicity. Yet, the correction term (7.16) is also valid for general entropy variables v^i; see [Abg18]. Using (7.8), E is given by

$$\begin{aligned} E := & \left(\hat{g}_R^{\text{num}}(1) - \hat{g}_L^{\text{num}}(-1) \right) \\ & - \frac{2}{\Delta x_i} \sum_{k=0}^{K} \hat{u}_k^i \left(\left[f^i, (u^i)' \right]_{\mathbf{w}^*} - \left(f_R^{\text{num}} u^i(1) \right) - f_L^{\text{num}} u^i(-1) \right) \right). \end{aligned}$$

The correction term is consistent with zero and does not effect the conservation relation since

$$\sum_{k=0}^{K} r_k^i = \alpha \left(\sum_{k=0}^{K} (\hat{u}_k^i - \overline{u}) \right) = 0$$

holds. The correction term (7.16) is added to the right hand side of the DGDLS scheme (7.8) and results in an entropy stable scheme. This idea was already applied in [AMÖ18] to construct entropy stable FR schemes on polygonal meshes.

Future works will address both approaches to construct entropy stable (DLS based) discretizations of DG methods as well as their comparison.

7.1.5 Numerical results

In the subsequent numerical tests, we investigate conservation, L^2 stability, and approximation properties of the proposed DGDLS method on equidistant and scattered points. We will do so for a linear advection equation with constant velocity $\lambda = 1$ and a nonlinear inviscid Burgers' equation. Finally, we demonstrate the extension to systems of CLs and a variable coefficients problem in two spatial dimensions. The later numerical tests have all been performed using the SSPRK(3,3) method presented in Definition 3.39 with timestep size

$$\Delta t = \frac{C}{I(K+1)\lambda}$$

with $C = 0.1$ and where $\lambda = \max_u |f'(u)|$ is the fastest propagation speed.

Linear advection equation

Let us consider the linear advection equation with constant velocity $\lambda = 1$,

$$\partial_t u + \partial_x u = 0, \tag{7.17}$$

on $\Omega = [0,1]$ with a smooth IC $u_0(x) = \sin(4\pi x)$ and periodic BCs. By the method of characteristics, the (entropy) solution is given by

$$u(x,t) = u_0(x-t).$$

In the following, we evolve this solution until time $t = 1$, so that the solution $u(x,1)$ is equal to the IC $u_0(x)$.

Linear advection equation - Conservation and L^2 stability

We start with a numerical demonstration of conservation and L^2 stability of the DGDLS method on equidistant and scattered points. Figure 7.2 illustrates the behavior of the solution, including its mass and energy over time when equidistant collocation points are used. For this test, we have chosen $I = 5$ equidistant elements, a polynomial degree of $K = 3$, and the full upwind numerical flux (7.15).

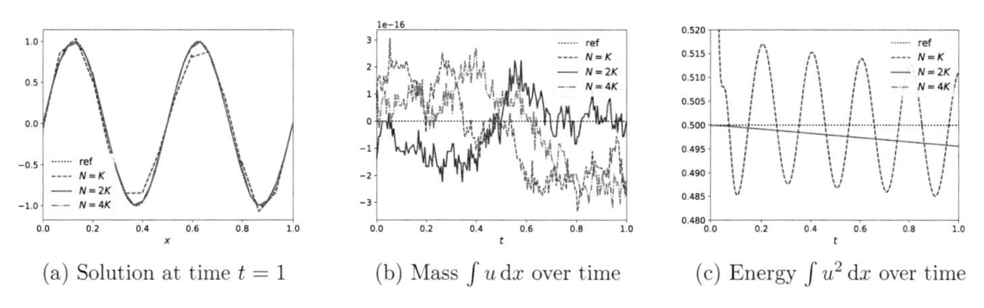

(a) Solution at time $t = 1$ (b) Mass $\int u \, dx$ over time (c) Energy $\int u^2 \, dx$ over time

Figure 7.2: Numerical solution, mass, and energy for $I = 5$, $K = 3$, and $N = K, 2K, 4K$. Linear advection equation and equidistant collocation points.

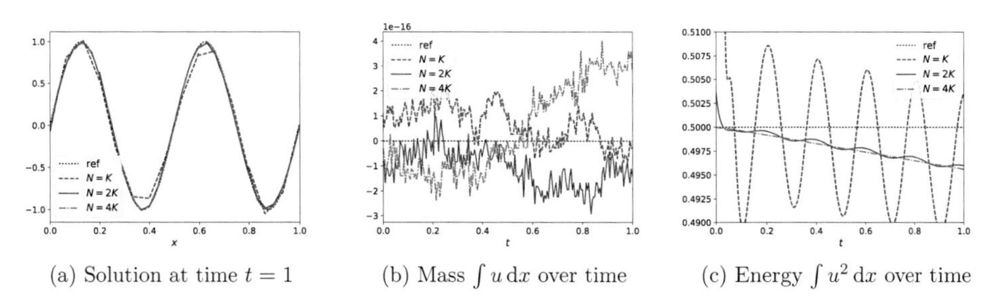

(a) Solution at time $t = 1$ (b) Mass $\int u \, dx$ over time (c) Energy $\int u^2 \, dx$ over time

Figure 7.3: Numerical solution, mass, and energy for $I = 5$, $K = 3$, and $N = K, 2K, 4K$. Linear advection equation and scattered collocation points.

Besides the reference solution, the figures show the numerical solutions for $N = K, 2K, 4K$, i. e. for an increasing number of equidistant points used in the DLS approximation and the LS-QR. All choices yield a conservative method, which can be observed in Figure 7.2(b). Yet, in

Figure 7.2(c), we observe the discretization to be (strongly) L^2 stable only for $N \geq 2K$. Due to an insufficiently high degree of exactness of the LS-QR, the numerical solution for $N = K$ yields spurious oscillation in the energy of the solution over time. As a consequence, the discretization for $N = 2K$ and $N = 4K$ can be observed to provide more accurate numerical solutions in Figure 7.2(a). We only observe slight differences between the numerical solutions for $N = 2K$ and $N = 4K$. Figure 7.3 provides a similar demonstration for a set of scattered collocation points. Here, the scattered collocation points are obtained by adding white uniform noise to the set of equidistant collocation points. Thus, the scattered collocation points are give by

$$\tilde{\xi}_0 = -1, \quad \tilde{\xi}_N = 1, \quad \tilde{\xi}_n = \xi_n + Z_n, \quad \xi_n = -1 + \frac{2n}{N}, \quad Z_n \in \mathcal{U}\left(-\frac{1}{40N}, \frac{1}{40N}\right), \tag{7.18}$$

for $n = 1, \ldots, N-1$, where the Z_n are independent, identically distributed, and further assumed to not be correlated with the ξ_n. Using $N = K, 2K, 4K$ scattered collocation points, again, all DLS discretizations yield a conservative method, see Figure 7.3(b). This time, however, slight oscillations can be observed in the energy profile displayed in Figure 7.3(c) even for $N = 2K$ collocation points. These oscillations only vanish when going over to a larger set of $N = 4K$ scattered collocation points. This behavior is again caused by an insufficiently high degree of exactness of the QR when scattered points are used. Yet, $N = 2K$ and $N = 4K$ provide similar accurate numerical solutions, as can be observed in Figure 7.3(a).

Linear advection equation - Accuracy and convergence

Next, we investigate the approximation properties of the proposed discretization for the same problem as before. Table 7.1 lists the L^2 errors for the DGDLS method on equidistant collocation points and an increasing number of degrees of freedom. Table 7.2 provides the same analysis for sets of scattered collocation points, which are again constructed by adding white uniform noise; see (7.18). The experimental orders of convergence (EOCs) have been computed by performing a LS fit for the parameters C and s in the model $y = C \cdot N^{-s}$, where y denotes the L^2 error for a fixed N.

We note from both tables that increasing the number of collocation points does not always yield more accurate numerical solutions when using a fairly small number of degrees of freedom, i. e. polynomials of degree $K = 1$ or only $I = 5$ elements. Yet, for all higher degrees $K \geq 2$ (and $I > 5$ for $K = 2$), we observe the numerical solutions to become more accurate when the number of collocation points is increased. Further, both tables provide a comparison of the DGDLS method with the usual DGSEM on a set of GLo points. The DGSEM can be considered as a special case of the proposed DGDLS method when GLo points are used and $N = K$ is chosen, i. e. using polynomial interpolation as well as an interpolatory (GLo-) QR. Of course, GLo points and their corresponding QR are known to be superior to equidistant or even scattered points, at least when the same number of points is used. Yet, when sufficiently increasing the number of collocation points, we often observe the DGDLS method to provide more accurate results on equidistant and even scattered collocation points than the DGSEM on GLo points.

Remark 7.3. In some cases, we observe the error to increase even though the number of elements I is increased, e. g. for $K = 4$ and $N = 4K$ when going over from $I = 10$ to $I = 20$ as well as for $K = 4$ and $N = 16K$ when going over from $I = 20$ to $I = 40$ in Table 7.2. The same observation can be made for the subsequent tables 7.4 and 7.5 concerning the inviscid Burgers' equation. This behavior is caused by the instability of the LS-QR when an insufficiently great number of collocation points N is used. Hence, we observe this problem to vanish when N is increased; see the case $N = 64K$ in tables 7.2 and 7.5.

		L^2 errors			
		DGSEM	DGDLS		
K	I	(GLo points)	$N=K$	$N=2K$	$N=4K$
1	5	5.8E-1	5.8E-1	6.4E-1	6.4E-1
	10	1.0E-1	1.0E-1	2.0E-1	1.9E-1
	20	2.6E-2	2.6E-2	3.5E-2	3.3E-2
	40	9.6E-3	9.6E-3	6.4E-3	5.9E-3
EOC:		2.5	2.5	1.7	1.8
2	5	6.6E-2	6.6E-2	1.0E-1	9.9E-2
	10	1.0E-2	1.0E-2	8.7E-3	7.9E-3
	20	1.3E-3	1.3E-3	1.0E-3	9.0E-4
	40	1.6E-4	1.6E-4	1.2E-4	1.1E-4
EOC:		2.7	2.7	3.5	3.6
3	5	1.1E-2	8.1E-2	1.0E-2	8.9E-3
	10	7.6E-4	2.0E-2	6.3E-4	5.4E-4
	20	4.9E-5	3.8E-4	4.0E-5	3.4E-5
	40	2.9E-6	3.7E-6	2.5E-6	2.1E-6
EOC:		3.8	2.1	3.9	4.0
4	5	1.3E-3	1.1E-2	1.2E-3	1.0E-3
	10	5.1E-5	4.3E-3	4.2E-5	3.4E-5
	20	2.3E-6	9.0E-4	1.5E-6	1.2E-6
	40	1.1E-7	9.1E-5	1.0E-7	9.7E-8
EOC:		4.1	1.5	4.1	4.1

Table 7.1: Linear advection equation and equidistant collocation points

		L^2 errors			
		DGSEM	DGDLS		
K	I	(GLo points)	$N=4K$	$N=16K$	$N=64K$
1	5	5.8E-1	6.4E-1	6.4E-1	6.4E-1
	10	1.0E-1	1.9E-1	1.9E-1	1.9E-1
	20	2.6E-2	3.3E-2	3.3E-2	3.3E-2
	40	9.6E-3	5.8E-3	5.7E-3	5.6E-3
EOC:		2.4	1.8	1.8	1.8
2	5	6.6E-2	9.9E-2	9.8E-2	9.8E-2
	10	1.0E-2	7.9E-3	7.7E-3	7.6E-3
	20	1.3E-3	9.0E-4	8.7E-4	8.6E-4
	40	1.6E-4	1.1E-4	1.0E-4	1.0E-4
EOC:		2.7	3.6	3.6	3.6
3	5	1.1E-2	8.9E-3	8.7E-3	8.7E-3
	10	7.6E-4	5.6E-4	5.2E-4	5.2E-4
	20	4.9E-5	3.7E-5	3.3E-5	3.3E-5
	40	2.9E-6	2.1E-6	2.0E-6	2.0E-6
EOC:		3.8	3.9	4.0	4.0
4	5	1.3E-3	1.1E-3	1.0E-3	1.0E-3
	10	5.1E-5	3.8E-4	4.3E-5	3.3E-5
	20	2.3E-6	3.9E-4	5.8E-6	2.0E-6
	40	1.1E-7	4.6E-6	6.4E-6	2.7E-7
EOC:		4.1	1.1	4.0	4.1

Table 7.2: Linear advection equation and scattered collocation points

Inviscid Burgers' equation

Let us now consider the nonlinear inviscid Burgers' equation,

$$\partial_t u + \partial_x \left(\frac{u^2}{2} \right) = 0,$$

on $\Omega = [0,1]$ with smooth IC $u_0(x) = 1 + \frac{1}{4\pi}\sin(2\pi x)$ and periodic BCs. For this problem a shock develops in the solution when the wave breaks at time

$$t_b = -\frac{1}{\min_{0 \leq x \leq 1} u_0'(x)} = 2.$$

In the subsequent numerical tests, we consider the solution at times $t=1$ and $t=3$. Moreover, a *local Lax–Friedrichs (LLF)* numerical flux

$$f_{LLF}^{\text{num}}(u_-, u_+) = \frac{1}{2}\left[f(u_-) + f(u_+)\right] - \frac{\lambda}{2} \cdot (u_+ - u_-) \quad \text{with} \quad \lambda = \max\{|u_-|, |u_+|\}$$

is used at the element interfaces. The reference solutions have been computed using characteristic tracing, solving the implicit equation $u(x,t) = u_0(x - tu)$ in smooth regions. The jump location, separating these regions, can be determined by the Rankine–Hugoniot condition. Revisit Chapter 2 for more details.

Inviscid Burgers' equation - Conservation and L^2 stability

Again, we start with a numerical investigation of conservation and L^2 stability. Note that our proof of L^2 stability in Chapter 7.1.4 only addresses the linear advection equation. Still, we observe similar results for the nonlinear Burgers' equation. These results are illustrated in Figure 7.4 for a set of equidistant collocation points and in Figure 7.5 for a set of scattered collocation points, again constructed by adding white uniform noise as described in (7.18).

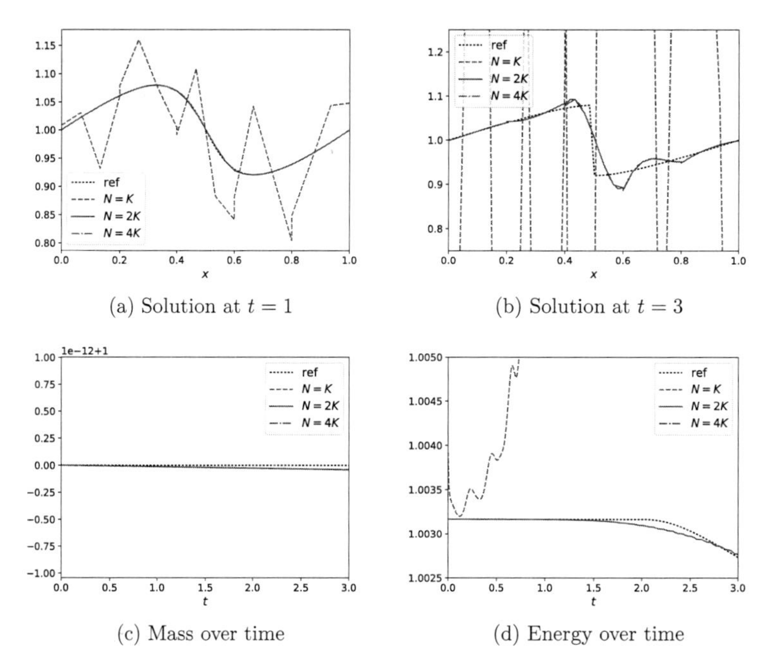

Figure 7.4: Numerical solution, mass, and energy for $I = 5$, $K = 3$, and $N = K, 2K, 4K$. Burgers' equation and equidistant collocation points.

Figures 7.4(a) and 7.5(a) show the solutions of the DGDLS method at time $t = 1$. At this time, no discontinuity has developed and the solution is still smooth. For the case of equidistant collocation points in Figure 7.4(a), the DGDLS method for $N = 2K, 4K$ is observed to provide reasonable numerical solutions. However, the numerical solution for $N = K$ shows heavy oscillations. The same can be observed for numerical solutions for $N = K$ as well as $N = 2K$ when scattered collocation points are used in Figure 7.5(a). Here, only the numerical solution for $N = 4K$ can be considered as reasonable. A similar observation can be made for both kinds of collocation points at time $t = 3$. In the case of scattered collocation points, the computation even broke down for $N = K$. Thus, Figure 7.5(b) instead illustrates the results for $N = 2K, 4K, 8K$. Note that all numerical results for $t = 3$, also the ones for equidistant collocation points in Figure 7.4(b), show at least some minor oscillations. This is a common problem for high order methods in the presence of discontinuities and might be overcome by post processing (assuming a stable computation until the final time has been reached) or additional shock capturing. Shock capturing might be performed, for instance, by AV methods [PP06, KWH11, GNJA+19, RGÖS18], modal filtering [Van91, HK08, MOSW13, GÖS18], FV subcells [HCP12, SM14, DZLD14, MO16], or other methods [GG19a, Gla19c]. Shock capturing in DLS

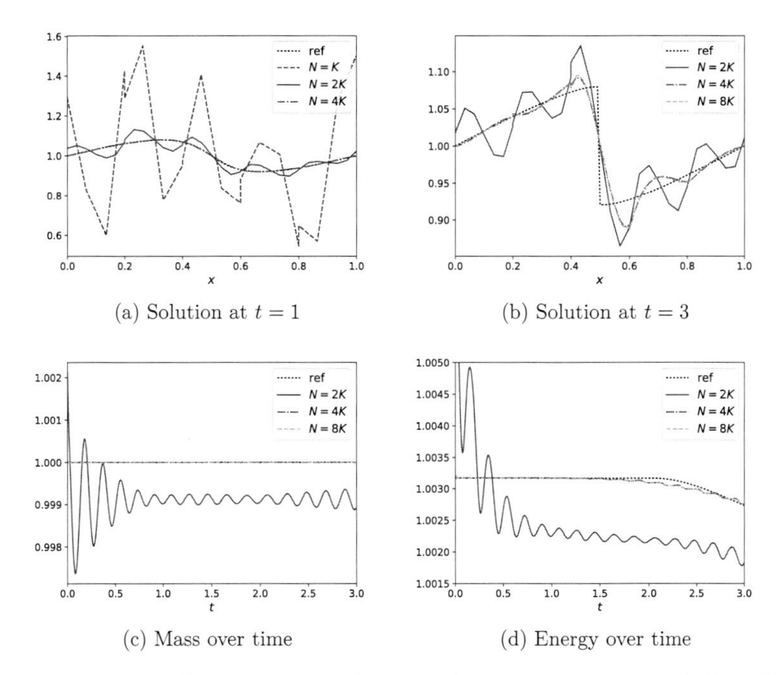

(a) Solution at $t = 1$

(b) Solution at $t = 3$

(c) Mass over time

(d) Energy over time

Figure 7.5: Numerical solution, mass, and energy for $I = 5$, $K = 3$, and $N = K, 2K, 4K$. Burgers' equation and scattered collocation points.

based high order methods might be investigated in future works. Finally, when a sufficiently large number of collocation points is used, we again observe the mass of the numerical solutions to nearly remain constant; see figures 7.4(c) and 7.5(c). Moreover, the energy is observed to nearly remain constant until the discontinuity occurs at time $t = 2$ and to decrease afterwards; see figures 7.4(d) and 7.5(d).

Inviscid Burgers' equation - Accuracy and convergence

We continue the above investigation of the DGDLS method by providing an additional error analysis. Tables 7.3 and 7.4 list the errors for DGDLS method at time $t = 1$ on equidistant and scattered points. We note that the DGDLS method for $N = K$ does not yield stable numerical solutions. Sufficiently increasing the number of collocation points in the DGDLS method, however, again results in more accurate numerical solutions. At least for equidistant collocation points, the DGDLS method even yields more accurate numerical solutions than the DGSEM on GLo points. Further, in both cases, we are able to recover or even exceed the EOCs of the DGSEM when a sufficiently large number of collocation points is used.

Extension to systems: The wave equation

The extension of the DGDLS method to systems of CLs is the same as for most discretizations of the DG method: We simply apply the discretization proposed in Chapter 7.1.3 to every component of the system separately. As a representative, we consider the second order scalar

		L^2 errors			
		DGSEM	DGDLS		
K	I	(GLo points)	$N = K$	$N = 2K$	$N = 4K$
1	5	1.3E-2	1.3E-2	1.1E-2	1.2E-2
	10	3.8E-3	3.8E-3	4.1E-3	3.8E-3
	20	1.1E-3	1.1E-3	9.3E-4	8.7E-4
	40	2.8E-4	2.8E-4	2.0E-4	1.8E-4
EOC:		1.8	1.8	1.5	1.7
2	5	1.7E-3	1.7E-3	3.4E-3	3.0E-3
	10	5.9E-4	5.9E-4	3.4E-4	3.5E-4
	20	6.7E-5	6.7E-5	5.0E-5	4.5E-5
	40	8.0E-6	8.0E-6	6.2E-6	5.5E-6
EOC:		1.7	1.7	3.3	3.0
3	5	1.0E-3	7.1E-2	4.9E-4	6.7E-4
	10	9.4E-5	4.2E-1	8.3E-5	7.5E-5
	20	5.8E-6	NaN	5.6E-6	4.7E-6
	40	3.6E-7	NaN	3.0E-7	2.6E-7
EOC:		3.4	-	2.6	3.1
4	5	1.9E-4	NaN	4.3E-4	3.1E-4
	10	1.0E-5	NaN	1.3E-5	1.1E-5
	20	3.7E-7	NaN	2.2E-7	2.2E-7
	40	1.9E-8	NaN	1.8E-8	1.4E-8
EOC:		4.2	-	4.3	4.4

Table 7.3: Burgers' equation at time $t = 1$ for equidistant collocation points

		L^2 errors			
		DGSEM	DGDLS		
K	I	(GLo points)	$N = 4K$	$N = 16K$	$N = 64K$
1	5	1.3E-2	1.2E-2	1.2E-2	1.2E-2
	10	1.3E-2	3.9E-3	3.7E-3	3.7E-3
	20	1.1E-3	8.6E-4	8.5E-4	8.4E-4
	40	2.8E-4	1.9E-4	1.8E-4	1.7E-4
EOC:		0.8	1.7	1.7	1.7
2	5	1.7E-3	3.0E-3	2.9E-3	2.9E-3
	10	5.9E-4	1.3E-3	3.5E-4	3.5E-4
	20	6.7E-5	1.3E-4	4.9E-5	4.4E-5
	40	8.0E-6	3.6E-4	6.9E-5	9.9E-6
EOC:		1.7	1.3	3.0	3.0
3	5	1.0E-3	9.1E-4	7.1E-4	6.9E-4
	10	9.4E-5	5.0E-4	9.9E-5	7.2E-5
	20	5.8E-6	4.4E-4	1.1E-5	4.9E-6
	40	3.6E-7	3.1E-4	7.7E-6	3.0E-6
EOC:		3.4	0.5	2.8	3.2
4	5	1.9E-4	7.5E-4	3.5E-4	3.0E-4
	10	1.0E-5	9.1E-4	6.0E-5	1.3E-5
	20	3.7E-7	2.2E-3	1.7E-4	9.2E-6
	40	1.9E-8	1.7E-4	6.6E-5	8.5E-6
EOC:		4.2	0.0	0.9	4.2

Table 7.4: Burgers' equation at time $t = 1$ for scattered collocation points

wave equation in one dimension,

$$\partial_{tt} u - c^2 \partial_{xx} u = 0,$$

on $\Omega = [0, 1]$ with periodic BCs. The wave equation can be rewritten as a first order system of CLs,

$$\partial_t u + c\partial_x v = 0,$$
$$\partial_t v + c\partial_x u = 0,$$

which is sometimes referred to as the one dimensional *acoustic problem* [GPR08]. Given ICs $u(x, 0) = u_0(x)$ and $v(x, 0) = v_0(x)$, the solution is given by

$$u(x,t) = \frac{1}{2}\left[u_0(x-ct) + u_0(x+ct)\right] + \frac{1}{2}\left[v_0(x-ct) - v_0(x+ct)\right],$$
$$v(x,t) = \frac{1}{2}\left[u_0(x-ct) - u_0(x+ct)\right] + \frac{1}{2}\left[v_0(x-ct) + v_0(x+ct)\right].$$

In the subsequent numerical tests, we choose $c = 1$ and consider the ICs

$$u_0(x) = e^{-20(2x-1)^2}, \quad v_0(x) = 0.$$

For the numerical flux, we have used an upwind flux

$$f^{\text{num}} = \frac{1}{2}\begin{pmatrix} (v^- + v^+) - (u^+ - u^-) \\ (u^- + u^+) - (v^+ - v^-) \end{pmatrix}.$$

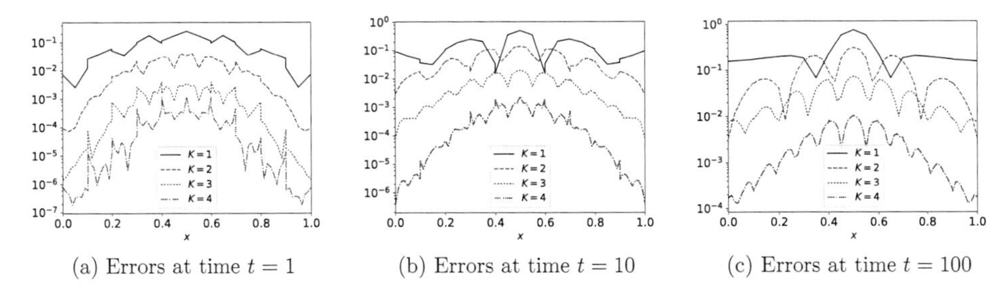

(a) Errors at time $t = 1$ (b) Errors at time $t = 10$ (c) Errors at time $t = 100$

Figure 7.6: Pointwise errors over time for the wave equation. DGDLS method for $I = 10$, $N = 2K$, and $K = 1, 2, 3, 4$. Equidistant collocation points.

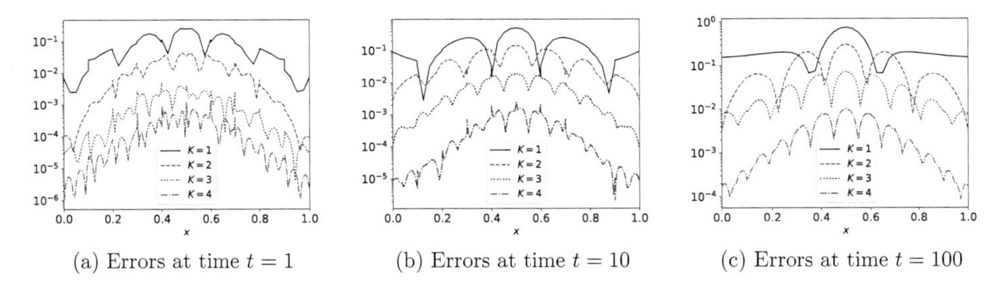

(a) Errors at time $t = 1$ (b) Errors at time $t = 10$ (c) Errors at time $t = 100$

Figure 7.7: Pointwise errors over time for the wave equation. DGDLS method for $I = 10$, $N = 4K$, and $K = 1, 2, 3, 4$. Scattered collocation points.

Figures 7.6 and 7.7 illustrate the pointwise errors

$$E(x, t) := |u(x, t) - u_{\text{num}}(x, t)| + |v(x, t) - v_{\text{num}}(x, t)|$$

for times $t = 1, 10, 100$. These have been computed using $I = 20$ elements and increasing polynomial degrees $K = 1, 2, 3, 4$. Moreover, in Figure 7.6, we have used $N = 2K$ equidistant collocation points. In Figure 7.7, on the other hand, we have used $N = 4K$ scattered collocation points. We note that the DGDLS method, regardless of whether equidistant or scattered collocation points are used, yields accurate results even for longtime simulations, especially when high polynomial degrees are used.

Extension to multiple dimensions: Linear advection equation with variable coefficients

Finally, we address the extension of the proposed DGDLS method to multiple dimensions by extending the ideas from the one dimensional case using a tensor product grid; see [Tra09, Chapter 7.1.6]. Let us consider the linear advection equation in two spatial dimensions,

$$u_t + (au)_x + (bu)_y = 0,$$

with variable coefficients $a = a(x, y)$ and $b = b(x, y)$ on $\Omega = [0, 1]^2 \subset \mathbb{R}^2$. Here, we choose $a(x, y) = x$ and $b(x, y) = 1$ together with the IC

$$u_0(x, y) = \sin(4\pi x) \left(1 - \frac{1}{2} \sin(2\pi y)\right)$$

and BCs

$$u(t, x, 0) = u(t, x, 1), \tag{7.19}$$

$$u(t, 0, y) = 0. \tag{7.20}$$

Note that (7.19) corresponds to periodic BCs at the upper and lower boundary of Ω, while (7.20) corresponds to a physical BC at the left boundary of Ω. The solution of the corresponding Cauchy problem can be calculated by the method of characteristics (see [Bre00, Chapter 3]) and is given by

$$\begin{aligned} u(t, x, y) &= \exp(-t) u_0(x \exp(-t), y) \\ &= \exp(-t) \sin(4\pi x \exp(-t)) \left(1 - \frac{1}{2}\sin(2\pi y)\right). \end{aligned} \tag{7.21}$$

Figures 7.8 and 7.9 illustrate different surface and contour plots.

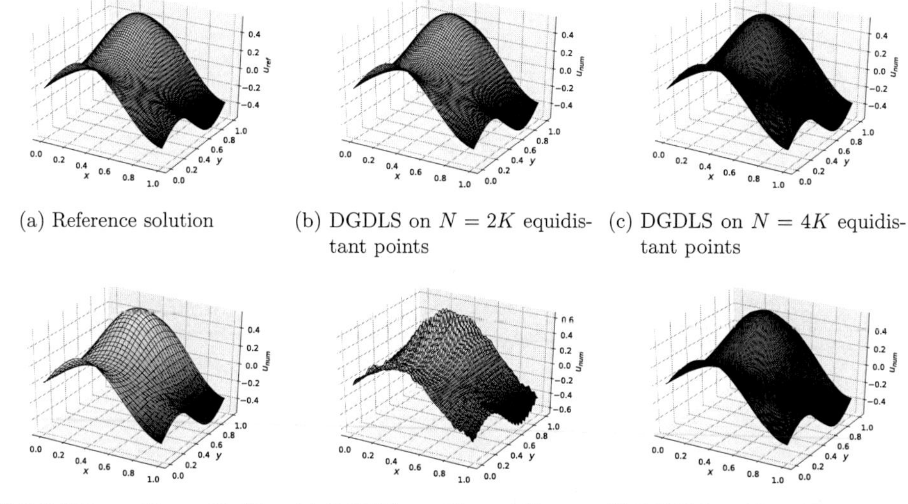

(a) Reference solution

(b) DGDLS on $N = 2K$ equidistant points

(c) DGDLS on $N = 4K$ equidistant points

(d) DGSEM on $N = K$ GLo points.

(e) DGDLS on $N = 2K$ scattered points

(f) DGDLS on $N = 4K$ scattered points

Figure 7.8: Surface plots of the solution at time $t = 1$. For the DGSEM and DGDLS method, a polynomial degree of $K = 3$ and $I = 20$ rectangular elements have been used in each direction.

Both figures demonstrate that the DGDLS method using $I = 20$ rectangular elements, a polynomial degree of $K = 3$, and $N = 2K$ equidistant collocation points in each direction yields numerical solutions which are in good agreement with the reference solution (7.21). When using $N = 2K$ scattered collocation points in each direction, the DGDLS method produces numerical solutions with slight oscillations; see figure 7.8(e) and 7.9(e). Yet, when going over to a larger number of $N = 4K$ scattered collocation points in each direction, the DGDLS method again yields numerical solutions which are in good agreement with the reference solution.

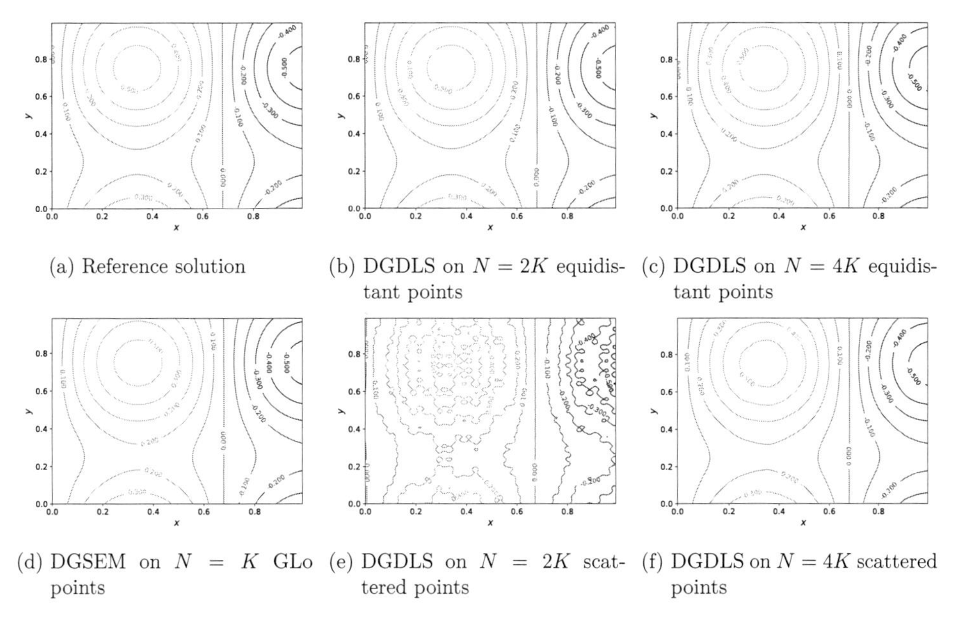

(a) Reference solution

(b) DGDLS on $N = 2K$ equidistant points

(c) DGDLS on $N = 4K$ equidistant points

(d) DGSEM on $N = K$ GLo points

(e) DGDLS on $N = 2K$ scattered points

(f) DGDLS on $N = 4K$ scattered points

Figure 7.9: Contour plots of the solution at time $t = 1$. For the DGSEM and DGDLS method, a polynomial degree of $K = 3$ and $I = 20$ rectangular elements have been used in each direction.

7.1.6 Concluding thoughts and outlook

In this chapter, we have proposed and investigated stable collocation-type discretizations of the DG method on equidistant and scattered collocation points. We have done so by utilizing DLS approximations instead of usual polynomial interpolation and LS-QRs, providing stable high order numerical integration even on equidistant and scattered points. As a consequence, we have been able to prove conservation and linear L^2 stability of the proposed DGDLS method. In several numerical tests we have observed that the DGDLS method on equidistant points is able to recover — sometimes even to exceed — the accuracy and EOC of the usual DGSEM on GLo points. Finally, the extension to the nonlinear viscous Burgers' equation, systems of CLs, long time simulations, and a variable coefficient problem in two dimensions using a tensor product approach have been demonstrated. Future work will address more general entropy stability and the extension to unstructured meshes, such as triangles and nonconvex polygons.

7.2 Stable radial basis function methods

Next, we investigate conservation and L^2 stability of RBF methods for hyperbolic CLs. We demonstrate that RBF methods based on the differential form of a hyperbolic CL might produce physically unreasonable solutions. We therefore propose RBF methods which are based on the weak form of a hyperbolic CL as an alternative. By including polynomials in the RBF approximation and using appropriate numerical fluxes, we are able to prove that our methods are conservative and L^2 stable. Numerical experiments validate our theoretical results. The material presented here resulted in the publication [GG19b].

7.2.1 Motivation

RBFs have become powerful tools in multivariate interpolation and approximation theory, since they are easy to implement, allow arbitrary scattered data, and can be spectrally accurate. With regard to the interpolation problem, we have already discussed them in Chapter 3.1.5. They are also often used to solve numerical PDEs [Kan90, Fas96, HM98, KH00, LF03, PD04], and are considered a viable alternative to more traditional methods, such as FD, finite element and spectral methods. Investigations into the stability of RBF methods are still underdeveloped and/or unsatisfactory, however. For instance, the L^2 (energy) stability has not been thoroughly studied. Moreover for time-dependent PDEs, differentiation matrices for RBF methods often have unstable eigenvalues [PD06], i. e. eigenvalues with positive real parts. Thus, in the presence of rounding errors, RBF approximations are less accurate [PD04, Sch95b, KH00] and may produce discretizations that become unstable in time unless fairly dissipative time integration methods are utilized [PD06, MP16].

In what follows, we seek to determine stable RBF methods for solving scalar one dimensional CLs

$$\partial_t u + \partial_x f(u) = 0 \tag{7.22}$$

on $\Omega = [a, b] \subset \mathbb{R}$ with appropriate ID $u(x, 0) = u_0(x)$ and BCs $u(a, t) = g_L(t)$ and $u(b, t) = g_R(t)$. Our approach involves using the weak form to solve (7.22) given by (see e. g. [Ran92])

$$\int_\Omega (\partial_t u) v \, \mathrm{d}x - \int_\Omega f(u)(\partial_x v) \, \mathrm{d}x + f(u)v\big|_{\partial\Omega} = 0 \tag{7.23}$$

with test function $v \in C^1(\Omega)$. Also see Chapter 2 or Chapter 6.1. In particular, recall that (7.23) is constructed from (7.22) by multiplying each term by v, integrating over Ω, and applying integration by parts. Observe that for (7.23) less regularity is required for the solution u. This is important since even for smooth ICs solutions of (7.22) can develop jump discontinuities [Lax73, Daf00]. Thus, by using (7.23) we permit the more general class of weak solutions, where (7.22) is satisfied in the sense of distribution theory. To distinguish the physically reasonable weak solution from all the other (many possible) weak solutions, (7.22) is augmented with an additional entropy condition

$$\partial_t \eta(u) + \partial_x \psi(u) \leq 0. \tag{7.24}$$

Here, η is an entropy function and ψ is a corresponding entropy flux satisfying $\eta' f' = \psi'$; see Chapter 2.7. A strict inequality in (7.24) reflects the presence of a physically reasonable shock wave. For scalar CLs in one dimension, the square entropy $\eta(u) = \frac{1}{2}u^2$ is often a valid entropy function. In this case, from the entropy inequality (7.24), we immediately get

$$\frac{\mathrm{d}}{\mathrm{d}t} \|u\|_{L^2}^2 = 2 \int_\Omega \partial_t \eta(u) \, \mathrm{d}x \leq -2\psi(u)\big|_{\partial\Omega} \tag{7.25}$$

for entropy solutions of (7.22). In particular, the entropy should not increase over time for an isolated physical system, and a physically reasonable weak solution of (7.22) should therefore satisfy

$$\frac{\mathrm{d}}{\mathrm{d}t} \|u\|_{L^2}^2 \leq 0 \tag{7.26}$$

for periodic BCs, i. e. the L^2 norm (*energy*) of the solution should not increase over time. We refer to (7.25) as L^2 or *energy stability*. Also see the previous Chapter 7.1. Together with the property of *conservation*, given by

$$\frac{\mathrm{d}}{\mathrm{d}t} \int_\Omega u \, \mathrm{d}x = -f(u)\big|_{\partial\Omega}, \tag{7.27}$$

L^2 stability is one of the most important design criteria for a numerical method to produce physically reasonable solutions. Note that (7.27) reduces to

$$\frac{\mathrm{d}}{\mathrm{d}t} \int_\Omega u \,\mathrm{d}x = 0 \tag{7.28}$$

in the case of periodic BCs.

Here, we show that it can be beneficial to write RBF methods in the weak form of CL (7.23) instead of the differential form (7.22), which is also referred to as the *strong form* in this chapter. Note that writing RBF methods in the strong form is the usual approach. We prove that RBF methods based on the weak form (called *weak RBF methods*), in fact, provide conservative (assuming constants are included in the RBF approximation, which has been explained in Chapter 3.1.5) as well as L^2 stable numerical solutions when appropriate numerical fluxes are used for the treatment of BCs. In contrast, we also demonstrate that usual RBF methods based on the strong form (referred to as *strong RBF methods*) violate conservation as well as L^2 stability and might produce physically unreasonable solutions. The treatment of BCs by numerical fluxes and weak enforcement further helps to overcome problems related to the common feature of RBF approximations to be relatively inaccurate at boundaries [FDWC02]. Our approach is closely related to the idea behind DG methods [CS91, CS89, CLS89, CHS90, CS98, HW07], for which a resembling but different L^2 stability analysis was performed in [JS94]. More details on L^2 stability for DG methods and related schemes can, for instance, be found in the previous Chapter 7.1 or in [Gas13, SN14, CS17, RGÖS18] and references therein. Yet, to the best of our knowledge, L^2 stability of RBF methods for hyperbolic CLs has not been investigated and proven before.

Outline

The rest of this chapter is organized as follows. In Chapter 7.2.2, we briefly describe the state of the art regarding stability results for RBF methods for time-dependent PDEs. Then, in Chapter 7.2.4 and 7.2.5 we describe how RBF methods can be incorporated into the weak form of a CL. We discuss two different realizations of the resulting weak RBF methods and, in particular, prove conservation and L^2 stability for these methods. However, the more efficient *weak RBF collocation method* (discussed in Chapter 7.2.5) is only proven to be L^2 stable for the linear advection equation. The relation of the proposed weak RBF methods to other methods — in particular to different DG methods — is described in Chapter 7.2.6. Next, Chapter 7.2.7 addresses implementation details for the introduced weak RBF methods, while Chapter 7.2.8 provides a numerical demonstration of the weak RBF method and a comparison with the usual strong RBF method. Some final remarks and an outlook on future research are offered in Chapter 7.2.9.

7.2.2 State of the art: Stability of radial basis function methods

Experience suggests that RBF approximations will produce discretizations that are unstable in time unless highly dissipative time stepping is used. In [PD06], it was shown that, under many conditions, differentiation matrices obtained with RBF collocation methods have eigenvalues with positive real parts. In particular, this was demonstrated for a simple one dimensional linear advection equation, suggesting its unsuitability for nonlinear hyperbolic CLs. A related observation was made in [FDWC02], where it was proposed that one source of instability might be inaccuracy of RBF approximations near boundaries. Indeed it was also proved in [PD06] that RBF collocation methods are time-stable (in the sense of eigenvalues for linear problems) for all conditionally positive definite RBFs and node distributions when *no BCs are needed*. Thus, RBFs are well suited for periodic domains, such as circles or unit spheres, but not, in applications where periodicity of the computational domain cannot be assumed. Moreover in Chapter 7.2.8 we demonstrate that conservation and L^2 stability are both violated by the standard RBF methods when applied to hyperbolic CLs, possibly leading to physically irrelevant solutions.

7.2.3 General idea

RBF methods typically use collocation to discretize (7.22). That is, u and f are both approximated by RBF interpolants with respect to the same set of centers x_k, $k = 1, \ldots, N$, in the computational domain $\Omega = [a, b]$. As discussed above, this yields unstable methods in the presence of BCs. Hence, our approach will be to use RBF methods based on the weak form of (7.22), given by (7.23). In what follows we describe two different RBF methods build from (7.23). In both cases the solution u is approximated by an RBF interpolant,

$$u_N(x) = \sum_{k=1}^{N} \alpha_k \phi\left(|x - x_k|\right) + \sum_{l=1}^{P} \beta_l q_l(x) \tag{7.29}$$

which as we noted earlier can include polynomials; also see Chapter 3.1.5. Remember that including polynomials of degree up to $P-1$ also adds the constraints

$$\sum_{k=1}^{N} \alpha_k q_l(x_k) = 0, \quad l = 1, \ldots, P, \tag{7.30}$$

to the RBF interpolant (7.29), where $\{q_l\}_{l=1}^P$ is a basis of the space of polynomials of degree at most $P-1$. For our purposes, the main advantage in including polynomials in the RBF interpolants is that the constraints in (7.30) enforce the RBF interpolants (7.29) to reproduce polynomials up to degree $P-1$:

$$u_N = u \quad \forall u \in \mathbb{P}_{P-1}(\mathbb{R})$$

For example, Figure 7.10 demonstrates that constant functions can be reconstructed exactly by RBF interpolants for $P \geq 1$.

This property will be crucial to prove conservation for the stable RBF methods proposed later. The method described in Chapter 7.2.4 uses the *analytical* flux function f applied to the RBF interpolant u_N. As a consequence, the resulting approximation $f(u_N) \approx f(u)$ still satisfies the interpolation condition but is no longer an RBF approximation. By contrast, the technique described in Chapter 7.2.5 utilizes the idea of collocation, where u and the flux $f(u)$ are both approximated by RBF interpolants.

7.2.4 Weak radial basis function analytical methods

Let u and v in the weak form (7.23) be replaced by RBF approximations $u_N, v_N \in V_{N,P}$ with

$$V_{N,P} := \left\{ \sum_{k=1}^{N} \alpha_k \phi\left(|x - x_k|\right) + \sum_{l=1}^{P} \beta_l q_l(x) \mid \boldsymbol{\alpha} \in \mathbb{R}^N, \boldsymbol{\beta} \in \mathbb{R}^P, \text{ and (7.30) holds} \right\}, \tag{7.31}$$

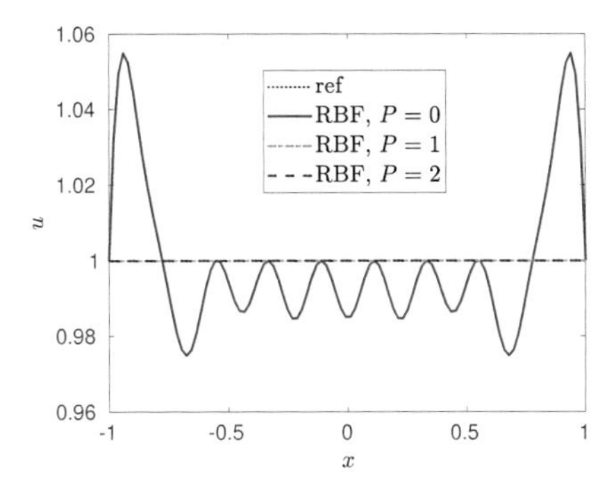

(a) RBF approximations of a constant function. Shape parameter: $\varepsilon = 6$.

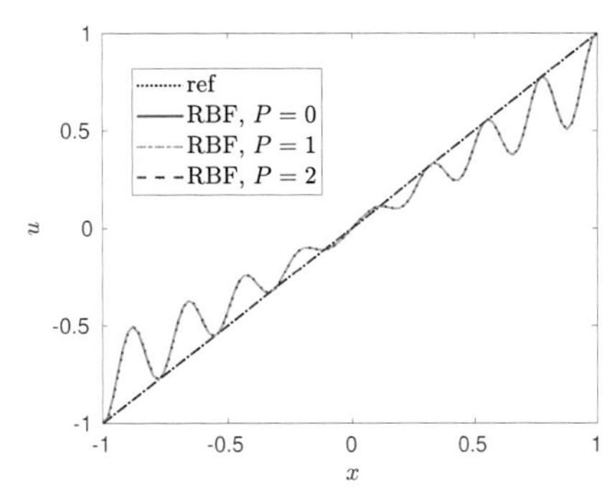

(b) RBF approximations of a linear function. Shape parameter: $\varepsilon = 10$.

Figure 7.10: RBF approximations including polynomials up to different degrees for a constant function $u(x) = 1$ and a linear function $u(x) = x$. In both cases Gaussian kernels have been used.

where $\boldsymbol{\alpha} := (\alpha_1, \ldots, \alpha_N)^T$ and $\boldsymbol{\beta} := (\beta_1, \ldots, \beta_P)^T$. Note that for $P = 0$ no polynomials are included in the RBF interpolant and the approximation space reduces to

$$V_{N,0} = \operatorname{span}\left\{\phi(|x - x_n|) \mid n = 1, \ldots, N\right\}.$$

Next observe that while one or both BCs may be given as part of (7.22), i.e. $u(a,t) = g_L(t)$ and $u(b,t) = g_R(t)$, it is also possible to assign these values with the RBF approximations evaluated there as $u(a,t) = u_L$ and $u(b,t) = u_R$ with $u_L := u_N(a)$ and $u_R := u_N(b)$. Hence to ensure well-defined boundary terms, we compute a single valued numerical flux at the boundaries as

$$f_L^{\text{num}} = f^{\text{num}}\left(g_L(t), u_L\right) \quad \text{and} \quad f_R^{\text{num}} = f^{\text{num}}\left(u_R, g_R(t)\right),$$

and therefore enforce the BCs in a weak sense. The numerical flux is chosen to be consistent ($f^{\text{num}}(u,u) = f(u)$), Lipschitz continuous, and monotone, that is f^{num} is nondecreasing in the first argument and nonincreasing in the second argument. Examples of commonly used numerical fluxes can be found in [CS89] and [Tor13]. We are now ready to define the *weak RBF analytical method* as

Definition 7.4 (Weak RBF analytical method)
Find $u_N \in V_{N,P}$ such that for all $v_N \in V_{N,P}$:

$$\int_\Omega (u_N)_t v_N \,\mathrm{d}x - \int_\Omega f(u_N)(v_N)_x \,\mathrm{d}x + \left(f_R^{\text{num}} v_R - f_L^{\text{num}} v_L\right) = 0 \tag{7.32}$$

Here, v_L and v_R respectively denote $v_N(a)$ and $v_N(b)$.

Note that in (7.32) all integrals as well as the flux $f(u_N)$ are assumed to be evaluated exactly. Next, we consider the properties of the weak RBF analytical method (7.32) for the one dimensional CL (7.22).

Conservation

We follow the same arguments as in Chapter 7.1.4. The rate of change of the total amount of the conserved variable u is given by

$$\frac{\mathrm{d}}{\mathrm{d}t} \int_\Omega u \,\mathrm{d}x = -f(u)\big|_{\partial\Omega}, \tag{7.33}$$

meaning that the total amount of change in u is due to the flux across the domain boundaries. In particular, conservation implies that

$$\frac{\mathrm{d}}{\mathrm{d}t} \int_\Omega u \,\mathrm{d}x = 0 \tag{7.34}$$

holds for periodic BCs. Numerical methods should satisfy (7.33) and (7.34) on a discrete level as well. Following the Lax–Wendroff theorem, this guarantees that discretely captured discontinuities propagate with the right speed. To prove conservation for the weak RBF analytical method, we choose $P \geq 1$ in order to include polynomials of degree $P-1$ in the approximation space $V_{N,P}$ defined by (7.31). Thus $1 \in V_{N,P}$, and since (7.32) holds for $v_N = 1$, we have

$$\frac{\mathrm{d}}{\mathrm{d}t} \int_\Omega u_N \,\mathrm{d}x = \int_\Omega (u_N)_t \,\mathrm{d}x = -\left(f_R^{\mathrm{num}} - f_L^{\mathrm{num}}\right),$$

which is the discrete counterpart to (7.33). Note that for periodic BCs, the numerical fluxes are given by $f_L^{\mathrm{num}} = f^{\mathrm{num}}(u_R, u_L)$ and $f_R^{\mathrm{num}} = f^{\mathrm{num}}(u_R, u_L)$, yielding

$$\frac{\mathrm{d}}{\mathrm{d}t} \int_\Omega u_N \,\mathrm{d}x = 0.$$

Observe that for periodic BCs, conservation of the continuous equation (7.22) is exact.

L^2 stability

Recall that the rate of change of the squared L^2 norm is given by

$$\frac{\mathrm{d}}{\mathrm{d}t} \|u\|_{L^2}^2 = 2 \int_\Omega u_t u \,\mathrm{d}x.$$

Hence by choosing $v_N = u_N$ in (7.32) we obtain

$$\begin{aligned} \frac{1}{2}\frac{\mathrm{d}}{\mathrm{d}t} \|u_N\|_{L^2}^2 &= \int_\Omega f(u_N)(\partial_x u_N) \,\mathrm{d}x - \left(f_R^{\mathrm{num}} u_R - f_L^{\mathrm{num}} u_L\right) \\ &= -\int_\Omega (\partial_x f(u_N)) u_N \,\mathrm{d}x + \left(f(u_R)u_R - f(u_L)u_L\right) - \left(f_R^{\mathrm{num}} u_R - f_L^{\mathrm{num}} u_L\right), \end{aligned}$$

with the second equality resulting from applying integration by parts. Note that for the square entropy $\eta(u) = \frac{u^2}{2}$ with corresponding entropy flux $\psi(u)$ satisfying $\eta' f' = \psi'$, we have

$$\partial_x \psi(u) = \psi'(u)\partial_x u = u f'(u) \partial_x u = u \partial_x f(u),$$

yielding

$$\frac{1}{2}\frac{\mathrm{d}}{\mathrm{d}t} \|u_N\|_{L^2}^2 = -\left(\psi(u_R) - \psi(u_L)\right) + \left(f(u_R)u_R - f(u_L)u_L\right) - \left(f_R^{\mathrm{num}} u_R - f_L^{\mathrm{num}} u_L\right). \tag{7.35}$$

Further, by defining

$$\gamma(u) := \int^u f(v) \,\mathrm{d}v,$$

the entropy flux $\psi(u)$ can be written as (see [JS94])

$$\psi(u) = \int^u f'(v) v \,\mathrm{d}v = f(u)u - \int^u f(v) \,\mathrm{d}v = f(u)u - \gamma(u) \tag{7.36}$$

so that (7.35) becomes

$$\begin{aligned}
\frac{1}{2}\frac{\mathrm{d}}{\mathrm{d}t}\|u_N\|_{L^2}^2 &= (\gamma(u_R)-\gamma(u_L)) - (f_R^{\text{num}}u_R - f_L^{\text{num}}u_L) \\
&= (\gamma(u_R)-\gamma(g_R)) - (\gamma(u_L)-\gamma(g_L)) \\
&\quad + (\gamma(g_R)-\gamma(g_L)) - (f_R^{\text{num}}u_R - f_L^{\text{num}}u_L),
\end{aligned}$$

where g_L and g_R are the BCs given as part of (7.22). By the mean value theorem, there exists a u_L^* between u_L and g_L as well as a u_R^* between u_R and g_R such that

$$\begin{aligned}
\gamma(u_L)-\gamma(g_L) &= (u_L-g_L)\,f(u_L^*), \\
\gamma(u_R)-\gamma(g_R) &= (u_R-g_R)\,f(u_R^*).
\end{aligned}$$

In this case we have

$$\begin{aligned}
\frac{1}{2}\frac{\mathrm{d}}{\mathrm{d}t}\|u_N\|_{L^2}^2 &= (u_R-g_R)\,f(u_R^*) - (u_L-g_L)\,f(u_L^*) + (\gamma(g_R)-\gamma(g_L)) \\
&\quad - (f_R^{\text{num}}u_R - f_L^{\text{num}}u_L) \\
&= (g_R-u_R)\,(f_R^{\text{num}} - f(u_R^*)) + (u_L-g_L)\,(f_L^{\text{num}} - f(u_L^*)) \\
&\quad + (\gamma(g_R)-\gamma(g_L)) - (g_R f_R^{\text{num}} - g_L f_L^{\text{num}}),
\end{aligned}$$

where the numerical fluxes are given respectively by

$$\begin{aligned}
f_L^{\text{num}} &= f^{\text{num}}(g_L, u_L) \\
f_R^{\text{num}} &= f^{\text{num}}(u_R, g_R).
\end{aligned}$$

Thus, by employing an *E-flux* (see [Osh84]) so that

$$(b-a)\,(f^{\text{num}}(a,b) - f(u)) \leq 0$$

for all u between a and b, we have

$$\frac{1}{2}\frac{\mathrm{d}}{\mathrm{d}t}\|u_N\|_{L^2}^2 \leq (\gamma(g_R)-\gamma(g_L)) - (g_R f_R^{\text{num}} - g_L f_L^{\text{num}})$$

Finally, utilizing (7.36) results in

$$\frac{\mathrm{d}}{\mathrm{d}t}\|u_N\|_{L^2}^2 \leq -2\psi(u_N)|_{\partial\Omega} + 2g_R\,(f(g_R) - f_R^{\text{num}}) - 2g_L\,(f(g_L) - f_L^{\text{num}}),$$

which is consistent with (7.25) since the numerical flux f^{num} is consistent with the flux f. In particular, the above inequality implies that

$$\frac{\mathrm{d}}{\mathrm{d}t}\|u_N\|_{L^2}^2 \leq 0$$

for periodic BCs. This yields a conservative and L^2 (energy) stable RBF method for general one dimensional scalar CLs.

7.2.5 Weak radial basis function collocation methods

Depending on the nonlinearity of f, the exact evaluation of $f(u_N)$ and resulting integrals may be impractical or even impossible. Remember that this has been part of the motivation for the

DGSEM; see Chapter 6.1. In the context of RBF methods, we therefore propose a collocation based alternative to the weak RBF analytic method given in Definition 7.4.

As before, we replace u and v with their RBF approximations $u_N, v_N \in V_{N,P}$ for $P \geq 1$, so that they also include constants, and possibly higher order polynomials. In the collocation case, $f(u)$ is approximated using an RBF interpolant $f_N \in V_{N,P}$ such that

$$f_N(x_k) = f(u_N(x_k)), \quad k = 1, \ldots, N.$$

We can now proceed as in the weak RBF analytical method. Specifically, we define:

Definition 7.5 (Weak RBF collocation method)
Find $u_N \in V_{N,P}$ such that for all $v_N \in V_{N,P}$:

$$\int_\Omega (\partial_t u_N) v_N \, \mathrm{d}x - \int_\Omega f_N (\partial_x v_N) \, \mathrm{d}x + (f_R^{\text{num}} v_R - f_L^{\text{num}} v_L) = 0 \tag{7.37}$$

Conservation

As in the weak RBF analytical case, conservation follows by including constants in the RBF interpolants, i. e. by choosing $P \geq 1$.

L^2 stability

For the weak RBF collocation method, we can only prove L^2 stability for the linear advection equation, given by

$$\partial_t u + \lambda \partial_x u = 0, \tag{7.38}$$

with constant velocity $\lambda > 0$. The case $\lambda < 0$ can be treated analogously. Here, the entropy flux is given as $\psi(u) = (\lambda/2)u^2$. By choosing $v_N = u_N$ in (7.37), we obtain

$$\begin{aligned}
\frac{1}{2}\frac{\mathrm{d}}{\mathrm{d}t} \|u_N\|_{L^2}^2 &= \int_\Omega (\partial_t u) u \, \mathrm{d}x \\
&= \lambda \int_\Omega u_N (\partial_x u_N) \, \mathrm{d}x - (f_R^{\text{num}} u_R - f_L^{\text{num}} u_L) \\
&= -\lambda \int_\Omega (\partial_x u_N) u_N \, \mathrm{d}x + \lambda \left(u_R^2 - u_L^2 \right) - (f_R^{\text{num}} u_R - f_L^{\text{num}} u_L),
\end{aligned}$$

where we have used integration by parts. Summing up the second and third equations above yields

$$\frac{\mathrm{d}}{\mathrm{d}t} \|u_N\|_{L^2}^2 = \lambda \left(u_R^2 - u_L^2 \right) - 2 (f_R^{\text{num}} u_R - f_L^{\text{num}} u_L), \tag{7.39}$$

which can be rewritten as

$$\frac{\mathrm{d}}{\mathrm{d}t} \|u_N\|_{L^2}^2 = -2\psi(u_N)\big|_{\partial\Omega} + 2\lambda u_R (u_R - f_R^{\text{num}}) - 2\lambda u_L (u_L - f_L^{\text{num}}).$$

Now, by employing a simple *upwind flux*, $f^{\text{num}}(a, b) = \lambda a$, we have

$$\begin{aligned}
f_L^{\text{num}} &= f^{\text{num}}(g_L, u_L) = \lambda g_L, \\
f_R^{\text{num}} &= f^{\text{num}}(u_R, g_R) = \lambda u_R,
\end{aligned}$$

and therefore

$$\frac{\mathrm{d}}{\mathrm{d}t} \|u_N\|_{L^2}^2 = -2\psi(u_N)\big|_{\partial\Omega} - 2\lambda u_L (u_L - g_L).$$

The above equation is consistent with (7.25). Note that for the linear advection equation (7.38) no shock waves arise and the inequalities (7.24) and (7.25) become equalities. For periodic BCs, (7.39) reduces to

$$\begin{aligned}
\frac{\mathrm{d}}{\mathrm{d}t} \|u_N\|_{L^2}^2 &= \lambda \left(u_R^2 - u_L^2\right) - 2\left(f^{\text{num}}(u_R, u_L)u_R - f^{\text{num}}(u_R, u_L)u_L\right) \\
&= -\lambda u_R^2 + 2\lambda u_L u_R - \lambda u_L^2 \\
&= -\lambda(u_R - u_L)^2 \\
&\leq 0.
\end{aligned}$$

Recall that for general CLs $\partial_t u + \partial_x f(u) = 0$, L^2 stability for the weak RBF analytical method in Definition 7.4 has been proven by utilizing the relation

$$\partial_x \psi(u_N) = (\partial_x u_N)\psi'(u_N) = (\partial_x u_N)\eta'(u_N)f'(u_N) = u_N \partial_x f(u_N), \tag{7.40}$$

for the square entropy $\eta(u) = \frac{u^2}{2}$. For the weak RBF collocation method in Definition 7.5, $f(u_N)$ in (7.40) is replaced by f_N and the final equality does not hold. Thus we are unable to prove L^2 stability for general nonlinear CLs.

7.2.6 Relationship to other methods

Let us compare the above approaches, i. e. the weak RBF analytical method and weak RBF collocation method to some commonly employed schemes.

Discontinuous Galerkin methods

DG methods, see [HW07] and references therein, are perhaps the most obviously comparable. DG methods use a partition of the domain Ω into smaller elements Ω_i with $\bigcup_i \Omega_i = \Omega$. In each element the problem is discretized in a weak form similar to (7.31), where the numerical solution u and the test functions v are typically replaced by polynomials in every element Ω_i. These polynomials are allowed to be discontinuous at the element interfaces and numerical fluxes are utilized to couple neighboring elements and to weakly enforce BCs. In this context, the proposed weak RBF method might be interpreted as (discontinuous) Galerkin method in which a single big element $\Omega_i = \Omega$ is used and the polynomial approximations are replaced with RBF interpolants. In a nodal approach this allows the use of more sophisticated sets of interpolation points, especially in higher dimensions (although these are not considered in this work). Note that by the Mairhuber–Curtis theorem polynomial interpolation in general is not well-defined in more than one dimension; see Chapter 3.1.4.

Spectral Galerkin tau methods

Spectral Galerkin methods solve the PDE in form of an integral equation as well, only *without* including the BCs in the integral equation. The BCs can, for instance, be enforced directly by choosing suitable trial functions to span the approximation space, e. g. by choosing $V_N = \text{span}\{\sin(\pi k x) \mid k = 1, \ldots, N\}$ in case of homogeneous Dirichlet BCs on $\Omega = [0, 1]$. The so-called *spectral Galerkin tau methods*, see [CHQZ06] and references therein, use trial functions that do not have to individually satisfy the BCs, but rather some additional equations are imposed to ensure the numerical solution satisfies BCs. To maintain a well-posed discretization, i. e., the number of equations being equal to the number of degrees of freedom, some of the integral integrations corresponding to the highest order test functions are dropped in favor of the BC equations. In the weak RBF method, on the other hand, these BC equations include numerical flux functions and are incorporated into the integral equations corresponding to the test functions. As a consequence, we do not need to remove any test functions from the integral equations, yielding higher order of accuracy.

Penalty-type boundary treatment in pseudospectral methods

As with strong RBF methods, classical pseudospectral methods typically are built from bases of Fourier, Chebyshev or Legendre polynomials, and require that the BCs are strongly (exactly) imposed; see [GH01] and references therein. Penalty methods, i. e. using a penalty term for treating BCS, have been used both for spectral methods in the weak [CQ82] and strong [FG88, FG91] forms. The basic idea behind penalty methods is that it suffices to impose the BCs to the order of the given scheme, which can be done by introducing a penalty term into the discretized equation. In particular, the BCs have to be satisfied exactly by the numerical solution only in the limit of infinite order. Depending on the method and problem under consideration it may be challenging to construct suitable penalty terms.

In the weak RBF method, such penalty terms are derived somewhat naturally by utilizing numerical flux functions. As a consequence, a large class of penalty terms may be available for practical use. Future work will address the development of stable RBF methods in strong form. As discussed above, a bottleneck for such an investigation will be the development of suitable penalty terms for the boundary treatment in a strong RBF method. This is consistent with the observation that classic strong RBF methods (in which BCs are imposed strongly), so far, have only been observed to be stable if no BCs were present [PD06].

7.2.7 Efficient implementation

We now address implementation details for both weak RBF analytical and weak RBF collocation methods for scalar CLs.

Numerical fluxes

There are several options for choosing numerical fluxes that result in L^2 stable weak RBF methods for one dimensional scalar CLs. For example:

- For the linear advection, $\partial_t u + \lambda \partial_x u = 0$, with constant velocity $\lambda \neq 0$, the general upwind flux,

$$f^{\text{num}}(a,b) = \begin{cases} \lambda a & \text{if } \lambda > 0, \\ \lambda b & \text{if } \lambda < 0, \end{cases}$$

  yields L^2 stability for both the analytical and collocation forms.

- For the nonlinear case, $\partial_t u + \partial_x f(u) = 0$, we can use an E-Flux as defined in [Osh84] (see also [CS89] and references therein). For example, the Godunov flux is given by

$$f^{\text{num}}(a,b) = \begin{cases} \min_{a \leq u \leq b} f(u) & \text{if } a \leq b, \\ \max_{a \geq u \geq b} f(u) & \text{if } a > b. \end{cases}$$

Weak RBF collocation methods

In the weak RBF collocation method, provided in Definition 7.5, u_N and f_N are collocated RBF interpolants with respect to the same set of centers x_n, $n = 1, \ldots, N$. It is important to note that (7.37) must hold for all $v \in V_N$.

Weak RBF collocation methods - Standard RBF approximations

To better understand the purpose of introducing polynomials in (7.31), we first consider the standard RBF approximation, that is when $P = 0$, yielding the space $V_{N,0} = V_N$ with a basis given by

$$\phi_n(x) := \phi\left(|x - x_n|\right), \quad n = 1, \ldots, N.$$

Then, by following (7.37), we obtain the system of equations

$$\int_\Omega (\partial_t u_N)\phi_n \,\mathrm{d}x - \int_\Omega f_N(\partial_x \phi_n)\,\mathrm{d}x + [f_R^{\mathrm{num}}\phi_n(b) - f_L^{\mathrm{num}}\phi_n(a)] = 0, \quad n = 1, \ldots, N.$$

Further expressing $u_N, f_N \in V_N$ with respect to the basis $\{\phi_k\}_{k=1}^N$ as in (7.29) with $P = 0$, the system becomes

$$\sum_{k=1}^{N} \left(\frac{\mathrm{d}}{\mathrm{d}t}\hat{u}_k\right) \int_\Omega \phi_k \phi_n \,\mathrm{d}x = \sum_{k=1}^{N} \hat{f}_k \int_\Omega \phi_k(\partial_x \phi_n)\,\mathrm{d}x - [f_R^{\mathrm{num}}\phi_n(b) - f_L^{\mathrm{num}}\phi_n(a)], \tag{7.41}$$

for $n = 1, \ldots, N$. We now define the corresponding *mass* and *stiffness matrices*, $M \in \mathbb{R}^{N\times N}$ and $S \in \mathbb{R}^{N\times N}$, as

$$M = \begin{pmatrix} \langle\phi_1, \phi_1\rangle & \ldots & \langle\phi_N, \phi_1\rangle \\ \vdots & & \vdots \\ \langle\phi_1, \phi_N\rangle & \ldots & \langle\phi_N, \phi_N\rangle \end{pmatrix}, \quad S = \begin{pmatrix} \langle\phi_1, \partial_x\phi_1\rangle & \ldots & \langle\phi_N, \partial_x\phi_1\rangle \\ \vdots & & \vdots \\ \langle\phi_1, \partial_x\phi_N\rangle & \ldots & \langle\phi_N, \partial_x\phi_N\rangle \end{pmatrix},$$

where the $\langle a, b\rangle$ denotes the inner product given by

$$\langle a, b\rangle := \int_\Omega a(x)b(x)\,\mathrm{d}x.$$

We also define the *restriction* and *boundary matrices*, $R \in \mathbb{R}^{2\times N}$ and $B \in \mathbb{R}^{2\times 2}$ as

$$R = \begin{pmatrix} \phi_1(a) & \ldots & \phi_N(a) \\ \phi_1(b) & \ldots & \phi_N(b) \end{pmatrix}, \quad B = \begin{pmatrix} -1 & 0 \\ 0 & 1 \end{pmatrix},$$

along with the vectors $\hat{\mathbf{u}} \in \mathbb{R}^N$, $\hat{\mathbf{f}} \in \mathbb{R}^N$ and $\boldsymbol{f}^{\mathrm{num}} \in \mathbb{R}^2$:

$$\hat{\mathbf{u}} = \begin{pmatrix} \hat{u}_1 \\ \vdots \\ \hat{u}_N \end{pmatrix}, \quad \hat{\mathbf{f}} = \begin{pmatrix} \hat{f}_1 \\ \vdots \\ \hat{f}_N \end{pmatrix}, \quad \boldsymbol{f}^{\mathrm{num}} = \begin{pmatrix} f_L^{\mathrm{num}} \\ f_R^{\mathrm{num}} \end{pmatrix}.$$

Note that the above notation is essentially the same as for DG methods, this time however formulated with respect to the modal degrees of freedom instead of the nodal ones; see Chapter 6.1.4. With the above definitions, we can now write (7.41) in vector matrix notation as

$$M\frac{\mathrm{d}}{\mathrm{d}t}\hat{\mathbf{u}} = S\hat{\mathbf{f}} - R^T B \boldsymbol{f}^{\mathrm{num}}.$$

Assuming that mass matrix M is invertible, we obtain the semidiscrete equation

$$\frac{\mathrm{d}}{\mathrm{d}t}\hat{\mathbf{u}} = M^{-1}S\hat{\mathbf{f}} - M^{-1}R^T B \boldsymbol{f}^{\mathrm{num}}, \tag{7.42}$$

which can be solved, for instance, by the SSPRK(3,3) method given in Definition 3.39. The existence of M^{-1} is addressed below. Finally, note that for the linear advection equation $\partial_t u + \lambda \partial_x u = 0$, (7.42) reduces to

$$\frac{\mathrm{d}}{\mathrm{d}t}\hat{\mathbf{u}} = \lambda M^{-1} S \hat{\mathbf{u}} - M^{-1} R^T B \boldsymbol{f}^{\mathrm{num}}, \tag{7.43}$$

for the weak RBF collocation method. An eigenvalue stability analysis, such as in [PD06], will be subjects of future work.

Weak RBF collocation methods - RBF approximations in $V_{N,P}$ with $P \geq 1$

If the RBF approximations do include polynomials ($P \geq 1$), the approximation space $V_{N,P}$ is given (7.31). In this case, it becomes harder to determine a basis of $V_{N,P}$. Here we proceed as follows: We choose the basis $\{b_n\}_{n=1}^N$ with

$$b_n(x) = \sum_{k=1}^N \alpha_k^{(n)} \phi_k(x) + \sum_{l=1}^P \beta_l^{(n)} q_l(x).$$

The coefficients $\boldsymbol{\alpha}^{(n)} = (\alpha_1^{(n)}, \dots, \alpha_N^{(n)})^T$ and $\boldsymbol{\beta}^{(n)} = (\beta_1^{(n)}, \dots, \beta_P^{(n)})^T$ are uniquely determined by the linear system

$$\begin{pmatrix} V & Q \\ Q^T & 0 \end{pmatrix} \begin{pmatrix} \boldsymbol{\alpha}^{(n)} \\ \boldsymbol{\beta}^{(n)} \end{pmatrix} = \begin{pmatrix} \mathbf{e}^{(n)} \\ 0 \end{pmatrix},$$

where $\mathbf{e}^{(n)}$ denotes the nth unit vector in $\mathbb{R}^N$, i. e. $\mathbf{e}^{(n)} = (\delta_{kn})_{k=1}^N$, and the matrix Q is given by

$$Q = \begin{pmatrix} q_1(x_1) & \dots & q_P(x_1) \\ \vdots & & \vdots \\ q_1(x_N) & \dots & q_P(x_N) \end{pmatrix};$$

also see Chapter 3.1.5. From here on, we can operate as described for the approximation space V_N without polynomials. Following (7.37), we obtain the system of equations

$$\int_\Omega (\partial_t u_N) b_n \,\mathrm{d}x - \int_\Omega f_N (\partial_x b_n) \,\mathrm{d}x + [f_R^{\mathrm{num}} b_n(b) - f_L^{\mathrm{num}} b_n(a)] = 0, \quad n = 1, \dots, N,$$

and expressing $u_N, f_N \in V_{N,P}$ with respect to the basis $\{b_n\}_{n=1}^N$ finally results in

$$\frac{\mathrm{d}}{\mathrm{d}t}\hat{\mathbf{u}} = M^{-1} S \hat{\mathbf{f}} - M^{-1} R^T B \boldsymbol{f}^{\mathrm{num}}. \tag{7.44}$$

This is the same matrix vector representation as in (7.42), only this time the matrices are defined by applying all operations to the basis $\{b_n\}_{n=1}^N$ instead of the RBF basis $\{\phi_n\}_{n=1}^N$, e. g.

$$M = \begin{pmatrix} \langle b_1, b_1 \rangle & \dots & \langle b_N, b_1 \rangle \\ \vdots & & \vdots \\ \langle b_1, b_N \rangle & \dots & \langle b_N, b_N \rangle \end{pmatrix}.$$

Weak RBF analytical methods

From Definition 7.4, we similarly can derive

$$\sum_{k=1}^{N} \left(\frac{\mathrm{d}}{\mathrm{d}t} \hat{u}_k \right) \int_\Omega b_k b_n \, \mathrm{d}x = \int_\Omega f(u_N)(\partial_x b_n) \, \mathrm{d}x - [f_R^{\mathrm{num}} b_n(b) - f_L^{\mathrm{num}} b_n(a)] \tag{7.45}$$

for $n = 1, \ldots, N$. However (7.45) does not have an equivalent form to (7.41) since the second integral cannot be decomposed using $\sum_{k=1}^{N} \hat{f}_k \langle b_k, \partial_x b_n \rangle$. Thus, by defining the flux stiffness vector $\mathbf{S_f} \in \mathbb{R}^N$ as

$$\mathbf{S_f} = \begin{pmatrix} \langle f(u_N), \partial_x b_1 \rangle \\ \vdots \\ \langle f(u_N), \partial_x b_N \rangle \end{pmatrix},$$

we can now write the weak RBF analytical method in vector matrix notation as

$$M \frac{\mathrm{d}}{\mathrm{d}t} \hat{\mathbf{u}} = \mathbf{S_f} - R^T B \boldsymbol{f}^{\mathrm{num}}.$$

Again assuming M is invertible we obtain

$$\frac{\mathrm{d}}{\mathrm{d}t} \hat{\mathbf{u}} = M^{-1} \mathbf{S_f} - M^{-1} R^T B \boldsymbol{f}^{\mathrm{num}}. \tag{7.46}$$

For the linear advection equation, $\partial_t u + \lambda \partial_x u = 0$, (7.46) reduces to (7.42), since in this case $f(u) = \lambda u$ yielding

$$\mathbf{S_f} = \begin{pmatrix} \langle \lambda u_N, \partial_x b_1 \rangle \\ \vdots \\ \langle \lambda u_N, \partial_x b_N \rangle \end{pmatrix} = \begin{pmatrix} \lambda \sum_{k=1}^{N} \hat{u}_k \langle b_k, (\partial_x \phi_1 \rangle \\ \vdots \\ \lambda \sum_{k=1}^{N} \hat{u}_k \langle b_k, \partial_x \phi_N \rangle \end{pmatrix} = \lambda S \hat{\mathbf{u}}.$$

That is, the weak RBF analytical and collocation methods coincide for linear advection.

Inverse of the mass matrix and orthonormal RBF bases

Both weak forms of RBF methods, (7.44) and (7.46), require the invertibility of the mass matrix M. Indeed in all of our numerical experiments, M is invertible, although in some cases it was found to be nearly singular. As is well known, a nearly singular mass matrix M may result in numerical instabilities.

To avoid this potential problem, one option might be to use a different basis to represent the approximation space $V_{N,P}$ given by (7.31). For example, we can represent the RBF interpolants (including polynomials) with respect to the orthonormal basis

$$\{\psi_n\}_{n=1}^{N} \subset V_{N,P} \quad \text{such that} \quad \langle \psi_k, \psi_n \rangle = \delta_{k,n}. \tag{7.47}$$

Such an orthonormal basis of RBFs including polynomials can, for instance, be constructed by the Stieltjes [Gau04] or (modified) Gram–Schmidt procedure [TBI97]. See Chapter 3.2 for more details (though we do not have a purely polynomial basis here). Note that when using (7.47), the mass matrix reduces to the identity matrix, i. e. $M = I = \mathrm{diag}(1, \ldots, 1)$.

As a second option, which was observed in our numerical experiments, we found that the mass matrix was well-conditioned, as long as the shape parameter ε is large enough. Hence we did not incorporate (7.47) in our implementations, although it might be useful for other applications.

7.2.8 Numerical results

Let us now demonstrate our theoretical findings for the weak RBF analytical and collocation methods. The subsequent numerical tests address the linear advection equation as well as the system of Euler equations. Further, for all test, we have again used the SSPRK(3,3) method (see Definition 3.39). For the time step Δt we use

$$\Delta t = C \cdot \frac{|\Omega|}{N \max |f'(u)|}$$

with $C = 0.1$. Here, $\max |f'(u)|$ is calculated for all u between $\min_{x \in \Omega} u_0(x)$ and $\max_{x \in \Omega} u_0(x)$. Note that for the linear advection equation $\partial_t u + \lambda \partial_x u = 0$, we simply have $\max |f'(u)| = |\lambda|$.

Linear advection equation

Let us again consider the linear advection equation in (7.38) for $\lambda = 1$ and $x \in \Omega = [-1, 1]$. We will test our algorithm on two different test cases, one with periodic and the other with inflow BCs given respectively by

$$u_1(x,0) = \sin(\pi x) + 1, \qquad u(-1,t) = u(1,t); \tag{7.48a}$$

$$u_2(x,0) = \exp\left(-20x^2\right), \qquad u(-1,t) = \exp(-20). \tag{7.48b}$$

We will compare solutions given by the weak RBF methods for $P = 0$ (no polynomials included) and $P = 1$ (constants included) to the standard RBF collocation method, which solves (7.38) in its strong form. The latter will be subsequently referred to as the *strong RBF collocation method* . Recall from Chapter 7.2.7 that in the linear advection case, the weak RBF analytical and collocation methods are the same.

Figure 7.11 illustrates the numerical results comparing the strong and weak RBF collocation methods using the inverse quadric kernel

$$\phi(r) = \frac{1}{(\varepsilon r)^2 + 1}$$

with shape parameter $\varepsilon = 7$ (also see Table 3.1) for solving $\partial_t u + \partial_x u = 0$ with ICs and BCs in (7.48a) at time $t = 2$ using both $N = 20$ and $N = 30$ equidistant centers. The solutions u_N are compared in figures 7.11(a) and 7.11(d), the corresponding momentum in time, $\int u_N(x,t)\,dx$, in figures 7.11(b) and 7.11(e), and the corresponding energy in time, $\int u_N^2(x,t)\,dx$, in figures 7.11(c) and 7.11(f).

Observe that in all cases, the weak RBF collocation method provides a more accurate solution, as well as physically reasonable behavior for both momentum and energy. Moreover, the momentum is conserved *only* when the RBF approximation includes the constant term ($P = 1$). This is consistent with our theoretical findings for periodic BCs that show that for $P \geq 1$, energy is noninceasing for the weak RBF collocation method. Our numerical solution also produces an energy that remains nearly constant in time, with the discrepancy to the true energy due to the difference between the exact IC and its approximation by the RBF interpolant. The strong RBF collocation method, by contrast, produces profiles for the momentum and energy that are not accurate, which is further observed in the numerical solution u_N. Moreover, although the L^2 norm (energy) is bounded in this example (until $t = 2$), this is not the case for all time, as is demonstrated in Figure 7.12. Hence we see that while the L^2 norm of the numerical solution of the weak RBF collocation method for $P = 0$ as well as

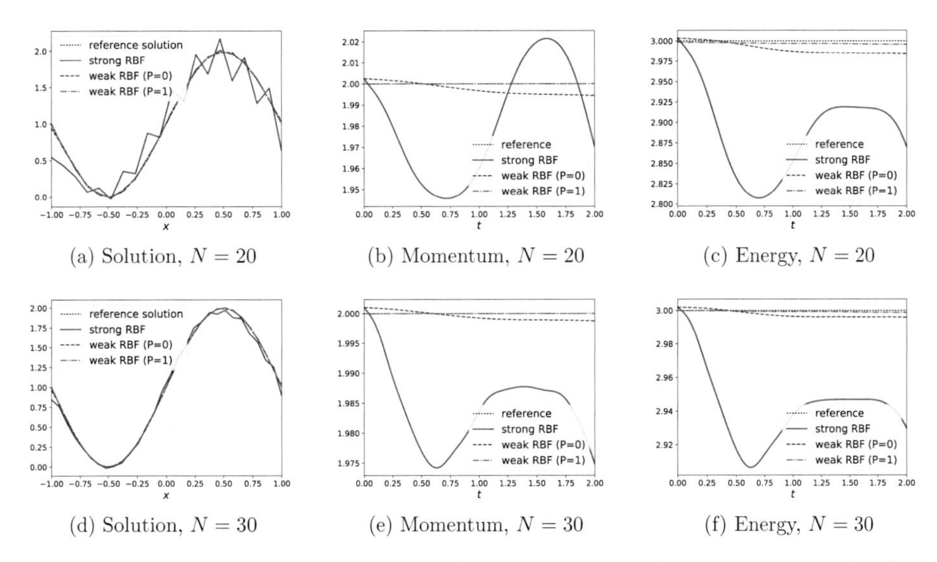

(a) Solution, $N = 20$ (b) Momentum, $N = 20$ (c) Energy, $N = 20$

(d) Solution, $N = 30$ (e) Momentum, $N = 30$ (f) Energy, $N = 30$

Figure 7.11: Numerical solutions as well as their momentum and energy over time for $\partial_t u + \partial_x u = 0$ with IC and BC given by (7.48a). Here we used the inverse quadric kernel with shape parameter $\varepsilon = 7$. The number of equidistant centers are (top) $N = 20$ and (bottom) $N = 30$.

$P = 1$ decreases and therefore remains bounded, the L^2 norm of the strong RBF collocation method becomes unstable at some later time t. The same observations hold true for other kernels, e. g. Gaussian and multiquadric, different shape parameters, and other ICs. See Table 3.1 in Chapter 3.1.5 for a discussion of the different RBFs.

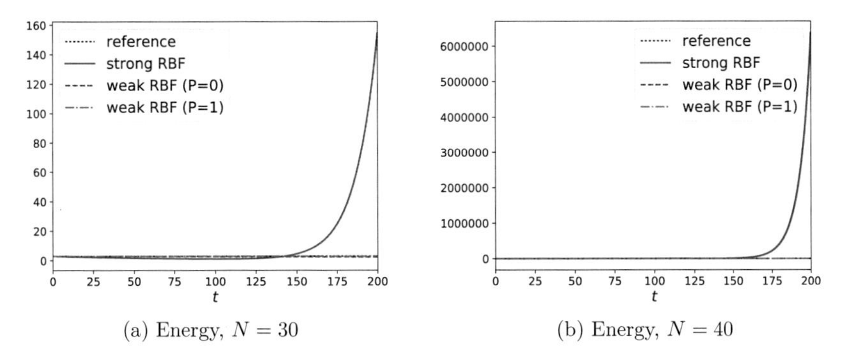

(a) Energy, $N = 30$ (b) Energy, $N = 40$

Figure 7.12: Energy of the numerical solutions for $\partial_t u + \partial_x u = 0$ with IC and BC given by (7.48a) over time period $0 \leq t \leq 200$. Once again we use an inverse quadric kernel with shape parameter $\varepsilon = 7$.

The weak RBF collocation method for $P = 0$ and $P = 1$ was more accurate in nearly all cases, as is reported in Table 7.5 for a Gaussian kernel, in Table 7.6 for a multiquadric kernel, and in Table 7.7 for an inverse quadric kernel. There, the second test case (7.48b) for nonperiodic inflow BCs was considered for time $t = 1$.

		$u_1(0,x) = \sin(\pi x)$ with periodic BC						$u_2(0,x) = \exp(-20x^2)$ with inflow BC					
		strong RBF		weak RBF				strong RBF		weak RBF			
				$P=0$		$P=1$				$P=0$		$P=1$	
ε	N	l^2-error	EOC	l^2-error	EOC	l^2-error	EOC	l^2-error	EOC	l^2-error	EOC	l^2-error	EOC
10	10	4.9e-1		1.1e-0		1.1e-0		4.8e-1		3.9e-1		3.9e-1	
	20	2.0e-1		4.1e-2		2.9e-2		5.1e-2		2.4e-2		1.7e-2	
	30	5.7e-2		1.9e-2		5.6e-3		1.7e-2		1.3e-2		1.0e-2	
	40	2.3e-2	1.6	5.8e-3	3.7	1.5e-3	3.8	3.9e-3	3.2	3.7e-3	3.6	3.0e-3	3.7
11	10	2.2e-1		1.3e-0		1.2e-0		4.8e-1		4.4e-1		4.4e-1	
	20	2.7e-1		3.9e-2		3.4e-2		6.3e-2		2.3e-2		1.7e-2	
	30	5.2e-2		2.5e-2		9.4e-3		2.8e-2		1.8e-2		1.4e-2	
	40	3.0e-2	0.7	8.7e-3	3.7	1.5e-3	3.7	6.8e-3	2.8	6.2e-3	3.5	5.0e-3	3.6
12	10	8.4e-2		1.1e-0		1.1e-0		4.8e-1		4.7e-1		4.6e-1	
	20	3.8e-1		3.4e-2		3.5e-2		1.0e-1		2.1e-2		1.6e-2	
	30	7.4e-2		3.1e-2		1.3e-2		4.0e-2		2.2e-2		1.7e-2	
	40	4.0e-2	0.2	1.2e-2	3.7	2.5e-3	3.7	1.1e-2	2.3	9.4e-3	3.6	7.5e-3	3.6

Table 7.5: l^2 error and EOC for the Gaussian kernel. The EOC was computed using a least squares fit for the parameters C and s in the model $y = C \cdot N^{-s}$, where y denotes the l^2-error for a fixed N.

		$u_1(0,x) = \sin(\pi x)$ with periodic BC						$u_2(0,x) = \exp(-20x^2)$ with inflow BC					
		strong RBF		weak RBF				strong RBF		weak RBF			
				$P=0$		$P=1$				$P=0$		$P=1$	
ε	N	l^2-error	EOC	l^2-error	EOC	l^2-error	EOC	l^2-error	EOC	l^2-error	EOC	l^2-error	EOC
18	10	4.8e-1		2.6e-2		2.8e-2		2.6e-1		1.1e-1		1.1e-1	
	20	2.4e-1		7.9e-3		7.6e-3		5.7e-2		9.7e-3		8.5e-3	
	30	1.0e-1		3.9e-3		3.6e-3		1.8e-2		4.5e-3		3.8e-3	
	40	3.3e-2	1.3	2.0e-3	1.7	1.9e-3	1.8	9.2e-3	2.2	2.2e-3	3.3	1.7e-3	3.4
19	10	4.9e-1		2.6e-2		2.8e-2		2.6e-1		1.1e-1		1.1e-1	
	20	2.4e-1		7.9e-3		7.7e-3		6.0e-2		9.7e-3		8.6e-3	
	30	1.1e-1		4.1e-3		3.7e-3		1.9e-2		4.7e-3		3.9e-3	
	40	4.1e-2	1.3	2.1e-3	1.7	2.0e-3	1.8	9.9e-3	2.2	2.4e-3	3.3	1.9e-3	3.4
20	10	5.0e-1		2.6e-2		2.8e-2		2.6e-1		1.1e-1		1.1e-1	
	20	2.4e-1		8.0e-3		7.8e-3		6.4e-2		9.8e-3		8.6e-3	
	30	1.2e-1		4.3e-3		3.8e-3		2.0e-2		4.8e-3		4.0e-3	
	40	4.8e-2	1.2	2.3e-3	1.6	2.1e-3	1.8	1.0e-2	2.1	2.5e-3	3.3	2.0e-3	3.4

Table 7.6: l^2 error and EOC for the multiquadric kernel.

Tables 7.5, 7.6, and 7.7 also demonstrate that the EOC increases in most cases using the weak RBF collocation method as opposed to when the usual strong RBF collocation method is used. Including constants in the weak RBF collocation method ($P = 1$) further enhances the accuracy and rate of convergence compared to the weak RBF collocation method without polynomials ($P = 0$). Moreover, although the shape parameter ε depends on the degree of

		$u_1(0,x) = \sin(\pi x)$ with periodic BC						$u_2(0,x) = \exp(-20x^2)$ with inflow BC					
		strong RBF		weak RBF				strong RBF		weak RBF			
				$P=0$		$P=1$				$P=0$		$P=1$	
ε	N	l^2-error	EOC	l^2-error	EOC	l^2-error	EOC	l^2-error	EOC	l^2-error	EOC	l^2-error	EOC
6	10	7.9e-1		4.7e-2		3.3e-2		3.0e-1		1.2e-1		1.2e-1	
	20	1.4e-1		1.5e-2		1.2e-2		1.2e-2		5.5e-3		3.2e-3	
	30	3.3e-2		7.2e-3		3.1e-3		3.6e-3		3.9e-3		2.5e-3	
	40	1.3e-2	2.6	3.8e-3	1.6	1.6e-3	1.7	1.4e-3	3.7	2.2e-3	3.6	1.7e-3	3.7
7	10	1.0e-0		7.8e-2		7.6e-2		3.4e-1		1.5e-1		1.5e-1	
	20	2.4e-1		1.6e-2		1.6e-2		2.1e-2		7.3e-3		4.7e-3	
	30	5.1e-2		8.9e-3		5.1e-3		6.1e-3		5.5e-3		3.7e-3	
	40	2.0e-2	2.2	4.8e-3	2.1	1.6e-3	2.3	2.5e-3	3.6	3.2e-3	3.6	2.4e-3	3.7
8	10	1.1e-0		1.3e-1		1.4e-1		3.8e-1		1.9e-1		1.9e-1	
	20	2.9e-1		1.8e-2		1.9e-2		3.9e-2		8.5e-3		5.9e-3	
	30	4.9e-2		1.0e-2		7.8e-3		9.7e-3		7.3e-3		5.1e-3	
	40	2.3e-2	2.1	6.0e-3	2.6	2.5e-3	2.8	4.2e-3	3.2	4.5e-3	3.6	3.3e-3	3.7

Table 7.7: l^2 error and EOC for the inverse quadric kernel.

freedom in the particular RBF method (i. e. the kernel used and number of centers), it does not depend on the IC. A sufficiently large shape parameter is needed to ensure numerical stability of the RBF collocation method, however, and in particular for the mass matrix to be well conditioned. Although this is not as significant when polynomials are included in the RBF approximation space, it is still advisable to increase the shape parameter with the number of centers N for the RBF collocation method. Finally, we point out that the best results were obtained using the weak RBF collocation method that includes polynomials in the basis, as was predicted by our theoretical results.

Euler equations

Finally, we address the extension to hyperbolic systems of nonlinear CLs. Let us consider the one dimensional Euler equations

$$\partial_t \boldsymbol{u} + \partial_x \boldsymbol{f}(\boldsymbol{u}) = 0$$

for $x \in \Omega = [-1,1]$, where $\boldsymbol{u}$ and $\boldsymbol{f}(\boldsymbol{u})$ respectively are the vector of conserved variables and fluxes, given by

$$\boldsymbol{u} = \begin{pmatrix} u_1 \\ u_2 \\ u_3 \end{pmatrix} = \begin{pmatrix} \rho \\ \rho u \\ E \end{pmatrix}, \quad \boldsymbol{f} = \begin{pmatrix} f_1 \\ f_2 \\ f_3 \end{pmatrix} = \begin{pmatrix} \rho u \\ \rho u^2 + p \\ u(E+p) \end{pmatrix}.$$

Here, ρ is the *density*, u is the *velocity*, p is the *pressure*, and E is the *total energy per unit volume*. The Euler equations are completed by addition of an equation of state (EOS) with general form

$$p = p(\rho, e),$$

where $e = E/\rho - u^2/2$ is the *specific internal energy*. Here, we consider ideal gases for which the EOS is given by

$$p = (\gamma - 1)\rho e$$

with γ denoting the *ratio of specific heats*. Also see Chapter 2.1 for more details on the Euler equations. For the subsequent numerical tests, we set $\gamma = 3$ and consider a *smooth isentropic flow* resulting from the Euler equations with smooth ICs

$$\rho(x,0) = \rho_0(x) = 1 + \frac{1}{2}\sin(\pi x), \quad u(x,0) = u_0(x) = 0, \quad p(x,0) = p_0(x) = \rho_0^\gamma(x),$$

and periodic BCs. A similar test problem has been proposed by Cheng and Shu [CS14] as well as Abgrall et al. [ABT19] in the context of (positivity-preserving) high order methods. Utilizing the method of characteristics, the exact density ρ and velocity u are given by

$$\rho(x,t) = \frac{1}{2}\left[\rho_0(x_1) + \rho_0(x_2)\right], \quad u(x,t) = \sqrt{3}\left[\rho(x,t) - \rho_0(x_1)\right],$$

where $x_1 = x_1(x,t)$ and $x_2 = x_2(x,t)$ are solutions of the nonlinear equations

$$x + \sqrt{3}\rho_0(x_1)t - x_1 = 0, \quad x - \sqrt{3}\rho_0(x_2)t - x_2 = 0.$$

Finally, the exact pressure p can be computed by the $p = C\rho^\gamma$ for smooth flows; see [Tor13, Chapter 3.1].

Figure 7.13 illustrates the numerical results at time $t = 0.1$ comparing the strong and weak RBF collocation methods using a Gaussian kernel with $\varepsilon = 11$ (Figures 7.13(a),7.13(b), and 7.13(c)), a multiquadric kernel with $\varepsilon = 19$ (Figures 7.13(d),7.13(e), and 7.13(f)), and an inverse quadric kernel with $\varepsilon = 7$ (figures 7.13(g),7.13(h), and 7.13(i)). In all tests, we used $N = 20$ equidistant centers. As for the linear advection equation, we observe that the weak RBF collocation method for $P = 0$ and $P = 1$ is more accurate. Finally, we note that using $P \geq 1$ did not improve the above results, and in some cases the accuracy was slightly reduced. This may be due to the fact that the above problem of a smooth isentropic flow does not rely on extra polynomials to recover the solution space. However, as observed in Figure 7.10, there may indeed be problems for which $P \geq 1$ are appropriate, especially given limited resolution or different shape parameters. This will be explored in future investigations.

7.2.9 Concluding thoughts and outlook

In this chapter, we investigated the conservation and L^2 stability properties of RBF methods. In the process we demonstrated that traditional RBF methods based on the differential (strong) form of hyperbolic CLs violate these properties and might therefore produce physically unreasonable solutions. As an alternative we proposed two novel RBF schemes based on the weak form of the hyperbolic CL. In the weak RBF analytical method, the flux $f(u_N)$ is assumed to be evaluated exactly, while in the corresponding weak RBF collocation method, the flux is approximated by an RBF interpolant f_N. The second formulation is clearly easier to implement and more efficient. We further proved that both methods are conservative assuming that (at least) constants are included in the RBF space, i. e. $P \geq 1$, and are also L^2 stable. In case of the weak RBF collocation method this was shown only for the linear advection equation, when appropriate numerical (E-) fluxes are included in the discretization. Thus, the weak RBF methods are able to provide numerical solutions with physically reasonable mass and energy profiles.

A drawback of this approach are potentially ill-conditioned mass matrices, which arise from the weak form of the CL. Yet, this can be overcome by choosing sufficiently large shape parameters. For more sophisticated applications requiring other kernels, however, it might be better to use orthonormal basis functions instead.

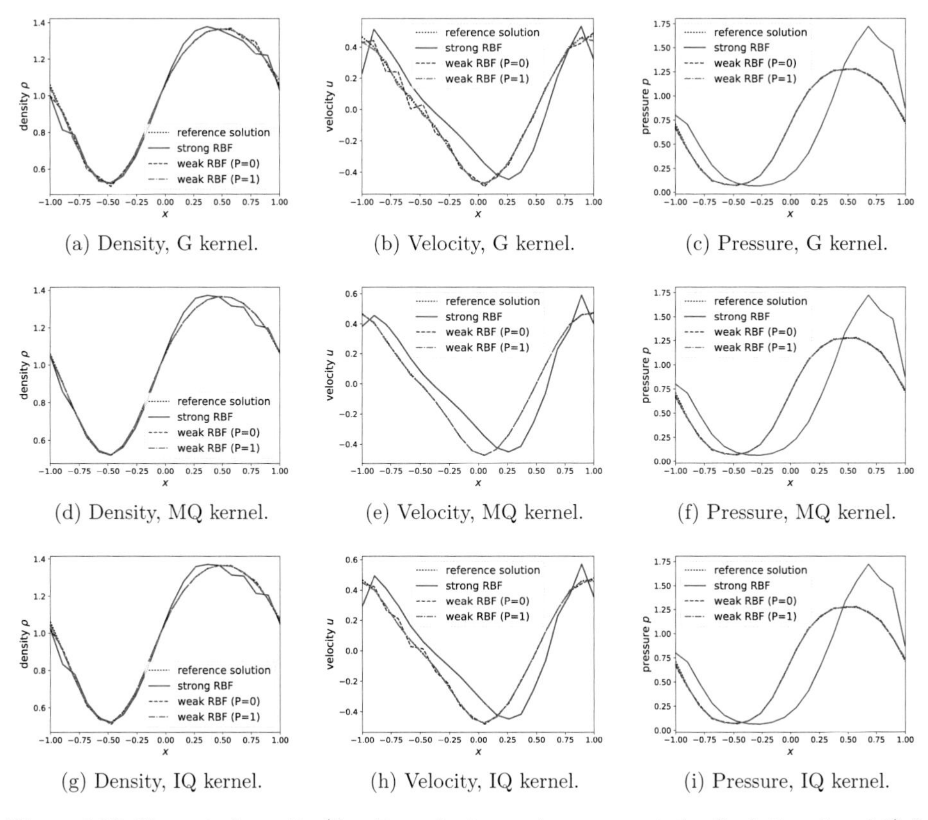

(a) Density, G kernel. (b) Velocity, G kernel. (c) Pressure, G kernel.

(d) Density, MQ kernel. (e) Velocity, MQ kernel. (f) Pressure, MQ kernel.

(g) Density, IQ kernel. (h) Velocity, IQ kernel. (i) Pressure, IQ kernel.

Figure 7.13: Numerical results (density, velocity, and pressure at the final time $t = 0.1$) for the Euler equations. In the first row, we used the Gaussian kernel with shape parameter $\varepsilon = 11$. In the second row, we used the multiquadric kernel with shape parameter $\varepsilon = 19$. In the third row, we used the inverse quadric kernel with shape parameter $\varepsilon = 7$. The number of equidistant centers is $N = 20$ in all tests.

Future work will focus on the application of the proposed weak RBF method to nonlinear problems and, in particular, on the adaptation of different methods [Tad90, KXR+04, HK08, KWH11, RGÖS18, GÖS18, GNJA+19, GG19a] from DG and related methods to further stabilize the weak RBF method in the presence of (shock) discontinuities. Finally, in addition to the L^2 stability analysis provided here, it would be useful to perform a (linear) eigenvalue stability analysis.

CHAPTER 8

ARTIFICIAL VISCOSITY METHODS

We have already observed in Chapter 2 that solutions of CLs might develop spontaneous (shock) discontinuities, even in finite time and for smooth ID. As a consequence, due to the famous Gibbs phenomenon, (polynomial) high order approximations to such discontinuous solutions show spurious oscillations and often yield the underlying numerical method to break down. In this chapter, we therefore address AV methods as a means to stabilize high order (discontinuous) SE methods, especially in the presence of (shock) discontinuities. In the process, we propose new viscosity distributions and present some original results concerning the discretization of AV methods using SBP operators in FR methods. The material presented here directly resulted in the publications [RGÖS18, GNJA$^+$19, GÖS18] and indirectly in the publication [ÖGR18, ÖGR19].

Outline

This chapter is organized as follows: We start by addressing the basic idea behind AV methods in Chapter 8.1 and revisiting the most commonly used state-of-the-art AV methods in Chapter 8.2. Our own contributions can be found starting from Chapter 8.3, where we provide a continuous study on conservation and stability properties of different AV terms. These results are also used to examine the previous state of the art AV methods and yield us to propose novel viscosity distributions in Chapter 8.4. Next, Chapter 8.5 addresses the strategy of modal filtering and its connection to the AV method. In particular, we note some serious pitfalls for modal filtering. Finally, in Chapter 8.6, we investigate the discretization of AV terms in the context of discontinuous SE methods using SBP operators. Concluding thoughts are provided in Chapter 8.7.

8.1 The idea behind artificial viscosity

At least for smooth solutions, SE methods — such as DG schemes (see Chapter 6.1) and FR methods (see Chapter 6.2) — are capable of reaching spectral orders of accuracy. Yet, special care has to be taken to the fact that solutions of hyperbolic CLs,

$$\partial_t \boldsymbol{u} + \partial_x \boldsymbol{f}(\boldsymbol{u}) = 0, \quad t > 0, \ x \in \mathbb{R}, \tag{8.1}$$

might form spontaneous discontinuities. These discontinuities do not only pose a challenge from a theoretical point of view but also for the numerical treatment of such solutions. Due to the Gibbs phenomenon [HH79], polynomial approximations of jump functions typically show spurious oscillations and can yield the underlying numerical scheme to break down. This phenomenon is illustrated in Figure 8.1 for the linear advection equation $\partial_t u + \partial_x u = 0$ with a square wave as IC.

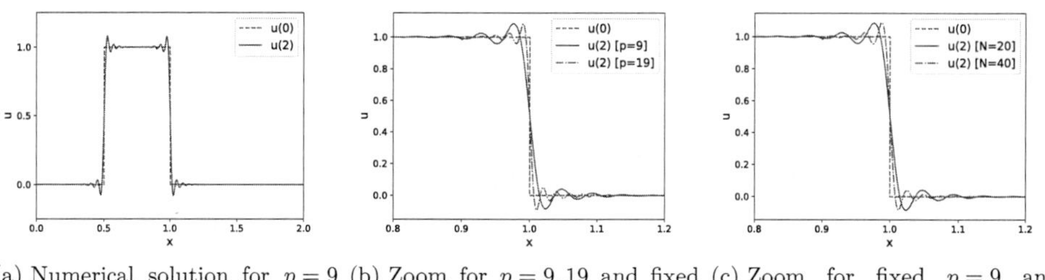

(a) Numerical solution for $p = 9$ and $N = 20$

(b) Zoom for $p = 9, 19$ and fixed $N = 20$

(c) Zoom for fixed $p = 9$ and $N = 20, 40$

Figure 8.1: Numerical solution for the linear advection using the DGSEM for the spatial discretization and the SSPRK(3,3) method to integrate in time. Here, p denotes the polynomial degree and N denotes the number of equidistant elements used in the DGSEM.

Figure 8.1 illustrates that in the presence of discontinuities sub-cell resolution can neither be enhanced by increasing the polynomial degree nor by refining the mesh. In both cases, the spurious oscillations are just closer to the discontinuity but will not vanish. A common approach therefore is to identify the elements lying in the shock region and to reduce the order of approximation there; see [BO99, BSB01] and references therein. Since this increases inter-element jumps and thus the amount of dissipation naturally added by a discontinuous SE (DG) method, at latest for piecewise constant approximations, the method should be able to handle any shock. Yet, it should be stressed that decreasing the order of approximation is equivalent to adding dissipation proportional to $\mathcal{O}(\Delta x)$. Clearly, the accuracy will be reduced. Thus, a widely accepted observation is that the solution, in fact, can be at most first order accurate near shocks. A common idea to bypass this problem is to adaptively refine the mesh in regions of discontinuity. Shocks, however, are lower dimensional objects and strongly anisotropic. An effective strategy for mesh adaptation therefore needs to incorporate some degree of directionality, especially in three dimensions. See, for instance, [DLGC03] and references therein.

Here, we therefore address another approach, which is strongly motivated by physics: *The AV method.* Originally, the idea of AV stems from the vanishing viscosity method to show existence of entropy solutions; see Chapter 2. There, entropy solutions are constructed as L^1 limits of solutions $\boldsymbol{u}_\varepsilon$ of the parabolic equation

$$\partial_t \boldsymbol{u}_\varepsilon + \partial_x \boldsymbol{f}(\boldsymbol{u}_\varepsilon) = \varepsilon\, \partial_{xx} \boldsymbol{u}_\varepsilon, \quad \varepsilon > 0,$$

for $\varepsilon \to 0$. Thus, the vanishing viscosity method can be used to characterize physically reasonable entropy solutions by identifying them as limits of solutions of equations in which a small dissipative mechanism has been added. In their pioneering work [vNR50], von Neumann and Richtmyer revised this idea to construct stable FD schemes for the equations of hydrodynamics by including AV terms. As they pointed out, when viscosity is taken into account, (shock) discontinuities are seen to be smeared out, so that the mathematical surfaces of discontinuity are replaced by thin layers in which pressure, density, temperature, etc. vary rapidly but continuously. Hence, the overall concept is to approximate (discontinuous) entropy solutions by smoother solutions of a parabolic equation and to apply the numerical method to this new equation, where shocks are replaced by thin but continuous layers now. Often, the smooth approximation $\boldsymbol{u}_\varepsilon$ is also called a *viscose profile* to the entropy solution $\boldsymbol{u}$.

8.2 State of the art

Since the early work of von Neumann and Richtmyer [vNR50] in the context of (classical) FD methods, many efforts have been made to adapt the idea of AV to several other schemes. Here, we focus on the realization and development of the AV method in discontinuous SE methods. In particular, we will address DG methods (see Chapter 6.1) and more recent FR schemes (see Chapter 6.2). In these methods, one usually considers AV extensions of the form

$$\partial_t \boldsymbol{u} + \partial_x \boldsymbol{f}(\boldsymbol{u}) = \partial_x \left(\varepsilon \partial_x \boldsymbol{u} \right), \tag{8.2}$$

where the *viscosity* $\varepsilon = \varepsilon(x,t) \geq 0$ is allowed to vary in space and time. In what follows, we revisit some of the most commonly used AV methods for DG methods. For more details we refer to the work [GNJA+19].

8.2.1 Persson and Peraire: Piecewise constant artificial viscosities

Considering a global AV, i. e. $\varepsilon(\cdot,t) \equiv const$, shocks might be spread over several elements if not even the whole domain. Hence, away from (shock) discontinuities, also other (especially small-scale) features of the original solution will be smeared. Breaking new ground in [PP06], Persson and Peraire therefore proposed a local AV in the framework of DG methods. By locally adapting the viscosity ε, AV is just added in the elements where (shock) discontinuities arise. Thus, the now spatially piecewise constant function ε is set to $\varepsilon = 0$ in elements where the solution is smooth and to $\varepsilon = \varepsilon_{\max} > 0$ in elements with (shock) discontinuities. Discontinuities that may appear in the original solution will spread over layers of thickness $\mathcal{O}(\varepsilon)$ in the solution of the modified equation. Hence, Persson and Peraire more precisely suggest that $\varepsilon_{\max}$ should be chosen as a function of the resolution available in the approximation space. If considering piecewise polynomials of degree at most K and elements with width h, the resolution scales like h/K. Thus, a value of $\varepsilon_{\max} \in \mathcal{O}(h/K)$ is taken near shocks. The whole procedure is illustrated in Figure 8.2(a).

8.2.2 Barter and Darmofal: Smoothing the artificial viscosity

In [BD10], Barter and Darmofal took up the work of Persson and Peraire [PP06] and further improved their ideas. Clear numerical shortcomings of a piecewise constant AV are stressed in their work. In particular, they note that element-to-element variations of the AV induce oscillations in state gradients, which pollute the downstream flow. Even though they seem to miss the violation of conservation in the continuous setting, which we address in Chapter 8.3, they clearly point out the missing smoothness of the viscosity as the crucial problem. Thus, in [BD10] a smooth AV was developed by employing an AV PDE model which is appended to the system of governing equations. Effectively, their idea to enhance smoothness of the AV is *diffusing the diffusivity*.

8.2.3 Klöckner, Warburton, and Hesthaven: Piecewise linear artificial viscosities

In [KWH11], Klöckner et al. numerically observed that there seems to be no advantage in having viscosities $\varepsilon(\cdot, t) \in C^k(\mathbb{R})$ for $k > 0$. [1] Following this observation, they formulated an algorithm to enforce continuity of the viscosity – in our opinion, more efficient than the one of Barter and Darmofal — by simple linear interpolation. Building up on a given piecewise constant viscosity, they propose the following steps:

1. At each vertex, collect the maximum viscosity occurring in each of the adjacent elements.
2. Propagate the resulting maximum back to each element adjoining vertex.
3. Use a linear ($\mathbb{P}_1$) interpolant to extend the values at the vertices into a viscosity on the entire element.

While the above algorithm works perfectly fine in regards to enforce continuity of the AV, it inherits a critical disadvantage, which is also shared by the approach of Barter and Darmofal [BD10]. As Sheshadri and Jameson already stated in [SJ14], enforcement of continuity of the AV ε across element boundaries increases the footprint of the added dissipation again. This is noticed immediately when consulting Figure 8.2, where the algorithm of Klöckner et al. is illustrated for a simple example.

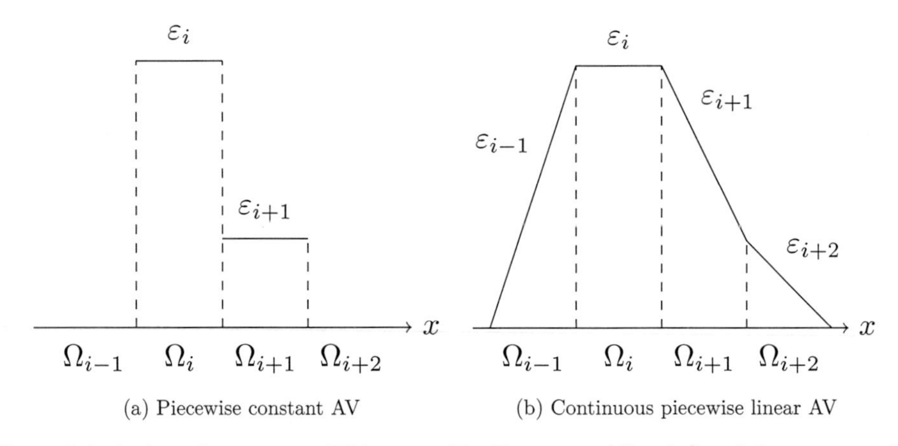

(a) Piecewise constant AV

(b) Continuous piecewise linear AV

Figure 8.2: A piecewise constant AV (proposed by Persson and Peraire) and the corresponding continuous piecewise linear AV constructed by the algorithm of Klöckner et al. are illustrated.

8.3 Conservation and stability properties

In what follows, we investigate conservation and (energy/entropy) dissipation properties of general AV extensions of the form

$$\partial_t \boldsymbol{u} + \partial_x \boldsymbol{f}(\boldsymbol{u}) = \varepsilon\, \partial_x (a \partial_x \boldsymbol{u}) \tag{8.3}$$

[1] It should be noted, however, that $\varepsilon(\cdot, t) \in C^k(\mathbb{R})$ for $k > 0$ is required for higher order artificial dissipation (AD) operators.

with a constant *viscosity strength* $\varepsilon \in \mathbb{R}_0^+$ and a *viscosity distribution* $a = a(x,t)$. Note that the above AV method includes the previous AV methods of the form (8.2) as the special case of $\varepsilon(x,t) = \varepsilon\, a(x,t)$. The investigation is done in the context of discontinuous SE methods, where the computational domain Ω is subdivided into a set of smaller elements Ω_i with $\Omega = \bigcup_{i=1}^{I} \Omega_i$, and provides us with new insights into the most commonly used AV methods described in Chapter 8.2. Finally, our results will also yield us to propose and discuss some novel viscosity distributions in the subsequent Chapter 8.4.

8.3.1 Conservation

Let us consider conservation properties of AV methods of the form (8.3). We start by noting that conservation of the continuous model (8.3) is only ensured for a viscosity distribution a which is continuous and compactly supported on Ω.

Theorem 8.1

Augmenting a CL $\partial_t \boldsymbol{u} + \partial_x \boldsymbol{f}(\boldsymbol{u}) = 0$ on Ω with a viscosity term $\varepsilon\, \partial_x(a \partial_x \boldsymbol{u})$, where $\varepsilon \in \mathbb{R}_0^+$ is constant and $a(\cdot,t)$ is continuous and compactly supported on Ω preserves conservation, i. e.

$$\frac{\mathrm{d}}{\mathrm{d}t} \int_\Omega \boldsymbol{u} \,\mathrm{d}x = -\boldsymbol{f}(\boldsymbol{u})\big|_{\partial\Omega}$$

holds componentwise for solutions $\boldsymbol{u}$ of (8.3).

Proof. Decomposing the domain Ω into disjoint, face-conforming elements Ω_i, we have

$$\begin{aligned}\frac{\mathrm{d}}{\mathrm{d}t} \int_\Omega u_n \,\mathrm{d}x &= -\int_{\Omega_i} \partial_x f_n(\boldsymbol{u}) \,\mathrm{d}x + \varepsilon \int_{\Omega_i} \partial_x (a \partial_x u_n) \,\mathrm{d}x \\ &= -f_n(\boldsymbol{u})\big|_{\partial\Omega_i} + \varepsilon a \partial_x u_n\big|_{\partial\Omega_i}\end{aligned}$$

for the nth component of $\boldsymbol{u} = (u_1, \dots, u_m)^T$. Now putting the elements together, the rate of change of the total amount of u_n is given by

$$\frac{\mathrm{d}}{\mathrm{d}t} \int_\Omega u_n \,\mathrm{d}x = -\sum_i f_n(\boldsymbol{u})\big|_{\partial\Omega_i} + \varepsilon \sum_i a \partial_x u_n\big|_{\partial\Omega_i}. \tag{8.4}$$

Assuming a sufficiently smooth solution $\boldsymbol{u}$ and recalling that a is continuous on Ω, this reduces to

$$\frac{\mathrm{d}}{\mathrm{d}t} \int_\Omega u_n \,\mathrm{d}x = -f_n(\boldsymbol{u})\big|_{\partial\Omega} + \varepsilon a \partial_x u_n\big|_{\partial\Omega}.$$

Finally, the assertion follows from a also being compactly supported in Ω. □

Note that the above proof would have also worked without decomposing Ω into smaller elements Ω_i. Yet, (8.4) allows us to immediately point out problems for the piecewise constant AV proposed by Persson and Peraire (see Chapter 8.2) regarding conservation on a continuous level. Note that for a piecewise continuous AV term of the form $\partial_x(\varepsilon \partial_x \boldsymbol{u})$ with $\varepsilon(\cdot,t) \equiv \mathit{const}$ on Ω_i for every $i = 1, \dots, I$, the above analysis again yields

$$\frac{\mathrm{d}}{\mathrm{d}t} \int_\Omega \boldsymbol{u} \,\mathrm{d}x = -\sum_i \boldsymbol{f}(\boldsymbol{u})\big|_{\partial\Omega_i} + \sum_i \varepsilon_i \partial_x \boldsymbol{u}\big|_{\partial\Omega_i},$$

where ε_i denotes the constant function $\varepsilon|_{\Omega_i}$ However, this time, only the first sum cuts down to the contributions at the boundaries of the whole computational domain $\partial\Omega$. For a general piecewise constant viscosity ε, the second sum is not ensured to vanish and we end up with

$$\frac{\mathrm{d}}{\mathrm{d}t}\int_\Omega \boldsymbol{u}\,\mathrm{d}x = -\boldsymbol{f}(\boldsymbol{u})\big|_{\partial\Omega} + \sum_i \varepsilon_i \partial_x \boldsymbol{u}\big|_{\partial\Omega_i}.$$

Note that especially for periodic BCs we have

$$\frac{\mathrm{d}}{\mathrm{d}t}\int_\Omega \boldsymbol{u}\,\mathrm{d}x = \sum_i \varepsilon_i \partial_x \boldsymbol{u}\big|_{\partial\Omega_i},$$

which typically is not equal to zero and therefore violates conservation.

8.3.2 Entropy dissipation

Next, we address the influence of AV terms on the dissipation of entropy. So far, we only required the viscosity distribution $a(\cdot,t)$ to be continuous and compactly supported on Ω. In the following theorem, we will show that $a(\cdot,t)$ should also be nonnegative for the total amount of entropy to not increase due to the addition of AV terms. Moreover, also the viscosity strength ε needs to be nonnegative.

Theorem 8.2

Let η and ψ be an convex entropy and corresponding entropy flux of the CL $\partial_t \boldsymbol{u} + \partial_x \boldsymbol{f}(\boldsymbol{u}) = 0$ on Ω; see Chapter 2.7. Moreover, let $\varepsilon \in \mathbb{R}_0^+$ and $a(\cdot,t)$ be continuous, compactly supported and nonnegative on Ω. Then, augmenting the CL $\partial_t \boldsymbol{u} + \partial_x \boldsymbol{f}(\boldsymbol{u}) = 0$ the viscosity term $\varepsilon\,\partial_x(a\partial_x \boldsymbol{u})$ preserves entropy stability, i.e.

$$\frac{\mathrm{d}}{\mathrm{d}t}\int_\Omega \eta(\boldsymbol{u})\,\mathrm{d}x \leq -\psi(\boldsymbol{u})\big|_{\partial\Omega}$$

holds for solutions $\boldsymbol{u}$ of (8.3).

Proof. Remember from Chapter 2.7 that entropies η and corresponding entropy fluxes ψ satisfy $\eta' f' = \psi'$. Hence, multiplying the viscosity extension (8.3) by the entropy variable η' (sometimes also called *entropy gradient*), we obtain

$$\partial_t \eta(\boldsymbol{u}) + \partial_x \psi(\boldsymbol{u}) = \varepsilon \eta'(\boldsymbol{u}) \partial_x (a \partial_x \boldsymbol{u}).$$

Next, since η is convex and a is nonnegative, we also have

$$\begin{aligned}
\partial_x \left(a \partial_x \eta(\boldsymbol{u})\right) &= \partial_x \left(a \eta'(\boldsymbol{u}) \partial_x \boldsymbol{u}\right) \\
&= (\partial_x a)\,\eta'(\boldsymbol{u})\partial_x \boldsymbol{u} + a\eta''(\boldsymbol{u})\left(\partial_x \boldsymbol{u}\right)^2 + a\eta'(\boldsymbol{u})\partial_{xx}\boldsymbol{u} \\
&\geq (\partial_x a)\,\eta'(\boldsymbol{u})\partial_x \boldsymbol{u} + a\eta'(\boldsymbol{u})\partial_{xx}\boldsymbol{u} \\
&= \eta'(\boldsymbol{u})\partial_x \left(a \partial_x \boldsymbol{u}\right).
\end{aligned}$$

Finally, this yields the entropy inequality

$$\partial_t \eta(\boldsymbol{u}) + \partial_x \psi(\boldsymbol{u}) \leq \varepsilon \partial_x \left(a \partial_x \eta(\boldsymbol{u})\right)$$

and the assertion follows from noting

$$\begin{aligned}
\frac{\mathrm{d}}{\mathrm{d}t}\int_\Omega \eta(\boldsymbol{u})\,\mathrm{d}x &\leq -\int_\Omega \partial_x \psi(\boldsymbol{u})\,\mathrm{d}x + \varepsilon \int_\Omega \partial_x \left(a \partial_x \eta(\boldsymbol{u})\right)\,\mathrm{d}x \\
&= -\psi(\boldsymbol{u})\big|_{\partial\Omega} + \varepsilon\, \underbrace{a \partial_x \eta(\boldsymbol{u})\big|_{\partial\Omega}}_{=0}.
\end{aligned}$$

□

Summarizing theorems 8.1 and 8.2, for AV terms to preserve basic conservation and entropy dissipation properties of the underlying CL, AV term of the form

$$\varepsilon \partial_x \left(a \partial_x \boldsymbol{u} \right)$$

should have a nonnegative viscosity strength ε and the viscosity distribution a should be continuous, compactly supported and nonnegative on the computational domain Ω.

8.4 New viscosity distributions

Building up on the previous investigation of AV methods with respect to conservation and entropy dissipation, in [GNJA+19] novel viscosity distributions have been proposed. These are derived by introducing certain functions from the fields of robust reprojection [GT06b] and spectral mollifiers [TT02, Tan06]. The resulting C_0^∞ AV method was demonstrated in [GNJA+19] to provide sharper profiles, steeper gradients, and a higher resolution of small-scale features compared to the most commonly used AV methods presented in Chapter 8.2. Even though this method is directly related to the present thesis and is able to improve on existing AV methods, we do not provide much more detail here. Instead, the construction of suitable viscosity distributions will be addressed again in Chapter 11. In our opinion, the ideas presented there are considerably more matured than the ones presented in [GNJA+19].

8.5 Modal filtering

Next, we address the strategy of modal filtering and its connection to the AV method. Let us consider the so-called *Legendre AV* defined on the reference element $\Omega_{\text{ref}} = (-1, 1)$ by

$$\varepsilon \partial_x \left(\left[1 - x^2 \right] \partial_x \boldsymbol{u} \right).$$

The Legendre AV method is a special case of the C_0^∞ AV methods mentioned in Chapter 8.4 and is applied in every element separately. Here, the viscosity strength ε might vary from element to element. Note that theorems 8.1 and 8.2 immediately ensure the preservation of conservation and entropy stability for the resulting Legendre AV method. By now, the Legendre AV method has become a widespread tool because it allows a reinterpretation resulting in an implementation which is often considered to be simpler, especially in modal SE methods where the solution is represented with respect to a basis of orthogonal Legendre polynomials. Here, a procedure first proposed by Majda, McDonough, and Osher in [MMO78] as well as by Kreiss and Oliger in [KO79] is utilized. Also see the monograph [GH01]. Applying a first order operator splitting in time, solving the AV extension

$$\partial_t \boldsymbol{u} + \partial_x \boldsymbol{f}(\boldsymbol{u}) = \varepsilon \partial_x \left(a \partial_x \boldsymbol{u} \right)$$

is divided into two steps:

$$\text{(1st) solve} \quad \partial_t \boldsymbol{u} + \partial_x \boldsymbol{f}(\boldsymbol{u}) = 0, \tag{8.5}$$

$$\text{(2nd) solve} \quad \partial_t \boldsymbol{u} = \varepsilon \partial_x \left(a \partial_x \boldsymbol{u} \right) \tag{8.6}$$

Applying a numerical solver to this approach, going forward in time is done by alternately integrating (8.5) and (8.6). This can be done, for instance, by an explicit RK method (see

Chapter 3.5). However, it should be stressed that in this approach a second equation (8.6) has to be solved in addition to the original CL (8.5). This would be a clear disadvantage regarding efficiency. Yet, when adjusting the AV method to the modal basis in which the numerical solution is expressed, equation (8.6) can be solved exactly. The exact solution is obtained by applying a modal exponential filter to the numerical solution of equation (8.5) then. In particular, the Legendre AV is used so often because the orthogonal Legendre polynomials P_k, $k \in \mathbb{N}_0$, arise as solutions of the corresponding Legendre differential equation

$$\frac{\mathrm{d}}{\mathrm{d}x}\left(\left[1-x^2\right]\frac{\mathrm{d}}{\mathrm{d}x}P_k(x)\right) = -\lambda_k P_k(x) \tag{8.7}$$

of a Sturm–Liouville type for eigenvalues $-\lambda_k = -k(k+1)$. Also see Example 3.24 in Chapter 3.2. Henceforth, let us consider a single component $u^i \in \mathbb{P}_K(\mathbb{R})$ of the numerical solution on Ω_i transformed to the reference element $\Omega_{\text{ref}} = (-1,1)$. Moreover, for sake of simplicity, we denote u^i by

$$u(x,t) = \sum_{k=0}^{K} \hat{u}_k(t)P_k(x), \quad x \in (-1,1),$$

omitting the element index i. Here, we have represented the numerical solution with respect to the orthogonal basis of Legendre polynomials. Thus, when we substitute the numerical solution into equation (8.6), we get

$$\begin{aligned}\sum_{k=0}^{K} \left(\partial_t \hat{u}_k\right)(t)\ P_k(x) &= \varepsilon \sum_{k=0}^{K} \hat{u}_k(t)\varepsilon\partial_x\left(a\partial_x P_k(x)\right) \\ &= -\varepsilon\sum_{k=0}^{K} \lambda_k\hat{u}_k(t)P_k(x)\end{aligned}$$

for $a(x) = 1-x^2$. Note that the last equation follows from the eigenvalue equation (8.7) for the Legendre polynomials. A simple comparison of the time dependent coefficients results in a system of $K+1$ decoupled ODEs

$$\frac{\mathrm{d}}{\mathrm{d}t}\hat{u}_k(t) = -\varepsilon\lambda_k\hat{u}_k(t), \quad k = 0,\dots,K,$$

with solutions given by

$$\hat{u}_k(t) = C_k \cdot e^{-\varepsilon\lambda_k t}, \quad C_k \in \mathbb{R}.$$

Hence, after one time step $\Delta t = t^{n+1} - t^n$, the coefficients $\hat{u}_k(t^{n+1})$ are given by

$$\begin{aligned}\hat{u}_k(t^{n+1}) &= C_k \cdot e^{-\varepsilon\lambda_k(\Delta t + t^n)} \\ &= e^{-\varepsilon\lambda_k\Delta t} \cdot C_k e^{-\varepsilon\lambda_k t^n} \\ &= e^{-\varepsilon\lambda_k\Delta t} \cdot \hat{u}_k(t^n).\end{aligned}$$

Thus, solving the parabolic equation (8.6) for the numerical solution u is equivalent to multiplying its modal coefficients $\hat{u}_k$ with the function $\sigma(k) = e^{-\varepsilon\Delta t\lambda_k}$. In order to speak of a modal filter [HK08], the exponent has to be rewritten as

$$\sigma\left(\frac{k}{K}\right) = e^{-\varepsilon K^2\Delta t\left(\frac{k}{K}\right)\left(\frac{k+1}{K}\right)} \approx e^{-\varepsilon K^2\Delta t\left(\frac{k}{K}\right)^2}. \tag{8.8}$$

Only now, $\sigma : [0,1] \to [0,1]$ can be seen as an *exponential filter* of *order* 2 with *filter strength* $\alpha := \varepsilon K^2\Delta t$. For instance, see the work [HK08] and references therein. Moreover — and directly connected to this thesis, in [GÖS18], modal filtering has been extended to SE methods, in particular SD methods, on triangular elements in two spatial dimensions.

A pitfall of modal filtering

By this time, modal filtering has become an established shock capturing tool and has been applied in a large number of works, [Hu96, YGH97, GH01, Boy01, GHW02, KCH06, HK08, MOS12, MOSW13, GÖS18, RGÖS18, ÖGR18]. However, in [GNJA+19] we observed the Legendre AV method as well as the corresponding modal filter applied in a Legendre basis to perform poorer than, for instance, the (continuous) piecewise linear AV method of Klöckner et al.; see Chapter 8.2. This finding of [GNJA+19] is also illustrated in the subsequent Figure 8.3.

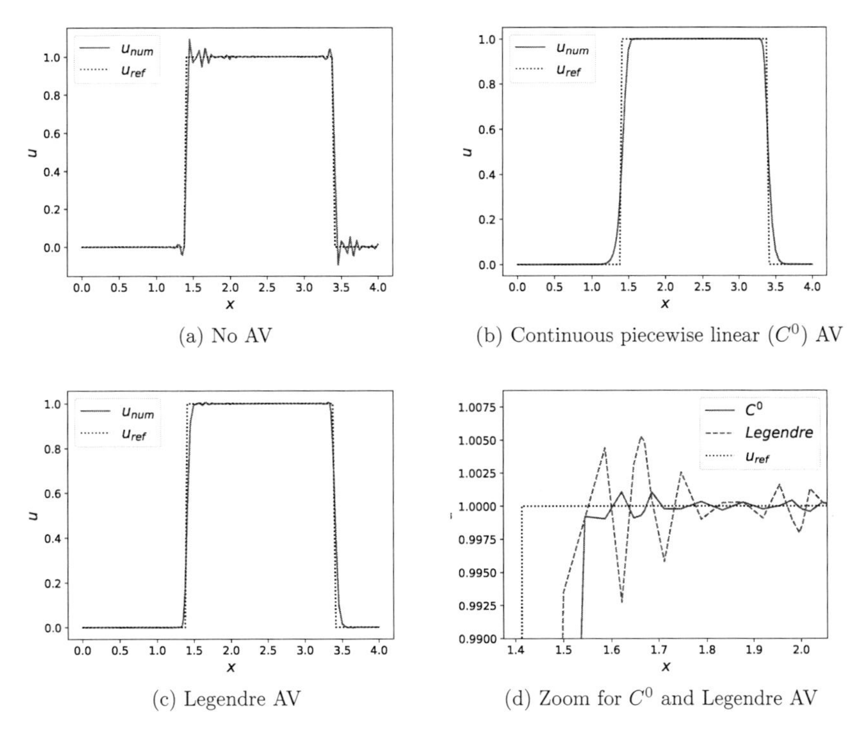

(a) No AV

(b) Continuous piecewise linear (C^0) AV

(c) Legendre AV

(d) Zoom for C^0 and Legendre AV

Figure 8.3: Numerical solutions (u_{num}) of the linear advection equation for different AV methods using $I = 12$ equidistant elements and polynomials of degree $K = 10$.

The numerical solutions were computed by a DGSEM (see Chapter 6.1.3 or [GNJA+19]). Moreover, the same viscosity strength was used for the continuous piecewise linear (C^0) AV method and the Legendre AV method. This viscosity strength was determined by using the modal decay based smoothness indicator as described in [GNJA+19] and references therein. Figure 8.3(b) demonstrates that there are nearly no oscillations present anymore in the numerical solution obtained by the C^0 AV method of Klöckner et al.. At the same time, the numerical solution obtained by the Legendre AV method still shows spurious oscillations (see Figure 8.3(c)). This is illustrated in greater detail in Figure 8.3(d), where the numerical solutions for the C^0 and Legendre AV method are compared around the first discontinuity. In [GNJA+19], the particular viscosity distribution of the Legendre AV was explored to be the determining factor for the spurious oscillations in the corresponding numerical solution. As illustrated in Figure 8.4, the Legendre viscosity rapidly vanishes away from the element center. Thus, if a shock discontinuity occurs close to an element boundary, nearly no dissipation is added there by the Legendre AV. This observation can be made whenever (shock) discontinuities arise close to element boundaries. This yields to the conclusion that the Legendre AV method as well as the corresponding exponential modal filters should be rejected.

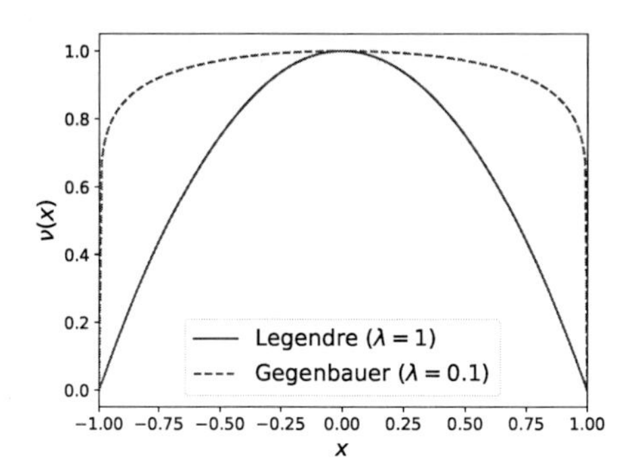

Figure 8.4: Legendre viscosity ($\lambda = 1$) and Gegenbauer viscosity for $\lambda = 0.1$

Naturally, the question arises if this bottleneck might be overcome by utilizing other viscosity distributions, which then again should correspond to modal filters in a proper basis of orthogonal polynomials. While the first attempt to construct more appropriate viscosity distributions was successfully tackled in [GNJA+19], the possibility of still finding corresponding modal filters has to be dismissed. The decisive property of the Legendre AV was that the orthogonal Legendre polynomials occur as eigenfunctions of the Sturm–Liouville operator

$$Ly = ([1 - x^2]y')'$$

with eigenvalues $-\lambda_k = -k(k+1)$. That is,

$$b(x)y'' + c(x)y' + d(x)y = \lambda y \tag{8.9}$$

with $b(x) = 1 - x^2$, $c(x) = 2x$, and $d(x) = 0$ for the Legendre AV. Yet, following a simple argument of Bochner [Boc29, Rou84], for every Sturm–Liouville operator featuring polynomials as eigenfunctions, the corresponding coefficients b, c and d must be polynomials of degree $2, 1$ and 0, respectively. Moreover, keep in mind that the resulting viscosity distribution is often desired to vanish at the element boundaries and should be nonnegative (for entropy stability). Thus, up to a positive constant, $b(x) = 1 - x^2$ is the very only choice. Note that every exponential 'modal filter',

$$\sigma_k = e^{-\alpha k(k+1+\alpha+\beta)},$$

applied in the associated basis of orthogonal Jacobi polynomials $P_k^{(\alpha,\beta)}$, $k \in \mathbb{N}_0$, is equivalent to an AV of the form

$$\partial_x \left([1 - x^2]\partial_x u\right) + (\beta - \alpha - (\alpha + \beta)x)\, \partial_x u,$$

by a first order operator splitting in time. This observation results from the Jacobi polynomials being eigenfunctions of the more general Sturm–Liouville operator

$$Ly = \left((1 - x^2)y'\right)' + (\beta - \alpha - (\alpha + \beta)x)\, y'.$$

Hence, while the dissipative term $\partial_x([1 - x^2]\partial_x u)$ is still the same, in this case also undesired dispersion of the form $(\beta - \alpha - (\alpha + \beta)x)\partial_x u$ would be added to the equation.

8.6 Discretization of artificial viscosity terms using summation by parts operators

Next, we address the discretization of AV terms in the context of discontinuous SE methods using SBP operators. Special focus is given to the DGSEM discussed in Chapter 6.1 and the FR method as in Chapter 6.2. Remember that both methods essentially result from subdividing the computational domain $\Omega \subset \mathbb{R}$ into smaller elements Ω_i and approximating the solution by a polynomial on every element Ω_i. Finally, since the resulting piecewise polynomial is allowed to be discontinuous at element interfaces, information is ensured to be passed between neighboring elements by utilizing numerical fluxes. Let us consider a hyperbolic conservation law (8.1) and let $\mathbf{u}_n \in \mathbb{R}^{K+1}$ denote the vector of (nodal or modal) coefficients of the nth component of $\boldsymbol{u} = (u_1, \dots, u_m)^T$ on a fixed element Ω_i transformed to the reference element. The same notation is used for the flux $\boldsymbol{f}$. For the DGSEM in its weak form, this results in the semidiscretization

$$M \frac{\mathrm{d}}{\mathrm{d}t} \mathbf{u}_n = D^T M \mathbf{f}_n - R^T B \boldsymbol{f}_n^{\mathrm{num}}, \quad n = 1, \dots, m,$$

on the reference element $\Omega_{\mathrm{ref}} = (-1, 1)$; see (6.9) in Chapter 6.1.4. For the FR method, on the other hand, we have the semidiscretization

$$\frac{\mathrm{d}}{\mathrm{d}t} \mathbf{u}_n = -D \mathbf{f}_n - C \left(\boldsymbol{f}_n^{\mathrm{num}} - R \mathbf{f}_n \right), \quad n = 1, \dots, m,$$

on the reference element $\Omega_{\mathrm{ref}} = (-1, 1)$; see (6.14) in Chapter 6.2.1. Note that for the canonical choice for the correction matrix,

$$C = M^{-1} R^T B, \tag{8.10}$$

both semidiscretization are equivalent when SBP operators are used. The later one means that the mass, differentiation, boundary and restriction matrices fulfill the SBP property

$$MD + D^T M = R^T B R \tag{8.11}$$

and therefore mimic integration by parts.

In what follows we show for FR methods that the conservation and (energy/entropy) stability properties from the continuous setting (see Chapter 8.3) also carry over to the semidiscrete setting when AV methods are discretized suitably by using SBP operators. Note that the DGSEM and FR method are equivalent when using the canonical correction matrix (8.10) and SBP operators. Thus, all subsequent results for the FR method also carry over to the DGSEM. The material presented here resulted in the related publication [RGÖS18]. Let us investigate the discretization of AV extensions for CLs in one spatial dimension,

$$\partial_t \boldsymbol{u} + \partial_x \boldsymbol{f}(\boldsymbol{u}) = \varepsilon \partial_x \left(a \partial_x \boldsymbol{u} \right), \tag{8.12}$$

with adequate IC and BCs. Henceforth, the viscosity strength ε is always assumed to be nonnegative. To investigate conservation and stability of the AV extension, it is sufficient to apply an operator splitting and study the equation

$$\partial_t \boldsymbol{u} = \varepsilon \partial_x \left(a \partial_x \boldsymbol{u} \right). \tag{8.13}$$

Then, using an FR method with the canonical correction matrix (8.10) and SBP operators, a direct discretization of (8.13) is given by

$$\frac{\mathrm{d}}{\mathrm{d}t} \mathbf{u}_n = \varepsilon D A D \mathbf{u}_n, \quad n = 1, \dots, m, \tag{8.14}$$

where the matrix A represents multiplication with the viscosity distribution a followed by some projection on $\mathbb{P}_K(\mathbb{R})$. Unfortunately, the direct discretization (8.14) is not even ensure to be conservative in all cases. Here, conservation is understood in a discrete sense and means that

$$\mathbf{1}^T M \frac{\mathrm{d}}{\mathrm{d}t} \mathbf{u}_n = 0, \quad n = 1, \ldots, m, \tag{8.15}$$

should hold. This can be noted from the following lemma.

Lemma 8.3
Let a vanish at the boundaries of Ω_{ref}, i. e. $a|_{\partial\Omega_{\text{ref}}} = 0$. Then, the discretization (8.14) *is conservative if and only if the projection used preserves boundary values ($RA = 0$).*

Proof. Multiplying (8.14) with $\mathbf{1}^T M$ from the left gives us

$$\mathbf{1}^T M \frac{\mathrm{d}}{\mathrm{d}t} \mathbf{u}_n = \varepsilon \mathbf{1}^T M D A D \mathbf{u}_n.$$

Next, applying the SBP property (8.11), we get

$$\mathbf{1}^T M \frac{\mathrm{d}}{\mathrm{d}t} \mathbf{u}_n = \varepsilon \mathbf{1}^T R^T B R A D \mathbf{u}_n - \varepsilon \mathbf{1}^T D^T M A D \mathbf{u}_n.$$

Note that the second term vanishes due to $D\mathbf{1} = \mathbf{0}$. Hence, this yields

$$\mathbf{1}^T M \frac{\mathrm{d}}{\mathrm{d}t} \mathbf{u}_n = \varepsilon \mathbf{1}^T R^T B R A D \mathbf{u}_n$$

and therefore the assertion. □

Thus, the resulting scheme is conservative if and only if $RA = 0$ holds, which means that the projection used preserves boundary values. This is the case for a nodal GLo basis including boundary points. However, a nodal GLe or a modal Legendre basis do not have this property. Yet, as described below, by a careful choice of the discrete terms these bases can be used as well.

Let us rewrite the direct discretization (8.14) by using the SBP property, yielding

$$\begin{aligned} \varepsilon D A D \mathbf{u}_n &= \varepsilon M^{-1} M D A D \mathbf{u}_n \\ &= \varepsilon M^{-1} \left(R^T B R A D \mathbf{u}_n - D^T M A D \mathbf{u}_n \right). \end{aligned}$$

Now, the basic idea is to enforce the boundary term $R^T B R A D \mathbf{u}_n$ to vanish. This results in the following discretization of (8.13):

$$\frac{\mathrm{d}}{\mathrm{d}t} \mathbf{u}_n = -\varepsilon M^{-1} D^T M A D \mathbf{u}_n, \quad n = 1, \ldots, m \tag{8.16}$$

In fact, for this discretization conservation is literally enforced and therefore always holds.

Lemma 8.4
The discretization (8.16) *is always conservative.*

Proof. Multiplying (8.16) with $\mathbf{1}^T M$ from the left gives us

$$\mathbf{1}^T M \frac{\mathrm{d}}{\mathrm{d}t} \mathbf{u}_n = -\varepsilon \mathbf{1}^T D^T M A D \mathbf{u}_n$$

and the assertion follows from noting $D\mathbf{1} = \mathbf{0}$. □

Next, let us address energy stability. We say that an AV discretization is *energy stable* if

$$\mathbf{u}_n^T M \frac{\mathrm{d}}{\mathrm{d}t} \mathbf{u}_n \leq 0, \quad n = 1, \ldots, m, \tag{8.17}$$

holds for the discretization of (8.13). This definition is motivated by the observation that the discrete squared L^2 norm

$$\|(\mathbf{u}_1, \ldots, \mathbf{u}_m)\|_M^2 := \sum_{n=1}^{m} \mathbf{u}_n^T M \mathbf{u}_n \tag{8.18}$$

does not increase in time then. The following lemma addresses energy stability of the AV discretization (8.16).

Lemma 8.5
The discretization (8.16) *is energy stable if and only if the matrix product MA is positive semidefinite.*

Proof. This time multiplying (8.16) with $\mathbf{u}_n^T M$ from the left hand side, we have

$$\mathbf{u}_n^T M \frac{\mathrm{d}}{\mathrm{d}t} \mathbf{u}_n = -\varepsilon \mathbf{u}_n^T D^T M A D \mathbf{u}_n.$$

Hence, denoting $D\mathbf{u}_n$ by $\mathbf{v}$ gives us

$$\mathbf{u}_n^T M \frac{\mathrm{d}}{\mathrm{d}t} \mathbf{u}_n = -\varepsilon \mathbf{v}^T M A \mathbf{v}.$$

It is now evident that energy stability holds if and only if the matrix product MA is positive semidefinite. □

Consulting Lemma 8.5 it is only natural to ask in which situations the matrix product MA is actually ensured to be positive semidefinite. In fact, it turns that this is ensured essentially in all situations of practical interest:

- For a **nodal basis** MA is ensured to be positive semidefinite when $a(x_k) \geq 0$, $k = 0, \ldots, K$, and the diagonal entries of the mass matrix M are all nonnegative. Note that for a nodal basis, the diagonal mass matrix M consists of the weights w_k, $k = 0, \ldots, K$, of a QR on the collocation points. For nonnegative quadrature weights $w_k \geq 0$ and $a(x_k) \geq 0$, the matrix product MA is therefore given by

$$\begin{aligned} MA &= \operatorname{diag}(w_0, \ldots, w_K) \operatorname{diag}(a(x_0), \ldots, a(x_K)) \\ &= \operatorname{diag}(w_0 a(x_0), \ldots, w_K a(x_K)), \end{aligned}$$

  where $w_k a(x_k) \geq 0$. Hence, MA is positive semidefinite. Note that this observation is independent of the collocation points corresponding to the nodal basis. Of course, the quadrature weights w_k have to be ensured to be nonnegative and the SBP property should be satisfied. Usual choices are the nodal GLo and the GLe bases.

- In a **modal basis**, e. g. using Legendre polynomials, the mass matrix M represents the L^2 scalar product, $a \geq 0$ is a polynomial, and multiplication is given by exact multiplication of polynomials followed by exact L^2 projection on $\mathbb{P}_K(\mathbb{R})$ (denoted by $\operatorname{proj}(\cdot)$). Then, for an arbitrary polynomial $v \in \mathbb{P}_K(\mathbb{R})$ with discrete representation $\mathbf{v} \in \mathbb{R}^{K+1}$, we have

$$\begin{aligned} \mathbf{v}^T M A \mathbf{v} &= \int v \operatorname{proj}(av) \, \mathrm{d}x \\ &= \int \operatorname{proj}(v) av \, \mathrm{d}x \\ &= \int vav \, \mathrm{d}x \\ &= \int av^2 \, \mathrm{d}x \\ &\geq 0. \end{aligned}$$

For the second equation, we have utilized that the orthogonal projection $\operatorname{proj}(\cdot)$ is selfadjoint. Hence, for $a \geq 0$, the matrix product MA is ensured to be be positive semidefinite also in this case.

We summarize our findings in the following theorem.

Theorem 8.6
Let a be a nonnegative polynomial. Then, the discretization (8.16) *is conservative (in the sense of* (8.15)*) and energy stable (in the sense of* (8.17)*) if*

- *a nodal basis with nonnegative diagonal mass matrix M,*
- *or a modal basis with exact L^2 scalar product and exact L^2 projection*

is used. Bases satisfying these conditions are nodal GLe and GLo bases as well as modal Legendre bases.

For a more general investigation of entropy stability, we refer to the later Chapter 11. Instead, we close this investigation of suitable discretizations of AV terms with the following remark.

Remark 8.7. In the case of a modal basis the polynomial $a \in \mathbb{P}_K(\mathbb{R})$ needs to be nonnegative for energy stability to hold for the discretization (8.16). Yet, it should be stressed that in the case of a nodal basis with diagonal mass matrix $M = \operatorname{diag}(w_0, \dots, w_K)$, in fact, it is sufficient that a is nonnegative only at the interpolation points.

8.7 Concluding thoughts

In this chapter, we have investigated different AV methods in the context of (discontinuous) SE methods. After revisiting some commonly used approaches, we started our investigation by collecting precise criteria on the viscosity strength $\varepsilon \in \mathbb{R}$ and the viscosity distribution $a : \Omega \to \mathbb{R}$ for the AV model

$$\partial_t \boldsymbol{u} + \partial_x \boldsymbol{f}(\boldsymbol{u}) = \varepsilon \partial_x \left(a \partial_x \boldsymbol{u} \right)$$

to preserve conservation and entropy stability properties. Essentially, we found that a should be compactly supported and continuous on Ω for conservation to hold in the continuous setting. Moreover, ε and a should both be nonnegative for the AV term to have a stabilizing effect on the CL (entropy is dissipated). We also extended this investigation to the semidiscrete setting, discussing the discretization of AV terms in the context of an FR method using SBP operators. It should be noted that these results also carry over to the DGSEM. Here, we observed that a direct (intuitive) discretization — as formulated in (8.14) — comes with some problems regarding conservation (as well as energy stability) when modal bases or nodal bases without including the boundary points as collocation points are used. Instead, we therefore proposed an alternative discretization of AV terms — see (8.16) — in which certain boundary terms are (artificially) enforced to vanish. As a consequence, we were able to prove that the resulting discretization is always conservative and in many situations also energy stable. These situation essentially include all methods of practical interest.

Furthermore, utilizing a strong connection to AV methods, we have also investigated the technique of modal filtering. In many works, modal filtering has been motivated as an alternative to AV methods which is fairly easy to implement. In fact, also certain timestep restrictions which can arise for AV methods (see Chapter 11), can be overcome by this method. Unfortunately, we observed serious pitfalls for modal filtering, at least when exponential modal filters are used. We discovered that, by a first order operator splitting in time, usual modal filters essentially correspond to a Legendre AV, possibly including additional dispersive terms (for more general modal Jacobi bases). At the same time, the Legendre AV has been observed to add an insufficient amount of dissipation near element boundaries, yielding it inferior to many other AV terms. In summary, we come to the conclusion that modal filtering (at least using exponential modal filters) is inferior to other techniques and can not be recommended as a means of shock capturing in (discontinuous) SE methods for CLs.

The ideas and investigation of AV methods started in this chapter will be continued in Chapter 11 for SBP based FD methods. In particular, we will address the question how the viscosity strength ε and the viscosity distribution a can be intelligently adapted to the smoothness of the (numerical) solution and the location of possible (shock) discontinuities. An outlook on future research will be given there.

CHAPTER 9

ℓ^1 REGULARIZATION AND HIGH ORDER EDGE SENSORS FOR ENHANCED DISCONTINUOUS GALERKIN METHODS

In this chapter, we investigate the use of ℓ^1 regularization for solving CLs based on high order DG approximations; see Chapter 6.1. Yet, the proposed procedure can also be applied to any other SEM. We first use the polynomial annihilation (PA) method to construct a HOES which enables us to flag "troubled" elements. The DG approximation is enhanced in these troubled regions by activating ℓ^1 regularization to promote sparsity in the corresponding jump function of the numerical solution then. The resulting ℓ^1 optimization problem is efficiently implemented using the alternating direction method of multipliers (ADMM). By enacting ℓ^1 regularization only in troubled cells, our method remains accurate and efficient, as no additional regularization or expensive iterative procedures are needed in smooth regions. We present results for the inviscid Burgers' equation as well as for a nonlinear system of CLs using the DGSEM as a solver. The subsequent material resulted in the publication [GG19a].

Outline

This chapter is organized as follows: After an introductory motivation of ℓ^1 regularization in the context of CLs in Chapter 9.1, Chapter 9.2 introduces ℓ^1 regularization and PA operators which are needed for the development of our method. In Chapter 9.3 we describe the application of ℓ^1 regularization by HOES to SE type methods. Further, a novel discontinuity sensor based on PA operators of increasing orders is proposed. Numerical tests for the inviscid Burgers' equation, the linear advection equation, and a nonlinear system of CLs are presented in Chapter 9.4. The tests demonstrate that we are able to better resolve numerical solutions when ℓ^1 regularization is utilized. We close this chapter with concluding thoughts in Chapter 9.5.

9.1 Why ℓ^1 regularization?

As mentioned before, over the last few decades, many shock capturing techniques have been developed for the robust and accurate treatment of (shock) discontinuities in CLs,

$$\partial_t \boldsymbol{u} + \partial_x \boldsymbol{f}(\boldsymbol{u}) = 0. \tag{9.1}$$

Such efforts date back more than 60 years to the pioneering work of von Neumann and Richtmyer, [vNR50] in which they add AV terms to (9.1) in order to construct stable FD schemes for the equations of hydrodynamics. See Chapter 8 for a discussion of AV methods in SE methods. However, augmenting (9.1) with additional AV terms requires care about their design and size. Otherwise, new time stepping constraints for explicit methods can considerably decrease computational efficiency; see [KWH11, equation (2.1)] or Chapter 11. Other interesting alternatives are based on modal filters [HK08, GÖS18, RGÖS18] or applying viscosity to the different spectral scales [Tad90]. We also mention those methods based on order reduction [CS89], mesh adaptation [DLGC03], and WENO concepts [SO88, SO89]. Yet, a number of issues regarding the fundamental convergence properties for these methods still remain unresolved. Moreover, even when the extension to multiple dimensions is straightforward, these schemes may be too computationally expensive for practical usage.

In this chapter, we therefore propose ℓ^1 regularization as a novel tool to capture shocks in SE methods by promoting sparsity in the jump function of the approximate solution. ℓ^1 regularization methods are frequently encountered in signal processing and imaging applications. Yet, they are still of limited use in solving PDEs numerically and only a few studies (see, e. g., [SCHO13, HLS15, Lav89, Lav91, SGP17b, GMP$^+$08]) have considered sparsity or ℓ^1 regularization of the numerical solution. A brief discussion of these investigations can be found in [SGP17b]. We note that while problems with discontinuous ICs were studied in [Lav89, Lav91, GMP$^+$08], problems that form shocks were not. The technique developed in [SCHO13] was designed to promote sparsity in the frequency domain, making it less amenable to problems emitting shocks, where the frequency domain is not sparse. Further, with the exception of [SGP17b], each of these investigations applied ℓ^1 regularization directly to numerical solution or to its residual, rather than incorporating it directly in the time stepping evolution.

Here, we follow the approach in [SGP17b], which incorporates ℓ^1 regularization directly into the time dependent solver. Specifically, we promote *the sparsity of the jump function* that corresponds to the discontinuous solution. The jump function approximation is performed using a (high order) PA operator [AGP16, AGY05, WAG15]. PA operators eliminate the unwanted "staircasing" effect, a common degradation of detail of the piecewise smooth solution arising from the classical TV regularization. More specifically, high order PA operators allow the resulting solution to be comprised of piecewise polynomials instead of piecewise constants. We solve the resulting ℓ^1 optimization problem by the ADMM [LYJZ13, San16, SGP17a]. A similar application of ℓ_1 regularization was used in [SGP17b] to numerically solve hyperbolic CLs, though only for the Lax–Wendroff scheme and Chebyshev and Fourier spectral methods. It was concluded in [SGP17b] that although the Lax–Wendroff scheme yielded sufficient accuracy for relatively simple problems, its lower order convergence properties made it difficult to resolve more complicated ones. The new technique fared better using Chebyshev polynomials, although their global construction made it difficult to resolve the local structures without oscillations or excessive smoothing.

One possible solution is to use SE methods as the underlying mechanism for solving the hyperbolic CL. SE methods have the advantage of being more localized and allow element-to-element variations in the optimization problem. In particular, for the method we develop here, ℓ^1 regularization is only activated in troubled elements, i. e. in elements where discontinuities are detected. This further enhances efficiency of the method. In the process, a novel discontinuity sensor based on PA operators of increasing orders is proposed, which is able to flag troubled elements. The discontinuity sensor steers the optimization problem and thus locally calibrates the method with respect to the smoothness of the solution. Numerical tests are performed for the DGSEM and the inviscid Burgers' equation as well as for a nonlinear system. It should be stressed that the proposed procedure also carries over to other classes of methods, with the obvious extension to SE type methods. The extension to other types of methods, such as FV methods, is also possible under slight modifications of the procedure. We should also stress that in our development of the ℓ^1 regularization method for solving PDEs, which admit shocks in their solutions, we *do not* require different methods to be used in smooth and nonsmooth regions. Such methods have been developed in, for instance, [DGLW16] and have been shown to be effective. Here we demonstrate that it is possible to avoid such additional complexities.

9.2 Preliminaries

Here, we briefly review all necessary concepts in order to introduce ℓ^1 regularization into the framework of DGSEMs. Remember that these yield a semidiscretization of the (strong) form

$$M\frac{\mathrm{d}}{\mathrm{d}t}\mathbf{u}_n = -MD\mathbf{f}_n - R^T B\left(\boldsymbol{f}_n^{\mathrm{num}} - R\mathbf{f}_n\right), \quad n = 1,\dots,m, \tag{9.2}$$

on the reference element $\Omega_{\mathrm{ref}} = (-1,1)$; see Chapter 6.1. Here, $\mathbf{u}_n$ once more denotes the nodal (or modal) values of the nth component of the (numerical) solution $\boldsymbol{u} = (u_1,\dots,u_m)^T$. Henceforth, we explain the proposed procedure and all necessary preliminaries for a scalar CL, i. e., $m = 1$. The extension to systems of CLs is achieved by simply applying the procedure to every component of a system.

9.2.1 ℓ^1 regularization

Let $u(\xi) = u(t,\xi)$ be the unknown solution on an element Ω_i transformed into the reference element Ω_{ref} and $u_K \in \mathbb{P}_K(\mathbb{R})$ a spatial polynomial approximation at fixed time t. Assume that some measurable features of u have sparse representation. Consequently, the approximation u_K is desired to have this sparse representation as well.

Let H be a regularization functional which measures sparsity. Then, the objective is to solve the constrained optimization problem

$$\underset{v\in\mathbb{P}_K(\mathbb{R})}{\arg\min}\, H(v) \quad \text{s.t.} \quad \|v - u_K\| = 0. \tag{9.3}$$

The equality constraint, referred to as the *data fidelity term*, measures how accurately the reconstructed approximation fits the given data with respect to some seminorm $\|\cdot\|$. Typically, the continuous L^2 norm $\|f\|_2^2 = \int |f|^2$ or some discrete counterpart is used. The *regularization term* $H(v)$ enforces the known sparsity present in the underlying solution u by penalizing missing sparsity in the approximation. The regularization functional H further restricts the approximation space to a desired class of functions, here $\mathbb{P}_K(\mathbb{R})$. Note that any L^p norm with $p \leq 1$ will enforce sparsity in the approximation. In this work, we choose H to be the ℓ^1 norm of certain transformations of v.

It should be stressed that if $\|\cdot\|$ is not just a seminorm but a strictly convex norm, for instance, induced by an inner product, the equality constraint immediately and uniquely determines the approximation. Thus, instead of (9.3), typically the related *denoising problem*

$$\underset{v\in\mathbb{P}_K(\mathbb{R})}{\arg\min}\, H(v) \quad \text{s.t.} \quad \|v - u_K\| < \sigma \tag{9.4}$$

with $\sigma > 0$ is solved, which relaxes the equality constraint on the data fidelity term. Equivalently, the denoising problem (9.4) can also be formulated as the *unconstrained (or penalized) problem*

$$\underset{v\in\mathbb{P}_K(\mathbb{R})}{\arg\min}\left(\|v - u_K\|_2^2 + \lambda H(v)\right) \tag{9.5}$$

by introducing a nonnegative *regularization parameter* $\lambda \geq 0$. λ represents the trade-off between fidelity to the original approximation and sparsity.

The unconstrained problem (9.5) is often solved with $H(v) = \mathrm{TV}(v)$, where $\mathrm{TV}(v)$ is the TV of v. Following [SGP17b], however, here we solve (9.5) with

$$H(v) = \|L_m[v]\|_1,$$

where L_m is a PA operator introduced subsequently. Using higher order PA operators will help to eliminate the staircase effect that occurs when using the TV operator (PA for $m = 1$) for H; see [SRG10].

9.2.2 Polynomial annihilation

PA operators were originally proposed in [AGY05]. One main advantage in using PA operators of higher orders ($m > 1$) for the regularization functional H is that they allow distinction between jump discontinuities and steep gradients, which is critical in the numerical treatment of nonlinear CLs. PA regularization is also preferable to TV regularization when the resolution is poor, even when the underlying solution is piecewise constant. Let $u(\xi^-)$ and $u(\xi^+)$ respectively denote the left- and right-hand-side limits of $u : \Omega_{\text{ref}} = [a, b] \to \mathbb{R}$ at ξ. We define the *jump function of* u as

$$[u](\xi) := u(\xi^+) - u(\xi^-) \tag{9.6}$$

and note that $[u](\xi) = 0$ at every $\xi \in \Omega_{\text{ref}}$ where u has no jump. If u is at least piecewise continuous, there will only be a finite set of points at which $[u](\xi) \neq 0$. We thus say that the jump function $[u]$ has a *sparse representation*. The *PA operator of order* m,

$$L_m[u](\xi) = \frac{1}{q_m(\xi)} \sum_{x_j \in S_\xi} c_j(\xi)\, u(x_j), \tag{9.7}$$

is designed in order to approximate the jump function $[u]$. Here,

$$S_\xi = \{x_0(\xi), \ldots, x_m(\xi)\} \subset \Omega_{\text{ref}}$$

is a set of $m+1$ local grid points around ξ, the *annihilation coefficients* $c_j : \Omega_{\text{ref}} \to \mathbb{R}$ are given by

$$\sum_{x_j \in S_\xi} c_j(\xi)\, p_l(x_j) = p_l^{(m)}(\xi), \quad l = 0, \ldots, m, \tag{9.8}$$

and $\{p_l\}_{l=0}^m$ is a basis of $\mathbb{P}_m(\mathbb{R})$. An explicit formula for the annihilation coefficients utilizing Newton's divided differences is given by ([AGY05])

$$c_j(\xi) = \frac{m!}{\omega_j(S_\xi)} \quad \text{with} \quad \omega_j(S_\xi) = \prod_{\substack{x_i \in S_\xi \\ i \neq j}} (x_j - x_i) \tag{9.9}$$

for $j = 0, \ldots, m$. Finally, the *normalization factor* q_m, calculated as

$$q_m(\xi) = \sum_{x_j \in S_\xi^+} c_j(\xi), \tag{9.10}$$

ensures convergence to the right jump strength at every discontinuity. S_ξ^+ denotes the set $\{x_j \in S_\xi \mid x_j \geq \xi\}$ of all local grid points to the right of ξ.

Here, the PA operator is applied to the reference element $\Omega_{\text{ref}} = (-1, 1)$ of the underlying nodal DGSEM using $K + 1$ collocation points $\{\xi_k\}_{k=0}^{K}$, where typically GLo points including the boundaries are used. Thus, we can construct PA operators up to order K by allowing the sets of local grid points S_ξ to be certain subsets of the $K + 1$ collocation points.

In [AGY05] it was shown that

$$L_m[u](\xi) = \begin{cases} [u](x) + \mathcal{O}(h(\xi)) & \text{if } x_{j-1} \leq \xi, x \leq x_j, \\ \mathcal{O}\left(h^{\min(m,k)}(\xi)\right) & \text{if } u \in C^k(I_\xi), \end{cases} \tag{9.11}$$

where $h(\xi) = \max\{|x_i - x_{i-1} \mid x_{i-1}, x_i \in S_\xi\}$ and I_ξ is the smallest closed interval such that $S_\xi \subset I_\xi$. Note that $h(\xi)$ depends on the density of the set of local grid points around ξ. Thus, L_m provides mth order convergence in regions where $u \in C^m$ and yields a first order approximation to the jump. It should be stressed that oscillations develop around points of discontinuity as m increases. The impact of the oscillations could be reduced by postprocessing methods, such as the minmod limiter, as was done in [AGY05]. However, as long as there is sufficient resolution between two shock locations, such oscillations do not directly impact our method. This is because we use the PA operator not to detect the precise location of jump discontinuities but rather to enforce sparsity.

Figures 9.1(b) and 9.1(c) demonstrate the PA operator for the discontinuous function illustrated in Figure 9.1(a), while Figures 9.1(e) and 9.1(f) do so for the continuous but not differentiable function shown in Figure 9.1(e), and Figures 9.1(h) and 9.1(i) illustrate the PA operator for the smooth function in Figure 9.1(g).

For each function we used eight GLo points to compute the PA operators for $m = 1, 3$ at the midpoints $\xi_{j+\frac{1}{2}} = (\xi_j + \xi_{j+1})/2$. As illustrated in Figure 9.1(b) and 9.1(c), the jump function approximation becomes oscillatory as m is increased from 1 to 3, especially in the region of the discontinuity. The maximal absolute value of the PA operator is also not decreasing in this case. On the other hand, for the continuous and smooth functions displayed in Figures 9.1(d) and 9.1(g), the maximal absolute value decreases significantly when going from $m = 1$ to $m = 3$. This is consistent with the results in (9.11) and will be used to construct the discontinuity sensor in Chapter 9.3. For a discussion on the convergence of the PA operator see [AGY05].

Remark 9.1. Here, we only consider one dimensional CLs. It should be stressed, however, that PA can be extended to multivariate irregular data in any domain. It was demonstrated in [AGY05] that PA is numerically cost efficient and entirely independent of any specific shape or complexity of boundaries. In particular, in [AGSX09] and [JAX11] the method was applied to high dimensional functions that arise when solving stochastic partial differential equations, which reside in a high dimensional space which includes the original space and time domains as well as additional random dimensions.

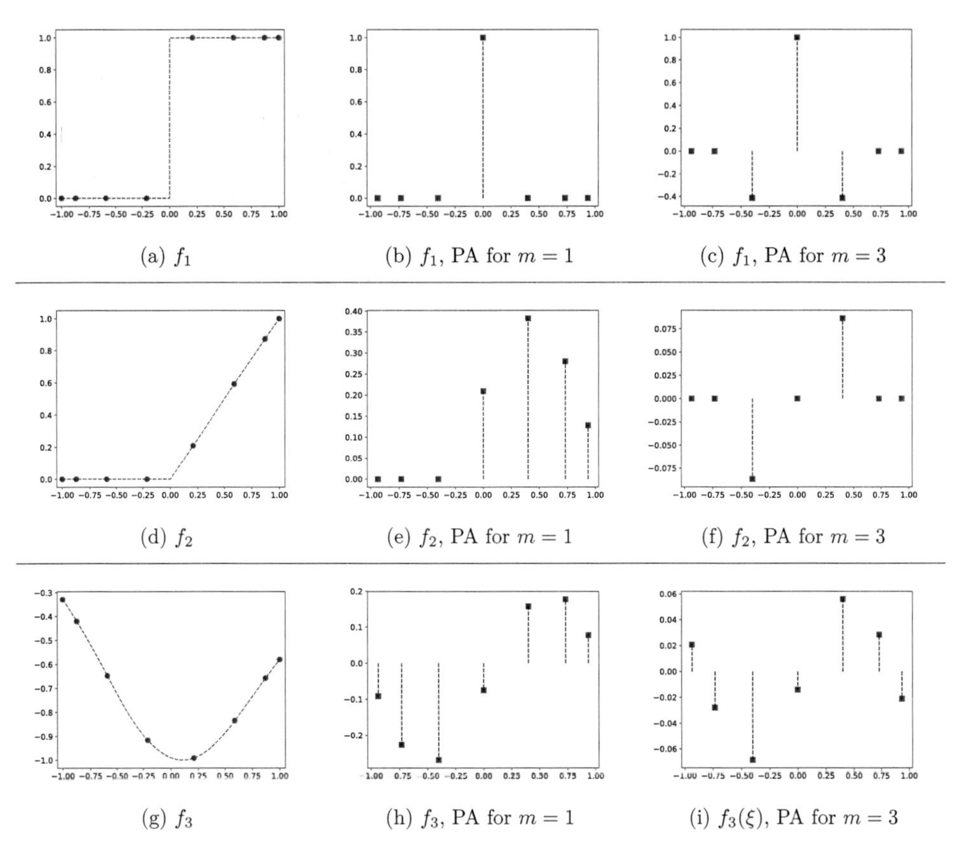

(a) f_1 (b) f_1, PA for $m = 1$ (c) f_1, PA for $m = 3$

(d) f_2 (e) f_2, PA for $m = 1$ (f) f_2, PA for $m = 3$

(g) f_3 (h) f_3, PA for $m = 1$ (i) $f_3(\xi)$, PA for $m = 3$

Figure 9.1: Three different functions f at eight GLo points. The PA operator is evaluated at their midpoints.

9.3 Application of ℓ^1 regularization to discontinuous Galerkin methods

Next, we bring the preliminaries from Chapter 9.2 into play and describe how the proposed ℓ^1 regularization using PA operators, i. e., $H(v) = \|L_m[v]\|_1$ in (9.5), can be incorporated into a DG method. While this kind of regularization functional was already investigated in [SGP17b], the work [GG19a] which resulted from this thesis was the first to extend these ideas to an SE method. In particular, this allows element-to-element variations in the optimization problem. It should be stressed that the subsequent procedure relies on a piecewise polynomial approximation in space. Yet, by appropriate modifications of the procedure, it is also possible to apply ℓ^1 regularization (with PA) to any other method.

9.3.1 Procedure

One of the main challenges in solving nonlinear CLs (9.1) is balancing high resolution properties and the amount of viscosity introduced to maintain stability, especially near shocks [vNR50, PP06, GNJA+19, Gla19b]. Applying the techniques presented in Chapter 9.2, we are now able to adapt the nodal DGSEM method described in Chapter 6.1.3 to include ℓ^1 regularization.

Our procedure consists of replacing the usual polynomial approximation $u_K \in \mathbb{P}_K(\mathbb{R})$ by a *sparse reconstruction*

$$u_K^{\text{spar}} = \arg\min_{v \in \mathbb{P}_K(\mathbb{R})} \left(\frac{1}{2} \|v - u_K\|_2^2 + \lambda \|L_m[v]\|_1 \right) \tag{9.12}$$

with *regularization parameter* λ in troubled elements after every (or every kth) time step by an explicit time integrator.

For the ADMM described in Chapter 9.3.5, it is advantageous to rewrite (9.12) in the usual form of an ℓ^1 regularized problem as

$$u_K^{\text{spar}} = \arg\min_{v \in \mathbb{P}_K(\mathbb{R})} \left(\|L_m[v]\|_1 + \frac{\mu}{2} \|v - u_K\|_2^2 \right), \tag{9.13}$$

where $\mu = \frac{1}{\lambda}$ is referred to as the *data fidelity parameter*. Note that (9.12) and (9.13) are equivalent. In the later numerical tests, the data fidelity parameter μ and the regularization parameter λ will be steered by a discontinuity sensor proposed in Chapter 9.3.3.

Remark 9.2. One of the main drawbacks in using ℓ^1 regularization for solving numerical PDEs, as well as for image restoration or sparse signal recovery, is in choosing the regularization parameter λ (or μ). Ideally, one would want to balance the terms in (9.12) or (9.13), but this is difficult to do without knowing their comparative size. Indeed, the ℓ^1 regularization term $\|L_m[v]\|_1$ heavily depends on the magnitudes of nonzero values in the sparsity domain, in this case the jumps. Larger jumps are penalized significantly more in the ℓ^1 norm than smaller values. Iterative *spatially varying* weighted ℓ^1 regularization techniques (see, e. g., [CWB08, GS19]) are designed to help reduce the size of the norm, since the remaining values should be close to zero in magnitude. Specifically, the jump discontinuities which are meant to be in the solution can "pass through" the minimization. In this way, with some underlying assumptions made on the accuracy of the fidelity term, one could argue that both terms are close to zero. Consequently, the choice of λ (or μ) should not have as much impact on the results, leading to greater robustness overall. For the numerical experiments in this investigation, we simply chose regularization parameters that worked well. We did not attempt to optimize our results and leave parameter selection to future work.

9.3.2 Selection of the regularization parameter λ

The ℓ^1 regularization should only be activated in troubled elements. In particular, we do not want to unnecessarily degrade the accuracy in the smooth regions of the solution. We thus propose to adapt the regularization parameter λ in (9.12) to appropriately capture different discontinuities and regions of smoothness. As a result, the optimization problem will be able to calibrate the resulting sparse reconstruction to the smoothness of the solution. More specifically, to avoid unnecessary regularization, we choose $\lambda = 0$ in elements corresponding to smooth regions. Note that this also renders the proposed method more efficient. On the other hand, when a discontinuity is detected in an element, ℓ^1 regularization will be fully activated by choosing $\lambda = \lambda_{\max}$ in (9.12), which corresponds to the amount of regularization necessary to reconstruct sharp shock profiles. While no effort was made to optimize or even adapt this parameter, we found that using $\lambda_{\max} = 4 \cdot 10^2$ in all of our numerical experiments yielded good results. A heuristic explanation for choosing $\lambda_{\max}$ in this way stems from the goal of balancing the size of $\|L_m[v]\|_1$ with the expected size of the fidelity term, which in this case means to be consistent with the order of accuracy of the underlying numerical PDE solver. As mentioned previously, choosing an appropriate λ will be the subject of future work.

Between these extreme cases, i. e., $\lambda = 0$ and $\lambda = \lambda_{\max}$, we allow the regularization parameter to linearly vary and choose λ as a function of the discontinuity sensor proposed in Chapter 9.3.3. As a consequence, we obtain more accurate sparse reconstructions while still maintaining stability in regions around discontinuities.

9.3.3 Discontinuity sensor

We now describe the discontinuity sensor which is used to activate the ℓ^1 regularization and to steer the regularization parameter λ in (9.12). The sensor is based on comparing PA operators of increasing orders. To the best of our knowledge, this is the first time PA operators are utilized for shock (discontinuity) detection in a PDE solver.[1]

At least for smooth solutions, DG methods are capable of spectral orders of accuracy. ℓ^1 regularization as well as any other shock capturing procedure [vNR50, PP06, GÖS18, RGÖS18, GNJA+19] should thus be just applied in (and near) elements where discontinuities are present. We refer to those elements as *troubled elements*.

Many shock and discontinuity sensors have been proposed over the last 20 years for the selective application of shock capturing methods. Some of them use information about the L^2 norm of the residual of the variational form [BR95, JJS95], the primary orientation of the discontinuity [Har06], the magnitude of the facial interelement jumps [BD10, FK07], or entropy pairs [GP08] to detect troubled elements. Others are not just able to detect troubled elements, but even the location of discontinuities in the corresponding element, such as the concentration method in [GT99, GT00, GT06a, GC08, ÖSW13, DGLW16, TW12]. The PA operator is also capable of detecting the location of a discontinuity [AGY05, AGY08, AGSX09] up to the resolution size, as in (9.11), using nonlinear postprocessing techniques. It should be stressed, however, that here we do not fully make use of this feature.

In what follows we present a novel discontinuity sensor based on PA operators. Let the *sensor value of order m* be

$$S_m = \max_{0 \le k \le K-1} \left| L_m[u_K] \left(\xi_{k+\frac{1}{2}} \right) \right|, \tag{9.14}$$

i. e., the greatest absolute value of the PA operator at the midpoints

$$\xi_{k+\frac{1}{2}} = \frac{\xi_k + \xi_{k+1}}{2}, \quad k = 0, \dots, K-1,$$

of the collocation points. Since L_m provides convergence to 0 of order m in elements where u has m continuous derivatives, we expect $S_3 < S_1$ to hold for an at least continuous function. Thus, ℓ^1 regularization should just be fully activated if

$$S(u) := \frac{S_3}{S_1} \ge 1, \tag{9.15}$$

[1] Of course (W)ENO schemes [SO88, SO89, QS05] compare slope magnitudes for determining troubled elements and choosing approximation stencils, so in this regard our method was inspired by WENO type methods.

i. e., if the sensor value does not decrease as m increases from 1 to 3. Sensor values of order 1 and 3 (rather than 2) are chosen since having symmetry of the grid points in S_ξ surrounding the point ξ yields a simpler form for implementation [AGY05]. Various modifications of the *PA sensor* (9.15) are possible and will be the topic of future research. In the following, the resulting PA sensor is demonstrated for the function displayed in Figure 9.2 on the interval $[0, 10]$. We decompose the interval into $I = 5$ elements and apply the PA operator and resulting PA sensor separately on each element.

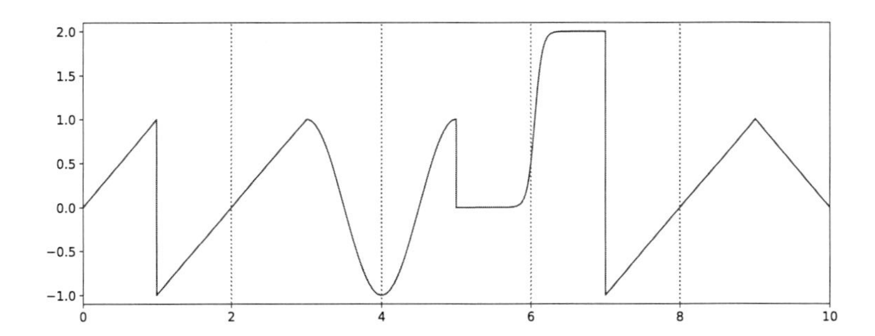

Figure 9.2: Test function with various different discontinuities as well as continuous but not differentiable points.

Table 9.1 lists the sensor values of order $m = 1$ and $m = 3$ on each element in Figure 9.2. The last row further shows whether the PA sensor (9.15) reacts or not. As can be seen in Table 9.1, the discontinuity sensor exactly identifies the troubled elements, simply by comparing the sensor values of order $m = 1$ and $m = 3$.

Element	$\Omega_1 = (0, 2)$	$\Omega_2 = (2, 4)$	$\Omega_3 = (4, 6)$	$\Omega_4 = (6, 8)$	$\Omega_5 = (8, 10)$
S_1	1.58	1.07	1.07	2.79	0.38
S_3	1.99	0.60	1.38	3.00	0.17
Discontinuity	yes	no	yes	yes	no

Table 9.1: Shock sensor for function displayed in Figure 9.2.

Even though we use this sensor only to detect troubled elements, in principle, this new sensor can also be used to more precisely determine the location and strength of the discontinuities. Such information can be used, for instance, for refined domain decomposition [AGSX09].

Finally, we decide for the regularization parameter λ to linearly vary between $\lambda = 0$ and $\lambda = \lambda_{\max}$ and thus utilize the parameter function

$$\lambda(S) = \begin{cases} 0 & \text{if } S \leq \kappa, \\ \lambda_{\max}(S - \kappa)/(1 - \kappa) & \text{if } \kappa < S < 1, \\ \lambda_{\max} & \text{if } 1 \leq S, \end{cases} \tag{9.16}$$

where $\kappa \in [0, 1)$ is a problem dependent *ramp parameter*. Observe that using $\lambda(S)$ is comparable to employing the weighted ℓ_1 regularization as discussed in Remark 9.2.

For the later numerical tests we also considered other parameter functions, some as discussed in [HCP12]. Yet the best results were obtained with (9.16). The same holds for other discontinuity sensors, such as the modal decay based sensor of Persson and Peraire [PP06, HCP12] and its refinements [BD10, KWH11] as well as the KXRCF sensor [KXR+04, QS05] of Krivodonova et al., which is built up on a strong superconvergence phenomenon of the DG method at outflow boundaries. For brevity, those results are omitted here.

Remark 9.3. We note that the PA sensor might produce false positive or false negative misidentifications in certain cases. A false negative misidentification might arise from a discontinuity where the solution is detected to be smooth. This is encountered by the ramp parameter κ, which is observed to work robustly for $\kappa = 0.8$ or $\kappa = 0.9$ in all later numerical tests. A false positive misidentification might arise from a smooth solution which is detected to be non-smooth (possibly discontinuous). In this case smooth parts of the solution with steep gradients will result in significantly greater values of $L_m[u]$ than parts of the solution with less steep gradients. As a result, the standard ℓ^1 regularization would heavily penalize these features of the solution, yielding inappropriate smearing of steep gradients in smooth regions. By making $\lambda = \lambda(S)$ dependent on the sensor value, this problem can be somewhat alleviated. Using a weighted ℓ^1 regularization, as suggested in Remark 9.2, should also reduce the unwanted smearing effect. Failure to detect a discontinuity would, after a number of time steps, yield instability. However it is unlikely that this would occur as the growing oscillations would more likely be identified as shock discontinuities.

9.3.4 Efficient implementation of the PA operator

While the PA operator was defined on the interior of the reference element $\Omega_{\text{ref}} = [-1, 1]$ in Chapter 9.2.2, the shock sensor proposed above only relies on the values of the PA operator at the K midpoints $\left\{\xi_{k+\frac{1}{2}}\right\}_{k=0}^{K-1}$ of the collocation points $\{\xi_k\}_{k=0}^{K}$. The same holds for the ℓ^1 regularization term $H(v) = \|L_m[v]\|_1$. The ℓ^1-norm of the PA transformation is therefore given by

$$\|L_m[v]\|_1 = \|L_m \mathbf{v}\|_1 = \sum_{k=0}^{K-1} \left|L_m[v]\left(\xi_{k+\frac{1}{2}}\right)\right|, \tag{9.17}$$

where the vector $\mathbf{v}$ once more consists of nodal degrees of freedom. We now aim to provide an efficient implementation of the PA operator L_m in form of a matrix representation, also denoted by L_m, which maps the nodal values $\mathbf{v}$ to the values of the PA operator at the midpoints. Even though we use the same notation for the PA operator and its matrix representation, it should always be clear by the context which object to use. Revisiting (9.7), the matrix representation of the PA operator L_m is given by

$$L_m = Q^{-1}C \in \mathbb{R}^{K\times(K+1)}$$

with

$$Q = \operatorname{diag}\left(q_m\left(\xi_{\frac{1}{2}}\right), \dots, q_m\left(\xi_{K-\frac{1}{2}}\right)\right),$$
$$q_m\left(\xi_{k+\frac{1}{2}}\right) = \sum_{x_j \in S^+_{\xi_{k+\frac{1}{2}}}} c_j\left(\xi_{k+\frac{1}{2}}\right)$$

and

$$C = \left(c_j\left(\xi_{k+\frac{1}{2}}\right)\right)_{k,j=0}^{K-1,K},$$

where

$$c_j\left(\xi_{k+\frac{1}{2}}\right) = \begin{cases} \dfrac{m!}{\omega_j\left(S_{\xi_{k+\frac{1}{2}}}\right)} & \text{if } x_j \in S_{\xi_{k+\frac{1}{2}}}, \\ 0 & \text{otherwise,} \end{cases}$$

$$\omega_j\left(S_{\xi_{k+\frac{1}{2}}}\right) = \prod_{\substack{x_i \in S_{\xi_{k+\frac{1}{2}}} \\ i \neq j}} (x_j - x_i)\,.$$

Utilizing all prior matrix vector representations, we can now give the discretization of the ℓ^1 regularized optimization problem (9.13) by

$$\mathbf{u}^{\text{spar}} = \operatorname*{arg\,min}_{\mathbf{v}\in\mathbb{R}^{K+1}} \left(\|L_m \mathbf{v}\|_1 + \frac{\mu}{2} \|\mathbf{v}-\mathbf{u}\|_M^2 \right). \tag{9.18}$$

Thus, we are able to solve the optimization problem directly for the nodal degrees of freedom of the sparse reconstruction u_p^{spar}. Alternatively, the fidelity term $\|v - u_K\|$ can also be approximated as

$$\|\mathbf{v}-\mathbf{u}\|_2^2 = \sum_{k=0}^{K} |v(\xi_k) - u(\xi_k)|^2\,,$$

i. e., by the Euclidean norm, yielding

$$\mathbf{u}^{\text{spar}} = \operatorname*{arg\,min}_{\mathbf{v}\in\mathbb{R}^{K+1}} \left(\|L_m \mathbf{v}\|_1 + \frac{\mu}{2} \|\mathbf{v}-\mathbf{u}\|_2^2 \right), \tag{9.19}$$

instead of (9.18). Future works will investigate the influence of the choice of the discrete norm on the performance of the ℓ^1 regularization. Here we decided to use the Euclidean norm and thus the minimization problem (9.19), making the computations in Chapter 9.3.5 more intelligible.

9.3.5 The alternating direction method of multipliers

Many techniques have been recently proposed to solve optimization problems in the form of (9.18). Following [SGP17b], we use the ADMM [LYJZ13, San16, SGP17a] in our implementation. The ADMM has its roots in [GM75] and details of its convergence properties can be found in [GM76, GLT89, EB92]. In the context of ℓ^1 regularization, ADMM is commonly implemented using the split Bregman method [GO09], which is known to be an efficient solver for a broad class of optimization problems. To implement the ADMM, it is first necessary to eliminate all nonlinear terms within the ℓ^1 norm. We thus introduce a slack variable

$$\mathbf{g} = L_m \mathbf{v} \in \mathbb{R}^K$$

and formulate (9.18) equivalently as

$$\operatorname*{arg\,min}_{\mathbf{v}\in\mathbb{R}^{K+1},\mathbf{g}\in\mathbb{R}^K} \left(\|\mathbf{g}\|_1 + \frac{\mu}{2} \|\mathbf{v}-\mathbf{u}\|_2^2 \quad \text{s. t.} \quad L_m \mathbf{v} = \mathbf{g} \right). \tag{9.20}$$

To solve (9.20), we further introduce Lagrangian multipliers $\sigma \in \mathbb{R}^K, \delta \in \mathbb{R}^{K+1}$ and solve the unconstrained minimization problem given by

$$\operatorname*{arg\,min}_{\mathbf{v}\in\mathbb{R}^{K+1},\mathbf{g}\in\mathbb{R}^K} \quad J_{\sigma,\delta}(\mathbf{v},\mathbf{g}) \tag{9.21}$$

with objective function

$$J_{\sigma,\delta}(\mathbf{v},\mathbf{g}) = \|\mathbf{g}\|_1 + \frac{\mu}{2}\|\mathbf{v}-\mathbf{u}\|_2^2 + \frac{\beta}{2}\|L_m\mathbf{v}-\mathbf{g}\|_2^2 - \langle L_m\mathbf{v}-\mathbf{g},\,\sigma\rangle_2 - \langle \mathbf{v}-\mathbf{u},\,\delta\rangle_2.$$

Here, $\beta > 0$ is an additional positive regularization parameter and recall that the data fidelity parameter μ is given by $\mu = \frac{2}{\lambda}$ for $\lambda > 0$; see (9.12) and (9.13). Note that if the Lagrangian multipliers σ, δ are updated a sufficient number of times, the solution of the unconstrained problem (9.21) will converge to the solution of the constrained problem (9.20). In the ADMM, the solution is approximated by alternating between minimizations of $\mathbf{v}$ and $\mathbf{g}$. A crucial advantage of this method is that, given the current value of $\mathbf{v}$ as well as the Lagrangian multipliers, the optimal value of $\mathbf{g}$ can be exactly determined by the *shrinkage*-like formula [GO09]

$$(\mathbf{g}_{r+1})_i = \operatorname{shrink}\left((L_m\mathbf{v})_i - \frac{1}{\beta}(\sigma_k)_i, \frac{1}{\beta}\right), \tag{9.22}$$

where

$$\operatorname{shrink}(x,\gamma) = \frac{x}{|x|}\cdot\max(|x|-\gamma,0). \tag{9.23}$$

Given the current value $\mathbf{g}_{r+1}$, on the other hand, the optimal value of $\mathbf{v}$ is computed by the gradient descent method as

$$\mathbf{v}_{r+1} = \mathbf{v}_r - \alpha\nabla_{\mathbf{v}}J_{\sigma,\delta}(\mathbf{v},\mathbf{g}_{r+1}), \tag{9.24}$$

where

$$\nabla_{\mathbf{v}}J_{\sigma,\delta}(\mathbf{v},\mathbf{g}_{r+1}) = \mu(\mathbf{v}-\mathbf{u}) + \beta(L_m)^T(L_m\mathbf{v}-\mathbf{g}_{r+1}) - (L_m)^T\sigma_r - \delta_r,$$

and the step size $\alpha > 0$ is chosen to provide a sufficient descent in direction of the gradient. Finally, the Lagrangian multipliers are updated after each iteration by

$$\begin{aligned}\sigma_{r+1} &= \sigma_r - \beta(L_m\mathbf{v}_{r+1} - \mathbf{g}_{r+1}),\\ \delta_{r+1} &= \delta_r - \mu(\mathbf{v}_{r+1}-\mathbf{u}).\end{aligned} \tag{9.25}$$

Algorithm 1 is borrowed from [SGP17a, SGP17b] and compactly describes the above ADMM.

Algorithm 1 ADMM

1: Determine parameters μ, β, and *tol*
2: Initialize $\mathbf{v}_0, \mathbf{g}_0, \sigma_0$, and δ_0
3: **for** $r = 0$ to R **do**
4:     **while** $\|\mathbf{v}_{r+1} - \mathbf{v}_r\| > tol$ **do**
5:         Minimize J for $\mathbf{g}$ according to (9.22)
6:         Minimize J for $\mathbf{v}$ according to (9.24)
7:     Update the Lagrangian multipliers according to (9.25)

Figure 9.3 demonstrates the effect of ℓ^1 regularization to a polynomial approximation with degree $K = 13$ of a discontinuous sawtooth function $u(x) = \operatorname{sign}(x) - x$ on $[-1, 1]$.

Due to the Gibbs phenomenon, the original polynomial approximation u_K in Figure 9.3(a) shows spurious oscillations [HH79]. As a consequence, the approximation of the corresponding jump function by the PA operator for $m = 3$ in Figure 9.3(b) also shows undesired oscillations away from the detected discontinuity at $x = 0$. Note that the neighboring undershoots around $x = 0$ are inherent in the PA operator for $m = 3$. Yet, the oscillations left and right from these undershoots stem from the Gibbs phenomenon and thus are parasitical. Applying ℓ^1

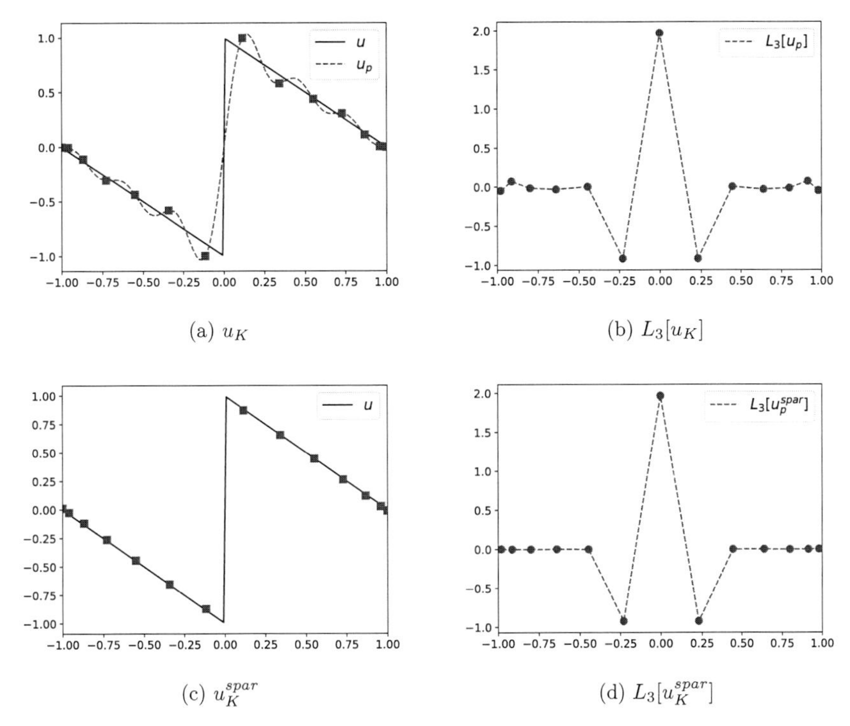

(a) u_K (b) $L_3[u_K]$

(c) u_K^{spar} (d) $L_3[u_K^{spar}]$

Figure 9.3: Demonstration of ℓ^1 regularization for the polynomial approximation (polluted by the Gibbs phenomenon) of a discontinuous function u. A polynomial degree of $K = 13$ has been used.

regularization to enhance sparsity of the PA transform, however, is able to correct these spurious oscillations, resulting in a sparse representation of the PA transformation in Figure 9.3(d). The corresponding nodal values, illustrated in Figure 9.3(c), now approximate the nodal values of the true solution accurately.

For the numerical test in Figure 9.3, we have chosen the parameters $R = 400$, $\mu = 0.005$ ($\lambda = 4 \cdot 10^2$), $\beta = 20$, $\alpha = 0.0001$, and $tol = 0.001$. The Lagrangian multipliers have been initialized with $\sigma_0 = \mathbf{0}$ and $\delta_0 = \mathbf{0}$.

9.3.6 Preservation of mass conservation

An essential property of standard (DG)SE methods for hyperbolic CLs is that they are conservative, i. e., that

$$\int_\Omega u(t^{n+1}, x)\,\mathrm{d}x = \int_\Omega u(t^n, x)\,\mathrm{d}x + f|_{\partial\Omega} \tag{9.26}$$

holds [Ran92, HW07, Gas13]. Any reasonable shock capturing procedure for hyperbolic CLs should preserve this property as well. In particular, in a troubled element Ω_i in which ℓ^1 regularization is applied,

$$\int_{\Omega_i} u_K^{\text{spar}}(x)\,\mathrm{d}x = \int_{\Omega_i} u_K(x)\,\mathrm{d}x \tag{9.27}$$

should hold in order for ℓ^1 regularization to preserve mass conservation of the underlying (DG) method. Unfortunately, (9.27) is violated when ℓ^1 regularization is applied too naively. This is demonstrated in Figure 9.4 for a simple discontinuous example $u(x) = \text{sign}(x - 0.5) + 1$ on $\Omega_{\text{ref}} = [-1, 1]$ with mass $\int_{\Omega_{\text{ref}}} u \, dx = 1$.

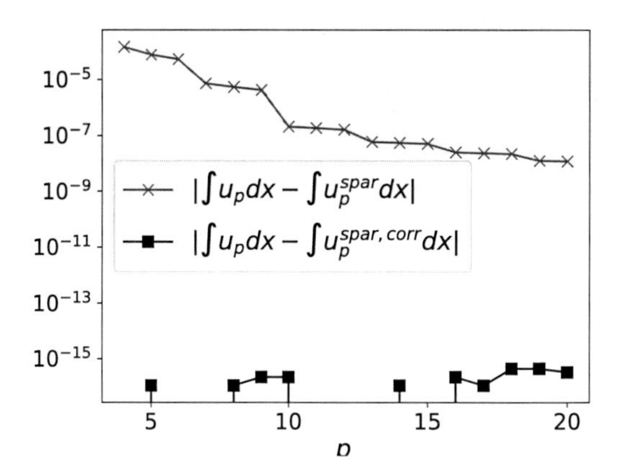

Figure 9.4: Absolute difference between the mass of u_K and u_K^{spar} (red crosses) and between the mass of u_K and $u_K^{\text{spar,corr}}$ (black squares) for $u(x) = \text{sign}(x-0.5)+1$ on $\Omega_{\text{ref}} = [-1, 1]$ and increasing polynomial degree $K = 4, \ldots, 20$ (denoted by p). In the cases where no (black) squares are visible the difference was below machine precision.

The (red) crosses illustrate the absolute difference between the mass of the polynomial interpolation u_K and its sparse reconstruction u_K^{spar}, i. e.

$$\left| \int_{\Omega_{\text{ref}}} u_K(x) \, dx - \int_{\Omega_{\text{ref}}} u_K^{\text{spar}}(x) \, dx \right|,$$

for increasing polynomial degrees $K = 4, \ldots, 20$. We observe that a naive application of ℓ^1 regularization might destroy mass conservation. Thus, in the following, we present a simple fix for this problem.

Remark 9.4. A generic approach — also to, for instance, ensure TVD and entropy conditions — is to add additional constraints to the optimization problem (9.12), e. g.,

$$\begin{aligned} u_K^{\text{spar}} &= \underset{v \in \mathbb{P}_K(\mathbb{R})}{\arg\min} \left(\frac{1}{2} \|v - u_K\|_2^2 + \lambda \|L_m v\|_1 \right) \\ \text{s. t.} & \int_{\Omega_{\text{ref}}} v(x) \, dx = \int_{\Omega_{\text{ref}}} u_K(x) \, dx \end{aligned} \tag{9.28}$$

for conservation of mass to be preserved. In the present case, where $u_K, v \in \mathbb{P}_K(\mathbb{R})$ are expressed with respect to basis functions $\{\varphi_k\}_{k=0}^K$ with zero average for $k > 0$, the conditions in (9.28) are easily met. Specifically,

$$u_K = \sum_{k=0}^{K} \hat{u}_k \varphi_k \quad \text{and} \quad v = \sum_{k=0}^{K} \hat{v}_k \varphi_k$$

with

$$\int_{\Omega_{\text{ref}}} \varphi_k \, dx = \|\varphi_0\|_2^2 \, \delta_{0k},$$

implies that the additional constraint in (9.28) reduces to

$$\hat{u}_0 = \hat{v}_0.$$

Basis functions $\{\varphi_k\}_{k=0}^K$ with zero average for $k > 0$ are, for instance, given by the orthogonal basis (OGB) of Legendre polynomials.

We now propose the following simple algorithm to repair mass conservation in the ℓ^1 regularization:

Algorithm 2 Mass correction

1: Compute $\hat{u}_0$
2: Compute u_K^{spar} according to (9.12)/(9.18)
3: Represent u_K^{spar} w. r. t. an OGB: $u_K^{\text{spar}} = \hat{u}_0^{\text{spar}} \varphi_0 + \cdots + \hat{u}_K^{\text{spar}} \varphi_K$
4: Replace $\hat{u}_0^{\text{spar}}$ by $\hat{u}_0$

The advantage of this additional step is demonstrated in Figure 9.4 as well, where the absolute difference between the mass of u_K and of its sparse reconstruction with additional mass correction, denoted by $u_K^{\text{spar,corr}}$, is illustrated by (black) squares. In contrast to the sparse reconstruction without mass correction, illustrated by (red) crosses, $u_K^{\text{spar,corr}}$ is demonstrated to preserve mass nearly up to machine precision ($\approx 10^{-16}$). Finally, we note that for the test illustrated in Figure 9.3(d), ℓ^1 regularization with and without mass correction resulted in the same approximations, due to u being an odd function. Thus, we omit those results. We present a flowchart in Figure 9.5 illustrating the proposed procedure for a fixed time $t < t_{\text{end}}$.

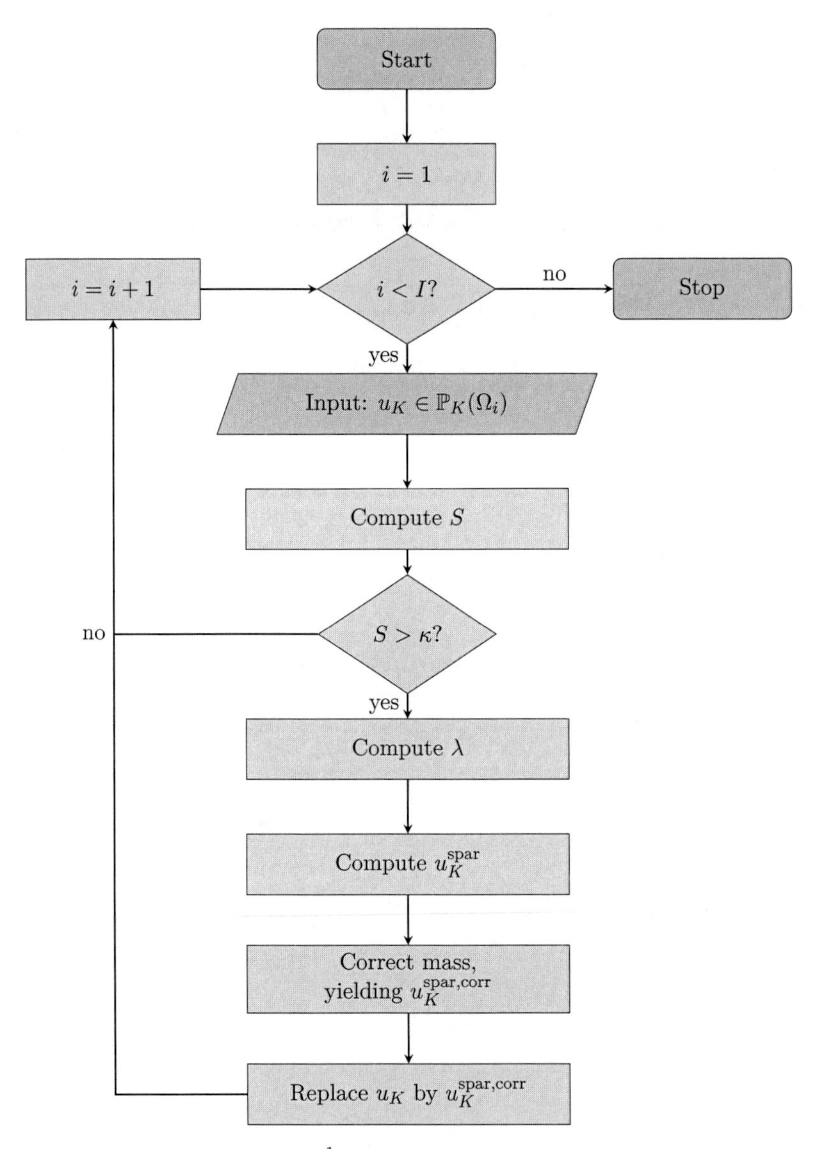

Figure 9.5: Flowchart describing ℓ^1 regularization with mass correction. The procedure is described for a fixed time and includes the loop over all I elements.

9.4 Numerical results

We now numerically demonstrate the application of ℓ^1 regularization (with and without mass correction) to the nodal DGSEM presented in Chapter 6.1.3 for the inviscid Burgers' equation, the linear advection equation, and a nonlinear system of CLs. Our results show that ℓ^1 regularization provides increased accuracy of the numerical solutions. In all numerical tests, we use a PA operator of third order and choose the same parameters as before, i. e., $\lambda_{\max} = 4 \cdot 10^2$, $R = 400$, $\beta = 20$, $\alpha = 0.0001$, and $tol = 0.001$. We have made no effort to optimize these parameters.

9.4.1 Inviscid Burgers' equation

We start our numerical investigation by considering the inviscid Burgers' equation

$$\partial_t u + \partial_x \left(\frac{u^2}{2} \right) = 0 \tag{9.29}$$

on $\Omega = [0, 2]$ with IC

$$u(x, 0) = u_0(x) = \sin(\pi x) \tag{9.30}$$

and periodic BCs. For this test case, a shock discontinuity develops in the solution at $x = 1$.

In the subsequent numerical tests, the usual local Lax–Friedrichs flux

$$f^{\text{num}}(u_-, u_+) = \frac{1}{2} \left(f(u_+) + f(u_-) \right) - \frac{\alpha_{\max}}{2} (u_+ - u_-) \tag{9.31}$$

with maximum characteristic speed $\alpha_{\max} = \max\{|u_+|, |u_-|\}$ is utilized. Further, we use the third order explicit SSP-RK method with three stages (SSPRK(3,3)) for time integration (see Definition 3.39 in Chapter 3.5) and choose the ramp parameter $\kappa = 0.8$ in (9.16).

Figure 9.6 illustrates the numerical solutions of the above test problem at time $t = 0.345$. In Figure 9.6(a) a polynomial degree of $K = 4$ was used on $I = 7$ equidistant elements, while in Figure 9.6(b) a polynomial degree of $K = 10$ was used on $I = 7$ equidistant elements and in Figure 9.6(c) a polynomial degree of $K = 18$ was used on $I = 7$ equidistant elements. All tests show spurious oscillations for the numerical solution without ℓ^1 regularization. These oscillations are significantly reduced when ℓ^1 regularization is applied to the numerical solution after every time step. As a consequence, Figures 9.6(d), 9.6(e), 9.6(f) illustrate how the pointwise error of the numerical solutions utilizing ℓ^1 regularization (with and without mass correction) is reduced compared to the numerical solution without ℓ^1 regularization. Again, considering an odd function, only very slight differences are observed between ℓ^1 regularization with and without mass correction. The reference solution u was computed using characteristic tracing.

We now extend the above error analysis for the ℓ^1 regularization. Table 9.2 lists the different common types of errors of the numerical solution for the above test problem (9.29), (9.30) and a varying polynomial degree p as well as a varying number of equidistant elements I. Here we consider the global errors with respect to the discrete M-norm, as defined in (6.8), approximating the continuous L^2 norm and the discrete 1-norm, given by

$$\|u_K\|_1 = \sum_{k=0}^{K} w_k |u_K(x_k)| = \langle \mathbf{1}, |\mathbf{u}| \rangle_M$$

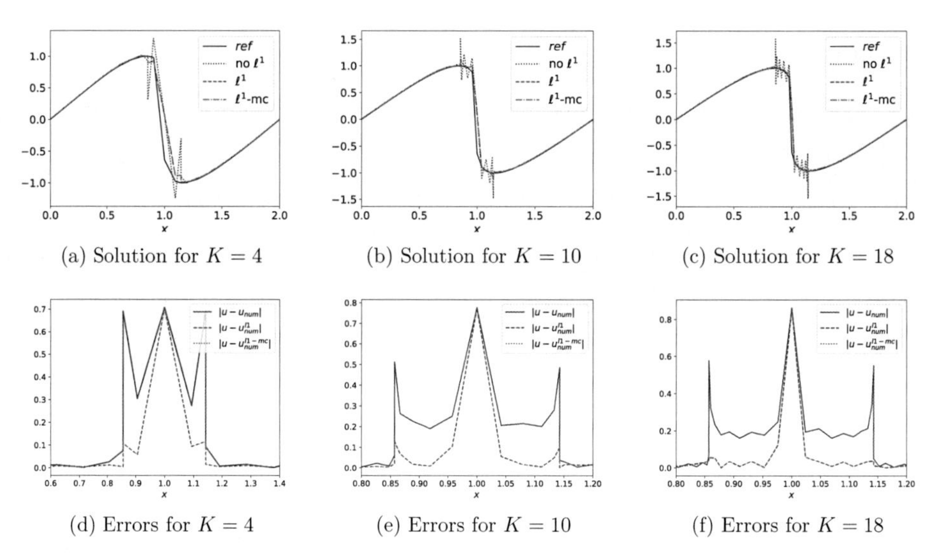

(a) Solution for $K = 4$ (b) Solution for $K = 10$ (c) Solution for $K = 18$

(d) Errors for $K = 4$ (e) Errors for $K = 10$ (f) Errors for $K = 18$

Figure 9.6: Numerical solutions (and their pointwise errors) at $t = 0.345$ by the DGSEM for $I = 7$ elements and polynomial degrees $K = 4, 10, 18$. Without ℓ^1 regularization (straight blue line), with ℓ^1 regularization (dashed red line), and with ℓ^1 regularization and mass correction (dash-dotted green line).

on the reference element. The discrete L^∞ norm is given by $\|u_K\|_\infty = \max_{k=0,\dots,K} |u_K(x_k)|$. For the discrete M norm and discrete 1 norm, the global norms are defined by summing up over the weighted local norms, i. e.,

$$\|u_{\text{num}}\|^2 = \sum_{i=1}^{I} \frac{|\Omega_i|}{2} \left\| u_K^i \right\|^2,$$

where u_K^i denotes the numerical solution (polynomial approximation) on the ith element Ω_i.

Table 9.2 demonstrates that for almost all these norms as well as combinations of polynomial degrees $K = 4, 5, 6, 7, 8, 9$ and number of elements $I = 15, 32, 63, 127$, the numerical solution with ℓ^1 regularization is more accurate than the numerical solution without ℓ^1 regularization. Further, we observe just a slight difference in accuracy for ℓ^1 regularization with and without mass correction. See, for instance, $K = 5$ in Table 9.2. We only utilize odd number of elements, so that the shock discontinuity arises in the interior of an element. By using an even number of elements, on the other hand, the shock would arise at the interface between two elements and the error analysis of the ℓ^1 regularization would be blurred by the dissipation added by the numerical flux.

In Table 9.2, all cases where the accuracy is increased or remains the same by applying ℓ^1 regularization are flagged with a checkmark. It should be stressed that no effort was made to optimize the parameters in the ℓ^1 regularization. In particular, the parameters in the ℓ^1 regularization have not been adapted to the specific choice of the polynomial degree K, the number of elements I, or the number of time steps used in the solver. We think that by further investigations of the parameters in the ℓ^1 regularization, all above errors could be further reduced.

Finally, special attention should be given to the cases flagged by an exclamation mark (!).

		$\|\cdot\|_M$-error			$\|\cdot\|_1$-error			$\|\cdot\|_\infty$-error		
K	I	no ℓ^1	ℓ^1	ℓ^1-mc	no ℓ^1	ℓ^1	ℓ^1-mc	no ℓ^1	ℓ^1	ℓ^1-mc
3	15	1.5e-2	3.3e-2	3.3e-2	1.2e-2	2.2e-1	1.2e-2	1.7e-1	3.3e-1	3.3e-1
	31	1.5e-2	2.0e-2	2.0e-2	1.1e-2	1.2e-2	1.2e-2	2.2e-1	3.2e-1	3.2e-1
	63	1.9e-2	2.4e-2	2.4e-2	1.1e-2	1.1e-2 ✓	1.1e-2	3.6e-1	5.6e-1	5.6e-1
	127	3.3e-2	2.7e-2 ✓	2.7e-2	1.3e-2	1.1e-2 ✓	1.1e-2	1.4e-0	8.3e-1 ✓	8.3e-1
4	15	7.0e-2	5.9e-2 ✓	5.9e-2	3.9e-2	2.7e-2 ✓	2.7e-2	7.6e-1	7.5e-1 ✓	7.5e-1
	31	5.5e-2	4.6e-2 ✓	4.6e-2	2.4e-2	1.7e-2 ✓	1.7e-2	8.7e-1	8.6e-1 ✓	8.6e-1
	63	4.3e-2	3.9e-2 ✓	3.9e-2	1.6e-2	1.3e-2 ✓	1.3e-2	1.0	1.0e-0 ✓	1.0e-0
	127	3.8e-2	3.6e-2 ✓	3.6e-2	1.2e-2	1.2e-2 ✓	1.1e-2	1.4	1.3e-0 ✓	1.3e-0
5	15	1.6e-2	1.5e-2 ✓	1.5e-2	1.2e-2	1.2e-2 ✓	1.2e-2	2.5e-1	1.9e-1 ✓	1.7e-1
	31	2.0e-2	1.4e-2 ✓	1.3e-2	1.3e-2	1.0e-2 ✓	1.0e-2	5.3e-1	2.6e-1 ✓	2.5e-1
	63	7.3e-2	1.9e-2 ✓	1.6e-2	2.2e-2	1.0e-2 ✓	1.0e-2	3.6e-0	5.3e-1 ✓	4.3e-1
	127	NaN	2.7e-2 !	2.2e-2	NaN	1.0e-2 !	1.0e-2	NaN	1.1e-0 !	9.1e-1
6	15	5.9e-2	5.2e-2 ✓	5.2e-2	3.2e-2	2.3e-2	2.3e-2	8.1e-1	8.0e-1 ✓	8.0e-1
	31	4.7e-2	4.3e-2 ✓	4.3e-2	2.0e-2	1.6e-2 ✓	1.6e-2	9.6e-1	9.5e-1 ✓	9.5e-1
	63	3.9e-2	3.7e-2 ✓	3.6e-2	1.2e-2	1.2e-2 ✓	1.2e-2	1.2e-0	1.1e-0 ✓	1.1e-0
	127	NaN	3.4e-2 !	3.2e-2	NaN	1.1e-2 !	1.1e-2	NaN	1.3e-0 !	1.3e-0
7	15	1.7e-2	1.8e-2	1.8e-2	1.4e-2	1.2e-2 ✓	1.2e-2	2.2e-1	2.7e-1	2.8e-1
	31	2.5e-2	1.7e-2 ✓	1.8e-2	1.5e-2	1.1e-2 ✓	1.1e-2	7.0e-1	3.8e-1 ✓	4.0e-1
	63	5.0e-1	2.1e-2 ✓	2.2e-2	1.4e-1	1.0e-2 ✓	1.0e-2	1.1e+1	6.9e-1 ✓	7.5e-1
	127	2.8e-2	2.5e-2 ✓	2.5e-2	1.0e-2	1.0e-2 ✓	1.0e-2	1.3e-0	1.2e-0 ✓	1.2e-0
8	15	5.9e-2	4.9e-2 ✓	4.9e-2	3.2e-2	2.1e-2 ✓	2.1e-2	9.9e-1	8.6e-1 ✓	8.6e-1
	31	4.5e-2	4.0e-2 ✓	4.0e-2	1.9e-2	1.4e-2 ✓	1.4e-2	1.0e-0	1.0e-0 ✓	1.0e-0
	63	3.8e-2	3.6e-2 ✓	3.7e-2	1.2e-2	1.2e-2 ✓	1.2e-2	1.4e-0	1.3e-0 ✓	1.3e-0
	127	1.5e-1	3.2e-2 ✓	3.1e-2	3.1e-2	1.1e-2 ✓	1.1e-2	6.9e-0	1.4e-0 ✓	1.4e-0
9	15	2.0e-2	1.9e-2 ✓	1.8e-2	1.5e-2	1.3e-2 ✓	1.3e-2	4.5e-1	3.0e-1 ✓	2.9e-1
	31	7.8e-2	1.9e-2 ✓	1.8e-2	2.5e-2	1.1e-2 ✓	1.1e-2	4.5e-0	4.5e-1 ✓	4.1e-1
	63	NaN	2.3e-2 !	2.2e-2	NaN	1.0e-2 !	1.0e-2	NaN	9.0e-1 !	8.3e-1
	127	NaN	2.7e-2 !	2.8e-2	NaN	1.0e-2 !	1.0e-2	NaN	1.4e-0 !	1.4e-0

Table 9.2: Errors of the numerical solutions without and with ℓ^1 regularization. In cases where the error value is NaN, the numerical solution broke down before the final time was reached. "no ℓ^1" refers to the underlying DG method without ℓ^1 regularization, "ℓ^1" refers to the DG method with ℓ^1 regularization, and "ℓ^1-mc" refers to the DG method with ℓ^1 regularization and additional mass correction.

In these cases, the numerical solver without ℓ^1 regularization broke down completely. Yet, when ℓ^1 regularization was applied, the same computations yielded fairly accurate numerical solutions.

9.4.2 Linear advection equation

The prior test case featured a shock discontinuity and might not fully reflect the behavior of the ℓ^1 regularized DG method in smooth regions. Note that for the underlying spectral DG method used in this work, given $K+1$ nodes, the optimal order of convergence is $K+1$ in sufficiently smooth regions. Especially in smooth regions, it is desirable that the convergence properties of the underlying method are preserved when using the ℓ^1 regularization method. Convergence of the ℓ^1 regularized DG method, however, also depends on the convergence of the PA operator. In all numerical tests presented in this work, a PA operator of order three, i. e., L_3, is used. Hence, if falsely activated (false positive; see Remark 9.3) by the discontinuity sensor (9.15), ℓ^1 regularization might affect the order of convergence of the underlying method in smooth regions. In our numerical tests, we observed the sensor to be fairly reliable for sufficient resolution, and ℓ^1 regularization did not get activated in smooth regions.

To investigate this potential drawback and demonstrate the reliability of the ℓ^1 regularized DG method in smooth regions, we now consider the linear advection equation

$$\partial_t u + \partial_x u = 0$$

on $\Omega = [0, 2]$ with IC

$$u(x, 0) = \sin(2\pi x)$$

and periodic BCs, which provides a smooth solution for all times. Table 9.3 lists comparative errors for the DG methods with and without the ℓ^1 regularization and with the additional mass conservation correction term at time $T = 2$.

Our results indicate that ℓ^1 regularization is only activated, and thus affects convergence of the underlying method, if the solution is heavily underresolved. This can be noted in Table 9.3 from $K = 3$ and $I = 2, 4$ as well as $K = 4, 5, 6$ and $I = 2$. For $K = 7$, even $I = 2$ provides sufficient resolution and the ℓ^1 regularization is not activated. For all numerical solutions, using at least $I = 8$ elements, ℓ^1 regularization does not affect accuracy and convergence of the underlying method in smooth regions. Finally, we note that in the cases where the numerical solution is heavily underresolved and ℓ^1 regularization is activated, ℓ^1 regularization with additional mass correction (l1-mc) provides slightly more accurate solutions than ℓ^1 regularization without mass correction (l1). Finally, we note that because the ℓ^1 regularization is only activated in elements containing discontinuities, the efficiency of our new method is comparable to the underlying DG method. Moreover, as in [SGP17b], we observed that for our specific set of test problems we were able to maintain stability for time step sizes larger than the standard CFL constraints suggest. Theoretical justification for this will be part of future investigations.

9.4.3 Systems of nonlinear conservation laws

We now extend our hybrid ℓ^1 regularized DG method to the nonlinear system of CLs

$$\partial_t \begin{pmatrix} u_0 \\ u_1 \end{pmatrix} + \frac{1}{2}\partial_x \begin{pmatrix} u_0^2 + u_1^2 \\ 2u_0 u_1 \end{pmatrix} = 0 \tag{9.32}$$

in the domain $\Omega = [0, 2]$. System (9.32) originates from a truncated polynomial chaos approach for Burgers' equation with uncertain IC [PIN09, PIN15, ÖGR18]. In this context, u_0 models the expected value of the numerical solution while u_1^2 approximates the variance. For the spatial

		$\|\cdot\|_M$-error			$\|\cdot\|_1$-error			$\|\cdot\|_\infty$-error		
K	I	no ℓ^1	ℓ^1	ℓ^1-mc	no ℓ^1	ℓ^1	ℓ^1-mc	no ℓ^1	ℓ^1	ℓ^1-mc
3	2	1.2e-0	1.2e-0	1.2e-0	1.5e-0	1.6e-0	1.6e-0	9.5e-1	9.8e-1	9.8e-1
	4	1.3e-1	4.4e-1	4.4e-1	1.4e-1	5.8e-1	5.8e-1	1.3e-1	4.1e-1	4.1e-1
	8	6.3e-3	6.3e-3	6.3e-3	7.0e-3	7.0e-3	7.0e-3	1.2e-2	1.2e-2	1.2e-2
	16	3.8e-4	3.8e-4	3.8e-4	3.8e-4	3.8e-4	3.8e-4	9.9e-4	9.9e-4	9.9e-4
4	2	3.4e-1	9.2e-1	9.2e-1	4.2e-1	9.7e-1	9.6e-1	3.8e-1	9.0e-1	8.8e-1
	4	7.8e-3	7.8e-3	7.8e-3	1.0e-2	1.0e-2	1.0e-2	1.2e-2	1.2e-2	1.2e-2
	8	4.2e-4	4.2e-4	4.2e-4	4.4e-4	4.4e-4	4.4e-4	1.2e-3	1.2e-3	1.2e-3
	16	1.3e-5	1.3e-5	1.3e-5	1.3e-5	1.3e-5	1.3e-5	4.4e-5	4.4e-5	4.4e-5
5	2	8.0e-2	1.0e-0	1.0e-0	1.0e-1	1.3e-0	1.3e-0	1.3e-1	8.4e-1	7.8e-1
	4	2.1e-3	2.1e-3	2.1e-3	2.3e-3	2.3e-3	2.3e-3	5.3e-3	5.3e-3	5.3e-3
	8	2.8e-5	2.8e-5	2.8e-5	2.9e-5	2.9e-5	2.9e-5	7.7e-5	7.7e-5	7.7e-5
	16	1.2e-6	1.2e-6	1.2e-6	1.5e-6	1.5e-6	1.5e-6	1.6e-6	1.6e-6	1.6e-6
6	2	2.1e-2	9.9e-1	9.9e-1	2.5e-2	1.1e-0	1.1e-0	5.1e-2	9.9e-1	9.9e-1
	4	7.5e-5	7.5e-5	7.5e-5	8.0e-5	8.0e-5	8.0e-5	1.8e-4	1.8e-4	1.8e-4
	8	6.0e-6	6.0e-6	6.0e-6	7.6e-6	7.6e-6	7.6e-6	7.5e-6	7.5e-6	7.5e-6
	16	7.3e-7	7.3e-7	7.3e-7	9.4e-7	9.4e-7	9.4e-7	7.4e-7	7.4e-7	7.4e-7
7	2	2.0e-3	2.0e-3	2.0e-3	2.0e-3	2.0e-3	2.0e-3	6.4e-3	6.4e-3	6.4e-3
	4	3.9e-5	3.9e-5	3.9e-5	4.9e-5	4.9e-5	4.9e-5	6.8e-5	6.8e-5	6.8e-5
	8	3.9e-6	3.9e-6	3.9e-6	5.0e-6	5.0e-6	5.0e-6	4.1e-6	4.1e-6	4.1e-6
	16	4.9e-7	4.9e-7	4.9e-7	6.3e-7	6.3e-7	6.3e-7	4.9e-7	4.9e-7	4.9e-7

Table 9.3: Errors of the numerical solutions without and with ℓ^1 regularization. "no ℓ^1" refers to the underlying DG method without ℓ^1 regularization, "ℓ^1" refers to the DG method with ℓ^1 regularization, and "ℓ^1-mc" refers to the DG method with ℓ^1 regularization and additional mass correction.

semidiscretization of (9.32) we follow [ÖGR18, RGÖS18], where a skew-symmetric formulation

$$\begin{aligned} \frac{\mathrm{d}}{\mathrm{d}t}\mathbf{u_k} &= -\frac{1}{3}\sum_{i,j=0}^{1}\langle\varphi_i\varphi_j\varphi_k\rangle\left(D\left(\mathbf{u_i}\circ\mathbf{u_j}\right)+U_j{}^*D\mathbf{u_i}\right) \\ &\quad - M^{-1}R^TB\left(\boldsymbol{f}_k^{\mathrm{num}}-\sum_{i,j=0}^{1}\langle\varphi_i\varphi_j\varphi_k\rangle\left(\frac{1}{3}R\left(\mathbf{u_i}\circ\mathbf{u_j}\right)+\frac{1}{6}\left(R\mathbf{u_i}\right)\circ\left(R\mathbf{u_j}\right)\right)\right) \end{aligned} \tag{9.33}$$

was proposed. Here, U_j denotes the diagonal matrix $\operatorname{diag}(u_j(\xi_0),\dots,u_j(\xi_K))$, $\langle\varphi_i\varphi_j\varphi_k\rangle$ denotes the triple product $\int\varphi_i\varphi_j\varphi_k\omega$, $\circ$ denotes the componentwise (Hadamard) product of two vectors, and $\{\varphi_k\}$ is a set of orthogonal polynomials (typically Hermite polynomials are used).

It was further proved in [ÖGR18] that (9.33) yields an entropy conservative semidiscretization when combined with the entropy conservative flux f_k^{num} presented in [ÖGR18]. Also see the short description in Chapter 6.2.3. An entropy stable semidiscretization is thus obtained by adding a dissipative term $-Q(\mathbf{u_{k+}}-\mathbf{u_{k-}})$ to the entropy conservative flux. Here we simply use a local Lax–Friedrichs type dissipation matrix

$$Q = \frac{\alpha}{2} I \quad \text{with} \quad \alpha = \max\{|\alpha(-)|, |\alpha(+)|\},$$

where $|\alpha(\pm)|$ is the largest absolute value of all eigenvalues of the Jacobian matrix $\boldsymbol{f}'(\boldsymbol{u}(\pm))$.

Even though the skew-symmetric formulation (9.33) combined with an appropriate numerical flux yields an entropy stable scheme, this test case demonstrates that additional regularization is still necessary to obtain reasonable numerical solutions. This was, for instance, stressed in [RGÖS18]. Here we demonstrate how ℓ^1 regularization enhances the numerical solution for the nonlinear system (9.32) with periodic BCs and IC

$$u_0(0,x) = 1 + \begin{cases} e \cdot \exp\left(-\frac{r^2}{r^2-(x-0.5)^2}\right) & \text{if } |x-0.5| < r, \\ 0 & \text{if } |x-0.5| \geq r, \end{cases}$$

$$u_1(0,x) = \begin{cases} e \cdot \exp\left(-\frac{r^2}{r^2-(x-0.5)^2}\right) & \text{if } |x-0.5| < r, \\ 0 & \text{if } |x-0.5| \geq r, \end{cases}$$

where $r = 0.5$.

For the more general case of systems of CLs, we propose a straightforward extension of our ℓ^1 regularization technique. Specifically, the PA sensor (9.15) is applied to every conserved variable u_k separately and, once a discontinuity is detected, ℓ^1 regularization (9.13) is performed for the respective variable. For this test case, the ramp parameter in (9.16) has been chosen as $\kappa = 0.9$.

Figure 9.7 illustrates the results of ℓ^1 regularization (with and without mass correction) for the above described test case and for an entropy stable numerical flux. In all subsequent tests $I = 100$ equidistant elements and a polynomial basis of degree $K = 6$ have been used. Further, for the ℓ^1 regularization, the same parameters as before have been used, i. e., $\lambda_{\max} = 4 \cdot 10^2$, $R = 400$, $\beta = 20$, $\alpha = 0.0001$, and $tol = 0.001$.

Note that while the numerical solution without ℓ^1 regularization shows heavy oscillations in both components, the numerical solution with ℓ^1 regularization provides a significantly sharper profile. Further, by consulting Figures 9.7(d) and 9.7(e), it should be stressed that only ℓ^1 regularization with additional mass correction is able to capture the exact shock location. Due to missing conservation, ℓ^1 regularization without mass correction results in a slightly wrong location for the shock. Finally, Figures 9.7(c) and 9.7(f) illustrate the energy of the different methods over time. We note from these figures that ℓ^1 regularization (with and without mass correction) slightly increase the energy in this test case.

In order to further emphasize the effect of ℓ^1 regularization (with and without mass correction), similar results using an entropy conservative numerical flux at the interfaces between elements are shown in Figure 9.8. Once again, ℓ^1 regularization is demonstrated to improve the numerical solution. The best results are obtained when ℓ^1 regularization is combined with mass correction. In particular, only ℓ^1 regularization with additional mass correction is able to accurately capture the shock at the right location. Finally, consulting Figures 9.8(c) and 9.8(f), ℓ^1 regularization decreases the energy overall. Yet, it is demonstrated once more that an entropy inequality is not satisfied by ℓ^1 regularization, i. e., the energy might increase as well as decrease by utilizing ℓ^1 regularization. Thus, future work will focus on incorporating energy stability (as well as other properties like TVD or positivity) by additional constraints in the minimization problem (9.12).

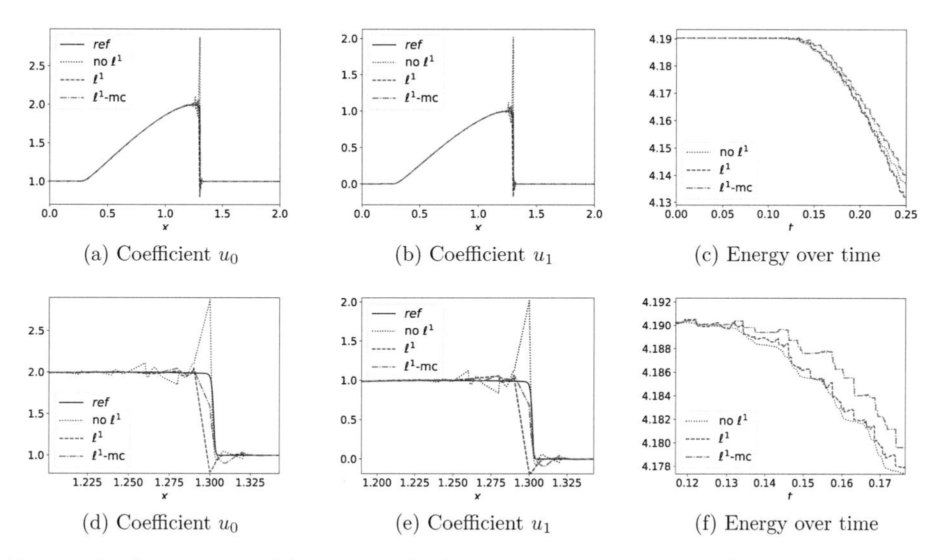

(a) Coefficient u_0 (b) Coefficient u_1 (c) Energy over time

(d) Coefficient u_0 (e) Coefficient u_1 (f) Energy over time

Figure 9.7: Components of the numerical solutions at $t = 0.25$ by the DG method for $I = 100$ elements and polynomial degree $K = 6$. Without ℓ^1 regularization (straight blue line), with ℓ^1 regularization (dashed red line), and with ℓ^1 regularization and mass correction (dash-dotted green line). In all tests an entropy *stable* numerical flux has been used.

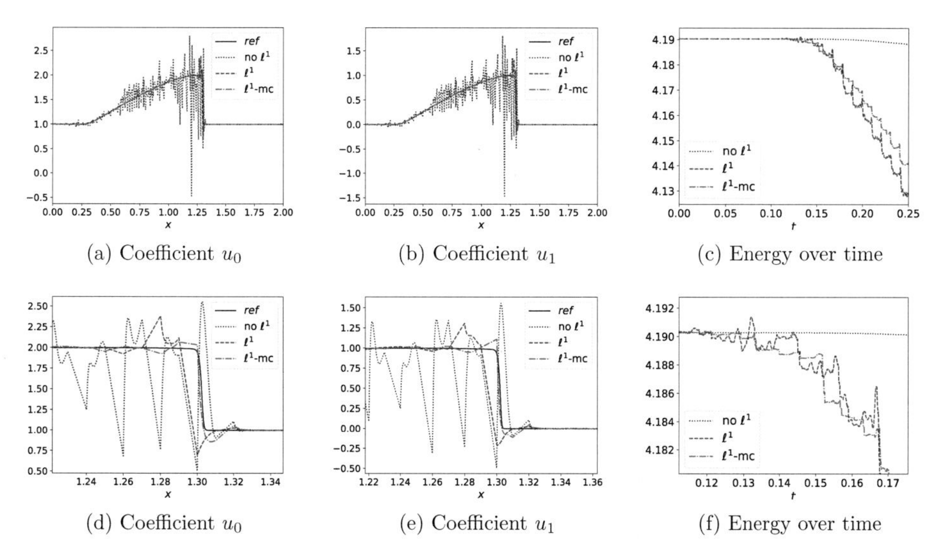

(a) Coefficient u_0 (b) Coefficient u_1 (c) Energy over time

(d) Coefficient u_0 (e) Coefficient u_1 (f) Energy over time

Figure 9.8: Components of the numerical solutions at $t = 0.25$ by the DG method for $I = 100$ elements and polynomial degree $K = 6$. Without ℓ^1 regularization (straight blue line), with ℓ^1 regularization (dashed red line), and with ℓ^1 regularization and mass correction (dash-dotted green line). In all tests an entropy *conservative* numerical flux has been used.

9.5 Concluding thoughts and outlook

In this chapter, we have presented a novel approach to shock capturing by ℓ^1 regularization using SE approximations. The approach presented here — and published in [GG19a] — not only is distinguished from previous studies [SCHO13, HLS15, Lav89, Lav91, GMP+08] by focusing on discontinuous solutions but further by promoting sparsity of the jump function instead of the numerical solution itself. By approximating the jump function with a (high order) PA operator, we help to eliminate the staircase effect that arises for classical TV operators. Our results demonstrate that it is possible to efficiently implement a method that yields increased accuracy and better resolves (shock) discontinuities. In particular, no additional time step restrictions are introduced, in contrast to AV methods (see Chapter 8) when no care is taken in their construction. This approach for solving numerical CLs was first used in [SGP17b], where the Lax–Wendroff scheme and Chebyshev and Fourier spectral methods were used as the numerical PDE solver. Our method improves upon the approach in [SGP17b] in two ways: First, we employ the SE approximation for solving the CL, which allows element-to-element variations in the optimization problem. In particular, ℓ^1 regularization is only activated in troubled elements, which enhances accuracy and efficiency of the method. Second, in the process we proposed a novel discontinuity sensor based on PA operators of increasing orders, which is able to flag troubled elements as well as to steer the amount of regularization introduced by the sparse reconstruction.

Numerical tests demonstrate the method using a nodal DGSEM for the inviscid Burgers' equation, the linear advection equation, and a system of nonlinear CLs. Our results show that the method yields improved accuracy and robustness.

No effort was made in the present study to optimize any of the parameters involved in solving the optimization problem. This will be addressed in future work, along with the possibility to include additional constraints (e. g., for entropy, TVD, and positivity constraints), since preliminary results presented here are encouraging. The generalization of the approach itself to higher dimensions is straightforward and has already been demonstrated in [SGP17b]. Of interest, however, would be the extension of the proposed approach to other classes of methods, such as FV methods. We believe ℓ^1 regularization might be an important ingredient to make high order methods viable in several research applications. Moreover, the PA operators introduced in this chapter (see Chapter 9.2.2) will also be used in Chapter 10 to steer another new shock capturing method and in Chapter 11 to construct novel AV operators in the context of (SBP based) FD methods.

CHAPTER 10

Shock capturing by Bernstein polynomials

In this chapter, we propose a second novel shock capturing procedure, appart from classical AV methods, for SE methods applied to scalar CLs

$$\partial_t u + \partial_x f(u) = 0, \quad x \in \Omega \subset \mathbb{R}, \tag{10.1}$$

with appropriate IC and BCs. We propose a procedure which consists of going over from the original (polluted) approximation to a convex combination of the original approximation and its Bernstein reconstruction. The coefficient in the convex combination, and therefore the procedure, is again steered by a discontinuity sensor (based on PA operators) and is only activated in troubled elements. Building up on classical Bernstein operators, we are able to prove that the resulting Bernstein procedure is TVD and preserves monotone (shock) profiles. Further, the procedure can be modified to not just preserve but also enforce certain bounds for the solution, such as positivity. In contrast to other shock capturing methods, e. g. AV methods, the new procedure does not reduce the time step or CFL condition and can be easily and efficiently implemented into any existing code. Numerical tests demonstrate that the proposed shock capturing procedure is able to stabilize and enhance SE approximations in the presence of shocks. Note that in this chapter, we make a slight change in notation and denote polynomial degrees by N instead of K. The material presented here resulted in the publication [Gla19c].

Outline

This chapter is organized as follows. In Chapter 10.1, we revise bases of Bernstein polynomials and their associated Bernstein approximation operator. This operator will later be used to obtain 'smoother' reconstructions of polynomial approximations near discontinuities and provides certain structure-preserving and approximation properties. Building up on these properties, Chapter 10.2 introduces the novel Bernstein procedure, which replaces polluted interpolation polynomial by convex combinations of the original (polluted) approximation and its Bernstein reconstruction. This process is steered by a polynomial annihilation sensor which is quite similar to the one presented in Chapter 9.3.3. Afterwards, Chapter 10.3 investigates some analytical properties of the proposed Bernstein procedure, such as its effect on entropy, total variation and monotone (shock) profiles of the numerical solution. We stress that the Bernstein reconstruction is proven to be total variation diminishing (nonincreasing). Finally, Chapter 10.4 provides numerical demonstrations for a series of different scalar test problems. We close this work with concluding thoughts in Chapter 10.5.

10.1 Bernstein polynomials and the Bernstein operator

Let us start by introducing Bernstein polynomials as well as some of their more important properties. On an interval $[a,b]$, the $N+1$ *Bernstein basis polynomials of degree* N are defined as

$$b_{n,N}(x) = \binom{N}{n} \frac{(x-a)^n (b-x)^{N-n}}{(b-a)^n}, \qquad n = 0, \dots, N,$$

and form a basis of $\mathbb{P}_N(\mathbb{R})$. Thus, every polynomial of degree at most N can be written as a linear combination of Bernstein basis polynomials,

$$B_N(x) = \sum_{n=0}^{N} \beta_n b_{n,N}(x),$$

called *Bernstein polynomial* or *polynomial in Bernstein form of degree* N. The coefficients are referred to as *Bernstein coefficients* or *Bézier coefficients*. We further define the linear *Bernstein operator of degree* N for a function $u : [a,b] \to \mathbb{R}$ by

$$B_N[u](x) = \sum_{n=0}^{N} u\left(a + \frac{n}{N}(b-a)\right) b_{n,N}(x). \tag{10.2}$$

$B_N[\cdot]$ maps a function u to a Bernstein polynomial of degree N with Bernstein coefficients

$$\beta_n = u\left(a + \frac{n}{N}(b-a)\right).$$

Without loss of generality, we can restrict ourselves to the intervals $[0,1]$ and $[-1,1]$. For sake of simplicity, the interval $[0,1]$ subsequently will be used for theoretical investigations. The interval $[-1,1]$, on the other hand, is typically used as a reference element in SE approximations. Thus, the proposed Bernstein procedure will be explained for this case.

10.1.1 Structure-preserving properties

Let us consider Bernstein polynomials on $[0,1]$. We start by noting that the Bernstein basis polynomials form a partition of unity, i. e.

$$\sum_{n=0}^{N} b_{n,N}(x) = 1 \tag{10.3}$$

for all $N \in \mathbb{N}$. This is a direct consequence of the binomial theorem; see [GRS19]. Thus, we can immediately note

Lemma 10.1
Let $B_N(x) = \sum_{n=0}^{N} \beta_n b_{n,N}(x)$ be a Bernstein polynomial of degree N with Bernstein coefficients $\beta_0, \dots, \beta_N$. Then

$$m \leq \beta_n \leq M \quad \forall n = 0, \dots, N \quad \implies \quad m \leq B_N(x) \leq M$$

holds for the Bernstein polynomial.

Proof. Let $m \leq \beta_n \leq M$ for all $n = 0, \dots, N$. We therefore have

$$m \sum_{n=0}^{N} b_{n,N}(x) \leq \sum_{n=0}^{N} \beta_n b_{n,N}(x) \leq M \sum_{n=0}^{N} b_{n,N}(x)$$

and the assertion follows from (10.3). □

In particular, Lemma 10.1 ensures that the Bernstein operator (10.2) preserves the bounds of the underlying function u. In fact, Lemma 10.1 not only ensures preservation of bounds by the Bernstein procedure, but also allows us to enforce such bounds. Moreover, the Bernstein operator preserves the boundary values of u, i. e.

$$B_N[u](0) = u(0) \quad \text{and} \quad B_N[u](1) = u(1)$$

hold. This makes the later proposed Bernstein procedure a reasonable shock capturing method not just in discontinuous SE methods, but also in continuous SE methods, where the numerical solution is required to be continuous across element interfaces. Further, we revise the formula

$$B_N^{(k)}[u](x) = N(N-1)\dots(N-k+1) \sum_{n=0}^{N-k} \Delta^k u\left(\frac{n}{N}\right) \binom{N-k}{n} x^n (1-x)^{N-n-k}$$

with forward difference operator

$$\begin{aligned} \Delta^k u\left(\frac{n}{N}\right) &= \Delta\left(\Delta^{k-1} u\left(\frac{n}{N}\right)\right) \\ &= u\left(\frac{n+k}{N}\right) - \binom{k}{1} u\left(\frac{n+k-1}{N}\right) + \dots + (-1)^k u\left(\frac{n}{N}\right) \end{aligned}$$

for derivatives of (10.2); see [Lor12, Chapter 1.4]. In particular, we have

$$B_N'[u](x) = N \sum_{n=0}^{N-1} \left[u\left(\frac{n+1}{N}\right) - u\left(\frac{n}{N}\right)\right] \binom{N-1}{n} x^n (1-x)^{N-1-n} \tag{10.4}$$

for the first derivative of $B_N[u]$. This formula will be important later in order to show that the Bernstein procedure is able to preserve monotone (shock) profiles.

10.1.2 Approximation properties

Bernstein polynomials were first introduced by Bernstein [Ber12a] in a constructive proof of the famous Weierstrass theorem, which states that every continuous function on a compact interval can be approximated arbitrarily accurate by polynomials [Wei85]; see Chapter 3.1.1. Hence, the sequence of Bernstein polynomials $(B_N[u])_{N \in \mathbb{N}}$ converges uniformly to the continuous function u. Assuming that u is bounded in $[0,1]$ and that the second derivative u'' exists at the point $x \in [0,1]$, we have

$$\lim_{N \to \infty} N\left(u(x) - B_N[u](x)\right) = -\frac{x(1-x)}{2} u''(x); \tag{10.5}$$

see [Lor12, Chapter 1.6.1]. In particular, at points x where the second derivative exists, the error of the Bernstein polynomial $B_N[u]$ therefore is of first order, i. e.

$$|u(x) - B_N[u](x)| \leq C N^{-1} u''(x)$$

for a $C > 0$. However, we should remember that solutions of scalar CLs – which we intend to approximate – might contain discontinuities. The structure of these solutions has been determined by Oleinik [Ole57, Ole64], Lax [Lax57], Dafermos [Daf77], Schaeffer [Sch73], Tadmor and Tassa [TT93], and many more. Most notably for our purpose, Tadmor and Tassa [TT93] showed for scalar convex CLs that if the initial speed has a finite number of decreasing inflection points then it bounds the number of future shock discontinuities. Thus, in most cases the solution consists of a finite number of smooth pieces, each of which is as smooth as the ID. Note that it is this type of regularity which is often assumed – sometimes implicitly – in the numerical treatment of hyperbolic CLs.

Hence, it appears to be more reasonable to investigate convergence of the sequence of Bernstein polynomials $(B_N[u])_{N\in\mathbb{N}}$ for only piecewise smooth functions u. Recall that we call a function u *piecewise* C^k on $[0,1]$ if there is a finite set of points $0 < x_1 < \dots < x_K < 1$ such that $u|_{[0,1]\setminus\{x_1,\dots,x_K\}} \in C^k$ and if the one-sided limits

$$u(x_k^+) := \lim_{x\to x_k^+} u(x) \quad \text{and} \quad u(x_k^-) := \lim_{x\to x_k^-} u(x)$$

exist for all $k = 1,\dots,K$; also see Chapter 2.4. In this case u is still integrable and the first order convergence of the Bernstein polynomials carries over in an L^p-sense.

Theorem 10.2

Let u be piecewise C^2 and let $(B_N[u])_{N\in\mathbb{N}}$ be the sequence of corresponding Bernstein polynomials. Then there is a constant $C > 0$ such that

$$\left(\int_0^1 |u(x) - B_N[u](x)|^p \,\mathrm{d}x\right)^{\frac{1}{p}} \leq CN^{-1}$$

holds for all $N \in \mathbb{N}$ and $1 \leq p < \infty$.

Proof. Let $0 < x_1 < \dots < x_k < 1$ be the points where u is not C^2. When further denoting $x_0 = 0$ and $x_{K+1} = 1$, we have

$$\int_0^1 |u(x) - B_N[u](x)|^p \,\mathrm{d}x = \sum_{k=0}^{K} \int_{x_k}^{x_{k+1}} |u(x) - B_N[u](x)|^p \,\mathrm{d}x$$

and, since u is piecewise C^2, there is a generic constant $C > 0$ such that

$$|u(x) - B_N[u](x)| \leq CN^{-1}u''(x) \tag{10.6}$$

holds for $x_k < x < x_{k+1}$. It should be noted however that u'' might be unbounded on the open interval (x_k, x_{k+1}). Therefore, let us choose $\varepsilon > 0$ such that $\varepsilon < \min_{k=0,\dots,K} |x_k - x_{k+1}|$ and let us consider the closed interval $[x_k + \varepsilon, x_{k+1} - \varepsilon]$. On these intervals u'' is bounded and (10.6) becomes

$$|u(x) - B_N[u](x)| \leq CN^{-1}$$

for $x_k + \varepsilon \leq x \leq x_{k+1} - \varepsilon$. Hence, we have

$$\begin{aligned}
&\int_{x_k}^{x_{k+1}} |u(x) - B_N[u](x)|^p \,\mathrm{d}x \\
&= \int_{x_k}^{x_k+\varepsilon} |u(x) - B_N[u](x)|^p \,\mathrm{d}x \\
&\quad + \int_{x_k+\varepsilon}^{x_{k+1}-\varepsilon} |u(x) - B_N[u](x)|^p \,\mathrm{d}x + \int_{x_{k+1}-\varepsilon}^{x_{k+1}} |u(x) - B_N[u](x)|^p \,\mathrm{d}x \qquad (10.7) \\
&= \int_{x_k}^{x_k+\varepsilon} |u(x) - B_N[u](x)|^p \,\mathrm{d}x \\
&\quad + (x_{k+1} - x_k - 2\varepsilon)C^pN^{-p} + \int_{x_{k+1}-\varepsilon}^{x_{k+1}} |u(x) - B_N[u](x)|^p \,\mathrm{d}x.
\end{aligned}$$

For the two remaining integrals, we remember that u and therefore all $B_N[u]$ are bounded. This yields

$$|u(x) - B_N[u](x)| \leq |u(x)| + |B_N[u](x)| \leq 2\,\|u\|_\infty$$

for all $x \in [0,1]$ and $N \in \mathbb{N}$. As a consequence, (10.7) reduces to

$$\int_{x_k}^{x_{k+1}} |u(x) - B_N[u](x)|^p \,\mathrm{d}x \leq (x_{k+1} - x_k - 2\varepsilon) C^p N^{-p} + 4\varepsilon\,\|u\|_\infty$$

and letting $\varepsilon \to 0$ results in

$$\int_{x_k}^{x_{k+1}} |u(x) - B_N[u](x)|^p \,\mathrm{d}x \leq (x_{k+1} - x_k) C^p N^{-p}.$$

Finally, we have

$$\begin{aligned}
&\int_{x_k}^{x_{k+1}} |u(x) - B_N[u](x)|^p \,\mathrm{d}x && \leq (x_{k+1} - x_k) C N^{-p} \\
\implies &\int_0^1 |u(x) - B_N[u](x)|^p \,\mathrm{d}x && \leq C N^{-p} \\
\implies &\left(\int_0^1 |u(x) - B_N[u](x)|^p \,\mathrm{d}x\right)^{\frac{1}{p}} && \leq C N^{-1},
\end{aligned}$$

which yields the assertion. □

One might reproach that this is an unacceptable order of convergence. We reply to this argument with a few selected counter-arguments.

Remark 10.3.

- For numerical solutions of hyperbolic CLs, it is almost universally accepted that near shocks, the solution can be first order accurate at most [PP06]. Thus, accuracy of the numerical solution won't decrease noticeably by reconstructing it as a Bernstein polynomial.

- In fact, high order methods often not even provide accuracy of first order but constant (or even decreasing) accuracy. This is due to the Gibbs–Wilbraham phenomenon [HH79, Ric91, GS97] for (polynomial) higher order approximations of discontinuous functions. Yet, for instance, Gzyl and Palacios [GP03] have shown the absence of the Gibbs–Wilbraham phenomenon for Bernstein polynomials. It is still an open problem which properties of the approximation cause the Gibbs–Wilbraham phenomenon, but it is our conjecture that the order of accuracy for smooth functions is the deciding factor.

- While Bernstein polynomials converge uniformly for every continuous function, for instance, polynomial interpolation in general only converges if u is at least (Dini–)Lipschitz continuous [Ber12b]. Even worse, Faber [Fab14] showed in 1914 that no polynomial interpolation scheme – no matter how the points are distributed – will converge for the whole class of continuous functions. For approximations by orthogonal projection, such as Fourier series, a similar result follows from divergence of the corresponding operator norms [Ber58, Pet90].

We conclude from Remark 10.3 that Bernstein polynomials, while appearing not attractive for approximating sufficiently smooth functions, provide some advantages for the approximation of just continuous or even discontinuous functions.

10.2 The Bernstein procedure

We are now ready to introduce our novel sub-cell shock capturing procedure by using Bernstein polynomials. The procedure is described for the reference element $[-1,1]$ and is based on replacing oscillatory high order approximations by their Bernstein reconstruction.

10.2.1 Related works

By now, Bernstein polynomials as polynomial bases have been successfully applied to solvers for PDEs in a number of works. Especially in [Kir11, AAD11], efficient SE operators have been constructed by using Bernstein polynomials as shape functions. Also in [BB15], Bernstein polynomials have been proposed as basis functions in a third order quasinodal DG method for solving flooding and drying problems with shallow water equations. Finally, [LKSM17, ADK+17] have investigated the potential of imposing discrete maximum principles for polynomials expressed in a basis of Bernstein polynomials. Yet, while Bernstein polynomials as basis functions have been studied in a few works by now, the associated approximation procedure resulting from the Bernstein operator is still of very limited use. In contrast to the above mentioned works, we do not only utilize Bernstein polynomials as basis functions, but we further propose to use the associated Bernstein operator as a building stone for new shock capturing procedures.

10.2.2 Bernstein reconstruction

Let us start by introducing the (modified) Bernstein reconstruction which is obtained by applying the Bernstein operator (10.2) to a polynomial approximation. Let $u \in \mathbb{P}_N([-1,1])$ be an approximate solution at time $t \geq 0$ with coefficients $\hat{\mathbf{u}} \in \mathbb{R}^{N+1}$ with respect to a (nodal or modal) basis $\{\varphi_n\}_{n=0}^N$ in the reference element $\Omega_{ref} = [-1,1]$. Then, the original approximation, for instance, obtained by interpolation or (pseudo) L^2-projection, is modified in the following way:

1. Compute the *Bernstein reconstruction of* u by

$$B_N[u](x) = \sum_{n=0}^{N} u\left(-1 + 2\frac{n}{N}\right) b_{n,N}(x). \tag{10.8}$$

2. Write $B_N[u]$ with respect to the basis $\{\varphi_n\}_{n=0}^N$ by a change of bases with transformation matrix T, i. e.

$$\mathbf{u}^{(B)} = T\mathbf{b}, \tag{10.9}$$

   where $\mathbf{b}$ is a vector containing the Bernstein coefficients $\beta_{n,N} = u\left(-1 + 2\frac{n}{N}\right)$.

To put it in a nutshell, the idea is to replace the coefficients $\hat{\mathbf{u}}$ in a troubled element by the coefficients $\mathbf{u}^{(B)}$ of the Bernstein reconstruction. Table 10.1 lists the condition numbers (see [TBI97]) of the transformation matrix T for the nodal Lagrange and modal Legendre bases. The nodal basis of Lagrange polynomials is considered with respect to the GLo points in $[-1,1]$. For all reasonable polynomial degrees ($N \leq 10$) and both bases we observe the condition number to be fairly small.

cond(T)										
N	1	2	3	4	5	6	7	8	9	10
Lagrange	1.0×10^0	2.3×10^0	$4.4. \times 10^0$	8.6×10^0	1.7×10^1	3.4×10^1	6.7×10^1	1.3×10^2	2.6×10^2	5.3×10^2
Legendre	1.0×10^0	1.9×10^0	2.9×10^0	4.3×10^0	5.4×10^0	7.7×10^0	1.0×10^1	1.6×10^1	2.4×10^1	4.1×10^1

Table 10.1: Condition numbers of the transformation matrix T for the Lagrange (with respect to GLo points) and Legendre bases. The spectral norm $||T|| = ||T||_2 := \sqrt{\lambda_{\max}((T^*) \cdot T)}$ has been used, where $\lambda_{\max}((T^*) \cdot T)$ is the largest eigenvalue of the positive-semidefinite matrix $(T^*) \cdot T$. This value is sometimes referred to as the largest singular value of the matrix T.

Concerning the now obtained Bernstein reconstruction, Lemma 10.1 does not only ensure preservation of bounds, but also allows us to enforce such bounds. Let us introduce the *modified Bernstein reconstruction with respect to the lower bound m and the upper bound M*

$$B_N^{(m,M)}[u](x) = \sum_{n=0}^{N} \beta_n b_{n,N}(x) \quad \text{with} \quad \beta_n = \begin{cases} u\left(-1+2\frac{n}{N}\right) & \text{if } m \leq u\left(-1+2\frac{n}{N}\right) \leq M, \\ m & \text{if } u\left(-1+2\frac{n}{N}\right) < m, \\ M & \text{if } u\left(-1+2\frac{n}{N}\right) > M. \end{cases} \tag{10.10}$$

To the best of our knowledge, the modified Bernstein operator (10.10) has not been defined anywhere else yet. The most beautiful property of the modified Bernstein operator is preservation – by default – of lower and upper bounds m and M. This follows directly from Lemma 10.1 and is summarized in

Theorem 10.4
Let $B_N^{(m,M)}$ be the modified Bernstein operator with respect to the lower bound m and the upper bound M given by (10.10) *and let u be some function. Then $B_N^{(m,M)}[u]$ fulfils*

$$m \leq B_N^{(m,M)}[u] \leq M.$$

We close this subsection by noting that, for instance, positivity (think about density in the Euler equations), can be easily ensured by setting $m = \varepsilon$, where $\varepsilon > 0$ is a suitable value larger than machine precision.

10.2.3 Proposed procedure

Finally, we propose a procedure on how to replace the original polynomial approximation by its (modified) Bernstein reconstruction. Let Ω_i be a troubled element with approximate solution $u \in \mathbb{P}_N([-1,1])$ and let the coefficients of u with respect to a (nodal or modal) basis $\{\varphi_n\}_{n=0}^N$ be given by $\hat{\mathbf{u}} \in \mathbb{R}^{N+1}$. The *Bernstein procedure* consists of two steps:

1. Compute the (modified) Bernstein reconstruction $B_N[u]$ of u.
2. Build an 'appropriate' convex combination $u^{(\alpha)}$ of the original approximation u and its Bernstein reconstruction $B_N[u]$, i. e.

$$u^{(\alpha)}(x) = \alpha u(x) + (1-\alpha) B_N[u](x) \tag{10.11}$$

with $\alpha \in [0,1]$.

With respect to the original basis $\{\varphi_n\}_{n=0}^N$, and therefore utilizing the transformation (10.9), the α *Bernstein reconstruction* $u^{(\alpha)} \in \mathbb{P}_N([-1,1])$ can be written as

$$u^{(\alpha)}(x) = \sum_{n=0}^{N} \left[\alpha \hat{u}_n + (1-\alpha) u_n^{(B)} \right] \varphi_n(x).$$

Hence, its coefficients are given by

$$\mathbf{u}^{(\alpha)} = \alpha \hat{\mathbf{u}} + (1-\alpha)\mathbf{u}^{(B)}.$$

This procedure can be incorporated easily into an already existing solver. Note that we have $u^{(1)} = u$ and $u^{(0)} = u^{(B)}$ in the extreme cases of $\alpha = 1$ and $\alpha = 0$. Thus, the α Bernstein reconstruction $u^{(\alpha)}$ (linearly) varies between the original (and potentially oscillating or boundary violating) approximation u and the more robust (modified) Bernstein reconstruction $B_N[u]$. This is illustrated in Figure 10.1 for the signum function $u(x) = \text{sign}(x)$, different parameters α, and $N = 1, 5, 9$.

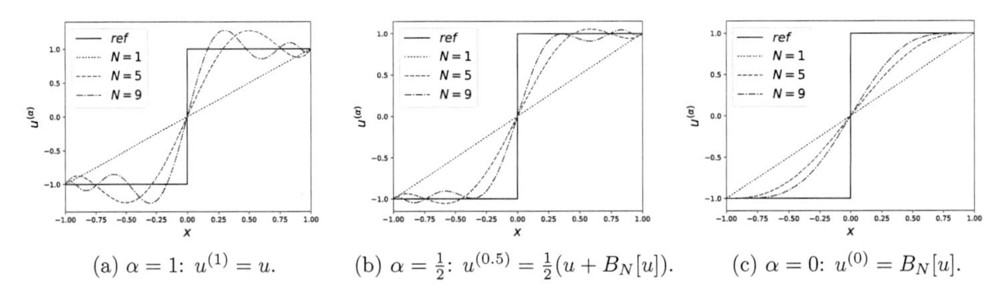

(a) $\alpha = 1$: $u^{(1)} = u$. (b) $\alpha = \frac{1}{2}$: $u^{(0.5)} = \frac{1}{2}(u + B_N[u])$. (c) $\alpha = 0$: $u^{(0)} = B_N[u]$.

Figure 10.1: Illustration of the Bernstein reconstruction $u^{(\alpha)}(x) = \alpha u(x) + (1-\alpha)B_N[u](x)$ for $\alpha = 1$ (original interpolation u), $\alpha = 0.5$, and $\alpha = 0$ ('full' Bernstein reconstruction $B_N[u]$).

Obviously, the order of the approximation is reduced to one in elements where $\alpha < 1$. Yet, in elements with $\alpha = 1$, which define the large majority of the domain, high accuracy is retained. Subsequently, we demonstrate how the parameter α can be adapted to the regularity of the underlying solution.

10.2.4 Selection of parameter α

The parameter α in (10.11) is adapted to adequately capture different discontinuities and regions of smoothness in the solution. The value of α adjusts in space and time to accurately capture strong variations in the solution. Hence, the proposed Bernstein procedure is able to calibrate the polynomial approximation to the regularity of the solution. It should be stressed that modification of the mesh topology, the number of degrees of freedom, node positions, or the type of SE method is utterly unnecessary. This makes the Bernstein procedure an efficient and easy to implement shock capturing method. As described above, the extreme values $\alpha = 1$ and $\alpha = 0$ yield the original approximation u and the Bernstein reconstruction $B_N[u]$, respectively. For intermediate parameter values $\alpha \in (0,1)$, the Bernstein reconstruction renders the convex combination $u^{(\alpha)} = \alpha u + (1-\alpha)u^{(B)}$ more and more robust. Thus, α allows us to adapt the amount of stabilization introduced by the Bernstein reconstruction. Here, we chose α being a function of a discontinuity sensor based on PA operators and similar to the one introduced in Chapter 9.3.3. We briefly describe the sensor below.

10.2.5 Discontinuity sensor

To detect troubled elements and to steer the parameter α, we propose to use a discontinuity sensor quite similar to the one presented in Chapter 9.3.3 and also based on PA operators. This sensor is an element-based function leading to a single scalar measure of the solutions' smoothness. It is a nonlinear functional

$$S : \Omega_{\text{ref}} \to \mathbb{R}, \quad s \mapsto S(s),$$

which depends on a *sensing variable* s. For systems, such as the Euler equations, the density or Mach number could be utilized for the sensing variable. Since we only consider scalar CLs in this chapter, s is simply chosen as the conserved quantity u. The sensor was first proposed in [GG19a] and is based on comparing PA operators L_m of increasing orders. As before, let us consider the sensor value of order m, given by

$$S_m = \max_{k=0,\dots,N-1} \left| L_m[s] \left(x_{k+\frac{1}{2}} \right) \right|, \tag{10.12}$$

i. e. by the greatest absolute value of the PA operator of order m at the mid points of the collocation points. Remember that if s has at least m continuous derivatives, the PA operator L_m provides convergence to 0 of order $m+1$, and we therefore expect the sensor value (10.12) to decrease for an increasing order m. In this case, a parameter value $\alpha = 1$ ($\alpha > 0$) is chosen, which means that the Bernstein procedure is not (fully) activated. Again, we only compare the sensor values of order $m = 1$ and $m = 3$ since it was endorsed in [AGY05] to use the same number of local grid points ξ_j on both sides of a point x. Of course, a variety of modification is possible for the PA sensor. Now, the parameter value $\alpha = 0$ is chosen only if

$$S(s) := \frac{S_3}{S_1} \geq 1 \tag{10.13}$$

holds, i. e. if the sensor value does not decrease when going over from order $m = 1$ to $m = 3$. Finally, we decide for the parameter α to linearly vary between $\alpha = 1$ and $\alpha = 0$. This is realized once more by the parameter function

$$\alpha(S) = \begin{cases} 1 & \text{if } S \leq \kappa, \\ \frac{1-S}{1-\kappa} & \text{if } \kappa < S < 1, \\ 0 & \text{if } 1 \leq S, \end{cases} \tag{10.14}$$

where $\kappa \in (0,1)$ is a problem dependent *ramp parameter*. Figure 10.2 illustrates the parameter function with respect to the values of the PA sensor S given by (10.13).

For the later numerical tests we also investigated different other parameter functions, of which some have been discussed in [HCP12]. Also see Chapter 9.3.3. Yet, we obtained the best results with (10.14). It should be noted that the above revisited PA sensor is only recommended for high orders $N \geq 4$. In our numerical test, we have observed some missidentifications for $N = 3$. For $N \leq 3$ other discontinuity sensors could be used instead. In particular, we would like to mention the modal decay based sensor of Persson and Peraire [PP06, HCP12] and its refinements [BD10, KWH11] as well as the KXRCF sensor [KXR$^+$04, QS05] of Krivodonova et al., which is build up on a strong superconvergence phenomena of the DG method at outflow boundaries. Future work will address a detailed comparison of different shock sensors.

Remark 10.5. The above sensor is simple to implement, but has its price, which is the introduction of the problem dependent and tunable parameter $\kappa \in (0,1)$ in (10.14). In practice,

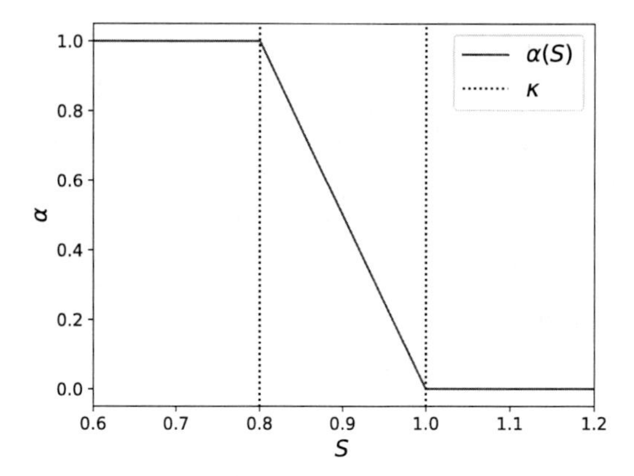

Figure 10.2: Parameter function $\alpha(S)$ as defined in (10.14).

we follow a strategy proposed by Guermond, Pasquetti, and Popov for their *entropy viscosity method for nonlinear CLs* [GPP11]. For a fixed polynomial degree N, the parameter κ is tuned by testing the method on a coarse grid. For all problems presented later in Chapter 10.4, the tuning has been done quickly on a coarse mesh of $I = 10$ elements. Further, we have observed the Bernstein procedure to be robust with respect to the parameter κ. This will be demonstrated in Chapter 10.4.1 for the linear advection equation.

10.3 Entropy, total variation, and monotone (shock) profiles

Next, we investigate some analytical properties of the proposed Bernstein procedure, such as its effect on entropy, TV and monotone (shock) profiles of the numerical solution. For sake of simplicity, all subsequent investigations are carried out on the reference element $\Omega_{\text{ref}} = [0, 1]$.

10.3.1 Entropy stability

Let $\eta \in C^1$ be a convex entropy function; see Chapter 2.7. We show that the proposed Bernstein reconstruction only yields a change in the total amount of entropy which is consistent with the total amount of the entropy of the original approximation u. This means that for increasing N the total amount of entropy of the Bernstein reconstruction $B_N[u]$ converges to the total amount of entropy of the function u. Yet, even though the change of entropy is consistent, it does not always yield a decrease of the entropy. This is demonstrated by

Example 10.6
Let $u(x) = x^2$ with Bernstein reconstruction $B_N[u](x) = x^2 + \frac{1}{N}x(1-x)$. For the usual L^2

entropy $\eta(u) = u^2$, we have

$$\int_0^1 \eta\left(u(x)\right) \,\mathrm{d}x = \frac{1}{5},$$

$$\int_0^1 \eta\left(B_N[u](x)\right) \,\mathrm{d}x = \int_0^1 x^4 \,\mathrm{d}x + \int_0^1 \frac{2}{N} x^3(1-x) + \frac{1}{N^2} x^2 (1-x)^2 \,\mathrm{d}x$$

$$= \int_0^1 \eta(u(x)) \,\mathrm{d}x + \frac{3N+1}{30N^2}.$$

Thus, the total amount of entropy is increased by applying the Bernstein procedure.

In fact, Example 10.6 is no exceptional case. We note that for every continuous and convex function u, the sequence of Bernstein reconstructions will converge to u from above, i. e.

$$B_N[u](x) \geq B_{N+1}[u](x) \geq u(x)$$

holds for all $N \geq 1$; see [Phi03, Theorem 7.1.8 and 7.1.9]. Hence, for $u \geq 0$, the L^2 entropy will be increased by the Bernstein procedure. Yet, we can further prove that the change of the total amount of entropy by the Bernstein reconstruction is consistent and vanishes for increasing N.

Theorem 10.7
Let $\eta \in C^1$ be a convex entropy function and let u be piecewise C^2. Then,

$$\lim_{N\to\infty} \int_0^1 \eta\left(B_N[u](x)\right) \,\mathrm{d}x = \int_0^1 \eta(u(x)) \,\mathrm{d}x$$

holds.

Proof. Since η is C^1 and u is piecewise C^2, equation (10.5) yields

$$\lim_{N\to\infty} \eta(B_N[u](x)) = \eta(u(x)) \quad \text{for all } x \in [0,1] \setminus \{x_1, \dots, x_k\},$$

where $0 < x_1 < \cdots < x_k < 1$ are the points where u'' does not exist. Hence, $\eta \circ B_N[u]$ converges almost everywhere to $\eta \circ u$. Further, u and all $B_N[u]$ are uniformly bounded, let us say by m and M, i. e.

$$m \leq u(x), B_N[u](x) \leq M$$

for all $x \in [0,1]$. Since η is continuous, η is also bounded on $[m, M]$ and there is a $v^* \in [m, M]$ such that

$$|\eta(v)| \leq |\eta(v^*)| =: C \quad \forall v \in [m, M].$$

Thus, we have

$$|\eta(B_N[u](x))| \leq C \quad \forall x \in [0,1], \ N \in \mathbb{N},$$

and the sequence $(\eta \circ B_N[u])_{N\in\mathbb{N}}$ is uniformly bounded. Finally, Lebesgue's dominated convergence theorem [Eva18, Chapter 1.3] yields

$$\lim_{N\to\infty} \int_0^1 \eta\left(B_N[u](x)\right) \,\mathrm{d}x = \int_0^1 \eta(u(x)) \,\mathrm{d}x$$

and therefore the assertion. □

Note that in the proposed Bernstein procedure, the Bernstein reconstruction is always computed for $u \in \mathbb{P}_N(\mathbb{R})$. Yet, Theorem 10.7 holds for general u which are piecewise C^2.

10.3.2 Total variation

A fundamental property of the (exact) solution of a scalar CL (10.1), assuming the ID u_0 has bounded TV, is that [Lax73, Tor13]

1. no additional spatial local extrema occur.
2. the values of local minima do not decrease and the values of local maxima do not increase.

As a consequence, the TV of the solution,

$$\mathrm{TV}(u(\cdot,t)) = \sup_{\substack{x_0<\dots<x_J \\ J\in\mathbb{N}}} \sum_{j=0}^{J-1} |u(x_{j+1},t) - u(x_j,t)|,$$

is a nonincreasing function in time, i. e.

$$\mathrm{TV}(u(\cdot,t_2)) \leq \mathrm{TV}(u(\cdot,t_1))$$

for $t_2 \geq t_1$. Also see Theorem 2.30 in Chapter 2.8. In the presence of (shock) discontinuities, in fact, the TV typically decreases [Har84]. We are therefore interested in designing shock capturing methods which mimic this behavior of being TVD.

The proposed Bernstein procedure is now shown to fulfill the TVD property in the sense that the Bernstein reconstruction $B_N[u]$ of a function u has a reduced (or the same[1]) TV, i. e.

$$\mathrm{TV}(B[u]) \leq \mathrm{TV}(u)$$

holds for the Bernstein reconstruction (10.8).

Theorem 10.8
Let $B_N[u] \in \mathbb{P}_N(\mathbb{R})$ be the Bernstein reconstruction of a function $u : [0,1] \to \mathbb{R}$. Then, the TV of $B_N[u]$ is less or equal to the TV of u, and the Bernstein procedure is TVD.

Proof. Since $B_N[u] \in \mathbb{P}_N(\mathbb{R})$ and by consulting (10.4), we have

$$\begin{aligned} \mathrm{TV}(B_N[u]) &= \int_0^1 |B_N'(x)| \,\mathrm{d}x \\ &\leq N \sum_{n=0}^{N-1} \left| u\left(\frac{n-1}{N}\right) - u\left(\frac{n}{N}\right) \right| \binom{N-1}{n} \int_0^1 x^n (1-x)^{N-n-1} \,\mathrm{d}x, \end{aligned}$$

where the integrals are given by

$$\int_0^1 x^n (1-x)^{N-n-1} \,\mathrm{d}x = N^{-1} \binom{N-1}{n}^{-1}.$$

Thus, inequality

$$\mathrm{TV}(B_N[u]) \leq \sum_{n=0}^{N-1} \left| u\left(\frac{n-1}{N}\right) - u\left(\frac{n}{N}\right) \right| \leq \mathrm{TV}(u)$$

follows, and therefore the assertion. □

[1] In all numerical tests, we actually observed the TV to decrease

10.3.3 Monotone (shock) profiles

Let $u : [0,1] \to \mathbb{R}$ be a piecewise smooth function with single discontinuity at $x_1 \in (0,1)$, representing a shock profile in a troubled element. It is desirable for the polynomial approximation of such a function to not introduce new (artificial) local extrema. Yet, typical polynomial approximations, such as interpolation and (pseudo) projections, are doing so by the Gibbs–Wilbraham phenomenon [HH79]. The Bernstein reconstruction, however, has been proved to not feature such spurious oscillations. This can, for instance, be noted from

Theorem 10.9 (Theorem 1.9.1 in Lorentz [Lor12])
Suppose that u is bounded in $[0,1]$ and let L^+, L^- respectively denote the right and left upper limits and l^+, l^- the right and left lower limits of u at a point x. Then

$$\frac{1}{2}\left(l^+ + l^-\right) \leq \liminf_{N\to\infty} B_N[u](x) \leq \limsup_{N\to\infty} B_N[u](x) \leq \frac{1}{2}\left(L^+ + L^-\right).$$

Note that Theorem 10.9 is fairly general. The absence of the Gibbs–Wilbraham phenomenon (without taking convergence into account) could have been noted by Lemma 10.1 already.

It should be stressed that we can not just rule out spurious (Gibbs–Wilbraham) oscillations for the Bernstein reconstruction of discontinuous (shock) profiles, but we are further able to ensure the preservation of monotonicity. Let u be a monotonic increasing (and possibly discontinuous) function on $[0,1]$. Then, the following lemma ensures that the Bernstein reconstruction $B_N[u]$ is monotonic increasing as well.

Lemma 10.10
Let $u : [0,1] \to \mathbb{R}$ be monotonic increasing and let $B_N[u]$ denote the Bernstein reconstruction of u. Then, $B_N[u]$ is monotonic increasing as well.

Proof. Note that for monotonic increasing u, we have

$$\Delta u\left(\frac{n}{N}\right) = u\left(\frac{n+1}{N}\right) - u\left(\frac{n}{N}\right) \geq 0.$$

Thus, by consulting (10.4), inequality

$$B_N'[u](x) \geq 0$$

follows and as a result also the assertion. □

Note that the same result holds for monotonic decreasing (shock) profiles.

10.4 Numerical results

Finally, we test the proposed Bernstein procedure incorporated into a nodal DGSEM as described in Chapter 6.1. For time integration, once more, we use the explicit SSP-RK method of third order using three stages (SSPRK(3,3)) given in Definition 3.39 in Chapter 3.5. Moreover, we use a usual time step size of

$$\Delta t = C \cdot \frac{|\Omega|}{I(2N+1)^2 \max|f'(u)|}$$

with $C = 0.1$ and where $\max |f'(u)|$ is calculated for all u between $\min_{x\in\Omega} u_0(x)$ and $\max_{x\in\Omega} u_0(x)$ in all subsequent numerical tests. Following [CS91] in parts, four different problems are investigated for which the exact solutions can be calculated. Most of them have already been addressed before. Note that we assume periodic BCs in all numerical tests. This restriction is utterly unnecessary for the proposed Bernstein procedure and is only made in order to compactly provide reference solutions, which refer to the exact entropy solutions. Further, for every problem the local Lax–Friedrichs (Rusanov) flux

$$f^{\text{num}}(u_-, u_+) = \frac{f(u_+) + f(u_-)}{2} - \frac{\lambda_{\max}}{2}(u_+ - u_-)$$

is applied, where $\lambda_{\max} = \max_{u\in[u_-,u_+]} |f'(u)|$ is a locally determined viscosity coefficient based on maximum characteristics speed [LeV02, Chapter 12.5].

10.4.1 Linear advection equation

Let us start by consider the linear advection equation

$$\partial_t u + \partial_x u = 0 \tag{10.15}$$

on $\Omega = [0, 1]$ with periodic BCs and a discontinuous IC

$$u(x,t) = u_0(x) := \begin{cases} 1 & \text{if } 0.4 \le x \le 0.8, \\ 0 & \text{otherwise.} \end{cases}$$

By the method of characteristics, the solution at time $t \ge 0$ is given by

$$u(x,t) = u_0(x - t);$$

see Chapter 2.3. The linear advection equation is the simplest PDE that can feature discontinuous solutions. Thus, the (shock capturing) method can be observed in a well-understood setting, isolated from nonlinear effects. Yet, the linear advection equation provides a fairly challenging example. Similar to contact discontinuities in Euler's equations, discontinuities are not self-steeping, i. e. once such a discontinuity is smeared by the method, it can not be recovered to its original sharp shape [KLT06].

Parameter study

We start by investigating the ramp parameter $\kappa \in (0, 1)$, which goes into the PA sensor (10.14) and steers the Bernstein procedure.

Figure 10.3 demonstrates the effect of the ramp parameter $\kappa \in (0, 1)$ on the results produced by the Bernstein procedure for a linear advection equation with discontinuous IC. The IC has thereby been evolved over time until $t = 1$. We observe the Bernstein procedure to be fairly robust with respect to the ramp parameter κ. On coarse meshes, such as $I = 10$ and $I = 20$, slight differences can be observed between different parameter values. Yet, these differences become less significant when the mesh is refined. As mentioned in Remark 10.5, the tuning of the ramp parameter has been done quickly on a coarse mesh of $I = 10$ elements for all problems presented.

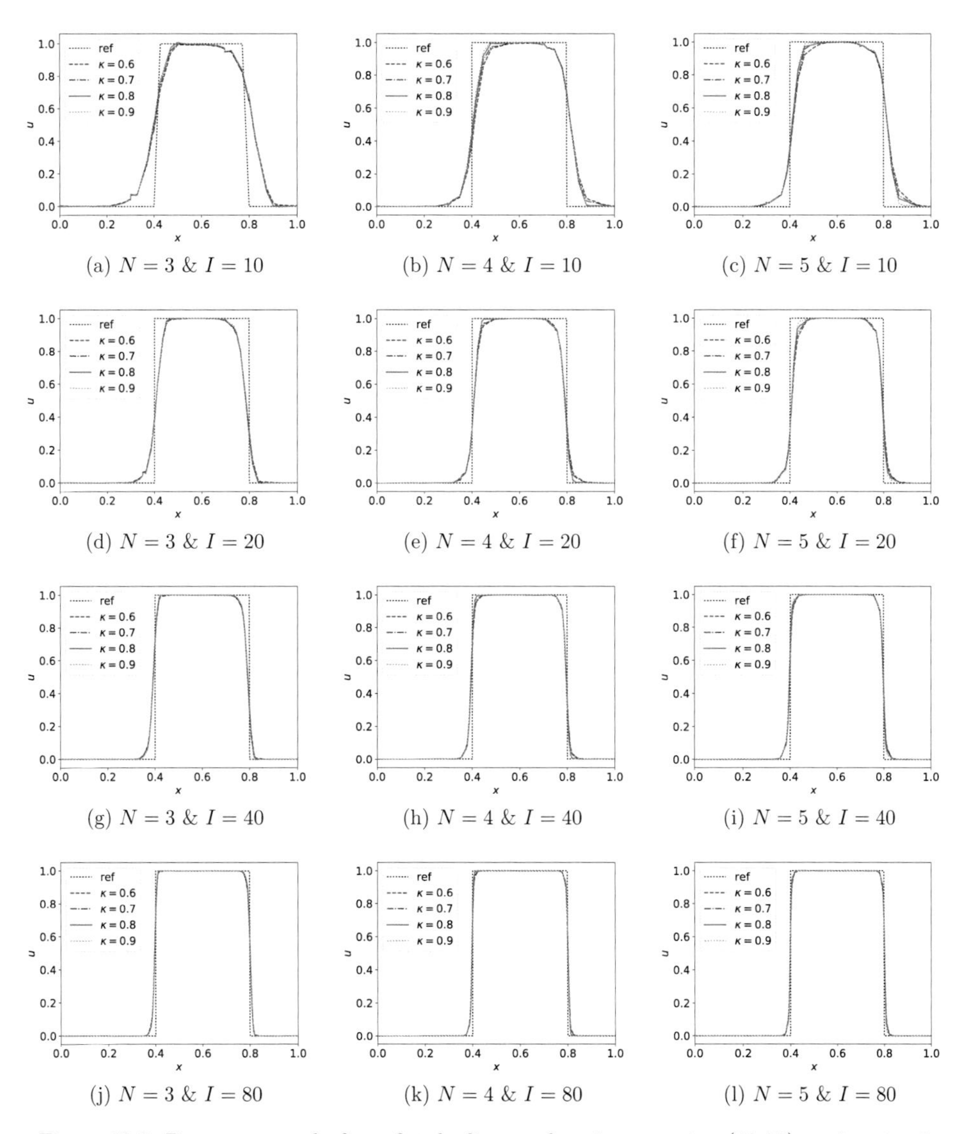

(a) $N = 3$ & $I = 10$ (b) $N = 4$ & $I = 10$ (c) $N = 5$ & $I = 10$

(d) $N = 3$ & $I = 20$ (e) $N = 4$ & $I = 20$ (f) $N = 5$ & $I = 20$

(g) $N = 3$ & $I = 40$ (h) $N = 4$ & $I = 40$ (i) $N = 5$ & $I = 40$

(j) $N = 3$ & $I = 80$ (k) $N = 4$ & $I = 80$ (l) $N = 5$ & $I = 80$

Figure 10.3: Parameter study for κ for the linear advection equation (10.15) at time $t = 1$.

Comparison with usual filtering in discontinuous Galerkin methods

Next, we investigate the approximation properties of a DGSEM enhanced with the Bernstein procedure for the linear advection equation with discontinuous IC. Figure 10.4 illustrates the results at time $t = 1$.

Further, we compare our results with the DGSEM without any filtering and with a usual

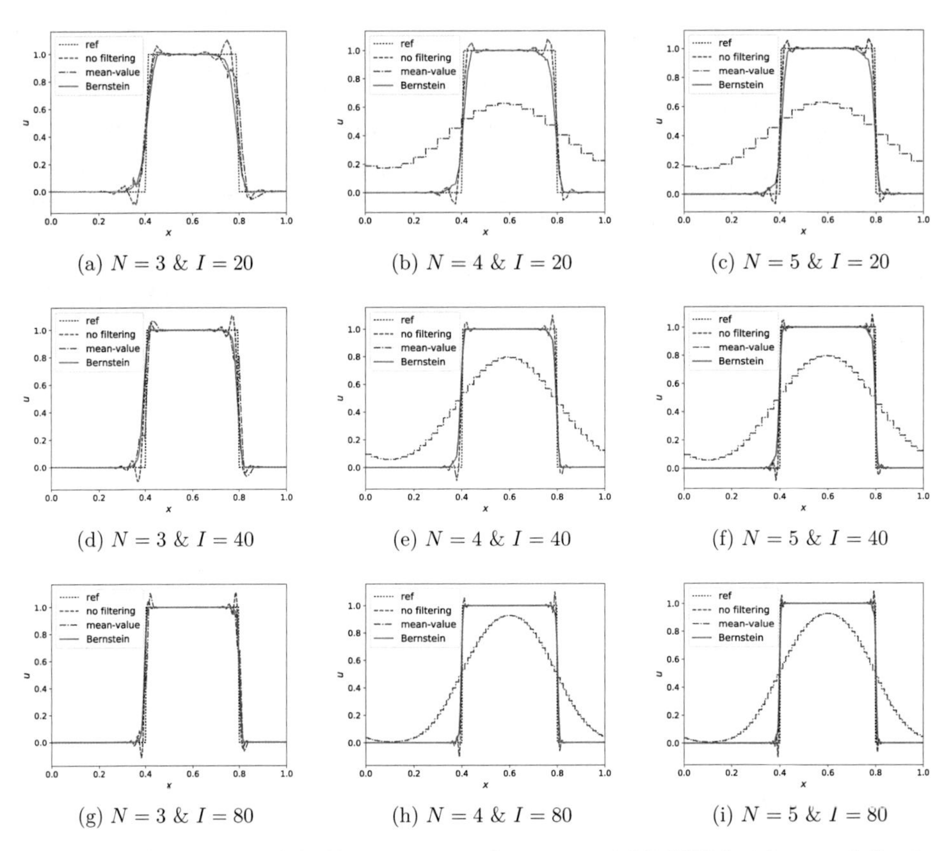

(a) $N = 3$ & $I = 20$ (b) $N = 4$ & $I = 20$ (c) $N = 5$ & $I = 20$

(d) $N = 3$ & $I = 40$ (e) $N = 4$ & $I = 40$ (f) $N = 5$ & $I = 40$

(g) $N = 3$ & $I = 80$ (h) $N = 4$ & $I = 80$ (i) $N = 5$ & $I = 80$

Figure 10.4: Comparison of the Bernstein procedure in a nodal DGSEM with a usual filtering technique and no filtering for the linear advection equation (10.15) at time $t = 1$.

filtering technique, where the DG solution $u \in \mathbb{P}_N(\Omega_i)$ is replaced by its mean,

$$\overline{u} = \int_{\Omega_i} u(x) \, \mathrm{d}x,$$

in a troubled element Ω_i. Troubled elements are detected by a critical value $S(u) \geq 1$, see (10.13). Note that the usual filtering is therefore expected to be applied in less elements than the Bernstein procedure, since the Bernstein procedure is already activated for $S(u) > \kappa$, see (10.14). Yet, we still observe the usual filtering to smear the numerical solution around discontinuities considerably. Over time, this smearing yields the numerical solution to nearly become constant. This can be observed in Figure 10.5, where the results are further evolved in time until $t = 10$.

At the same time, the results of the Bernstein procedure remain in their relatively sharp shape, even near discontinuities. Yet, no oscillations are observed, in contrast to the results produced by the DGSEM without any filtering. It should be stressed that this test case is especially challenging for the usual filtering by mean values, since the initial discontinuities have traveled through the domain several times until $t = 10$ is reached. The following tests might provide a fairer comparison. An extensive smearing of the numerical solution by the usual

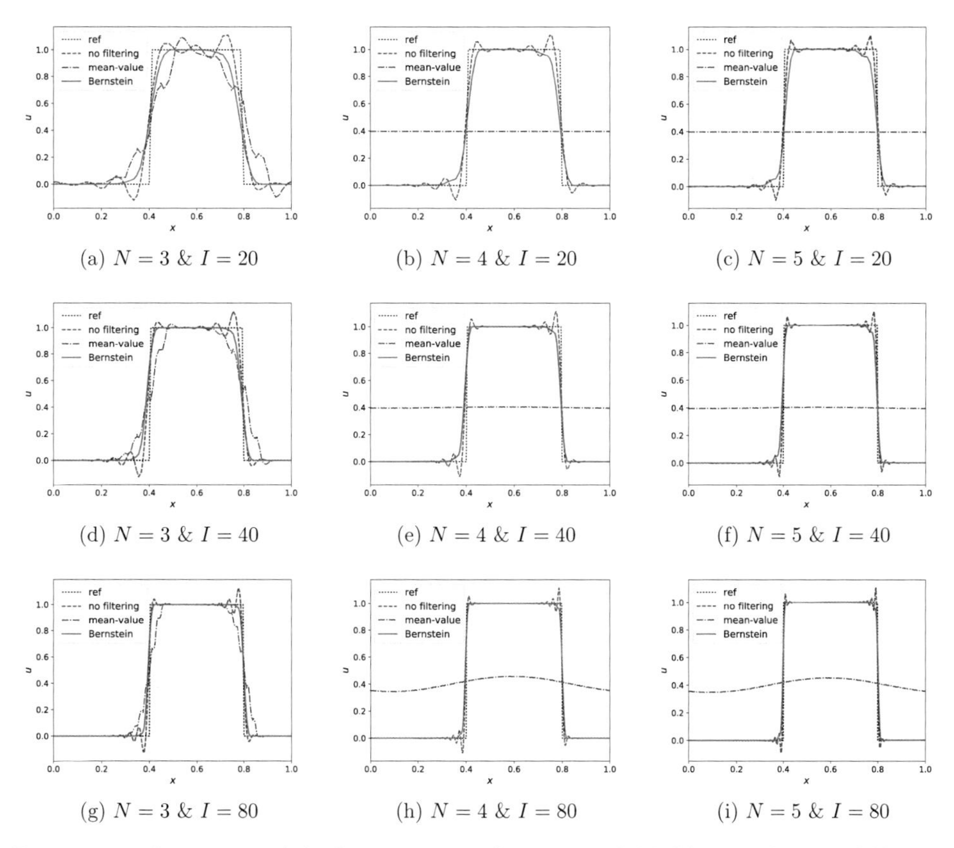

(a) $N = 3$ & $I = 20$ (b) $N = 4$ & $I = 20$ (c) $N = 5$ & $I = 20$

(d) $N = 3$ & $I = 40$ (e) $N = 4$ & $I = 40$ (f) $N = 5$ & $I = 40$

(g) $N = 3$ & $I = 80$ (h) $N = 4$ & $I = 80$ (i) $N = 5$ & $I = 80$

Figure 10.5: Comparison of the Bernstein procedure in a nodal DGSEM with a usual filtering technique and no filtering for the linear advection equation (10.15) at time $t = 10$.

filtering by mean values, especially compared to the Bernstein procedure, is always observed to some extent, however.

Remark 10.11. In Figure 10.4(g) (and Figure 10.9(d)) we note stronger oscillations for the DGSEM with mean value filtering than without any filtering. Typically, this is not expected. A reason for this behavior might be that the mean value filtering, when activated by the above sensor, is neither ensured to be TVD nor to preserve the relation between boundary values at element interfaces (while $u_- < u_+$ holds for the original approximation, mean value filtering in one or both elements connected by the interface might result in a reverse relation $\overline{u}_- > \overline{u}_+$). Such behavior could be prevented by local projection limiters which are steered by modified minmod function; see [CS91, CS89].

10.4.2 Inviscid Burgers' equation

Let us now consider the nonlinear inviscid Burgers' equation

$$\partial_t u + \partial_x \left(\frac{u^2}{2} \right) = 0 \tag{10.16}$$

on $\Omega = [0,1]$ with smooth IC

$$u(x,t) = u_0(x) := 1 + \frac{1}{4\pi} \sin(2\pi x)$$

and periodic BCs. For this problem a shock develops in the solution when the wave breaks at time

$$t_b = -\frac{1}{\min_{0 \leq x \leq 1} u_0'(x)} = 2.$$

In the subsequent numerical tests, we consider the solution at times $t = 2$ and $t = 3$. The reference solutions have been computed using characteristic tracing, solving the implicit equation $u(x,t) = u_0(x - tu)$ in smooth regions. The jump location, separating these regions, can be determined by the Rankine–Hugoniot condition.

Figure 10.6 illustrates the results at time $t = 2$, at which the shock waves starts to arise. As a consequence, only slight differences are observed at this state. In particular, we can observe that the Bernstein procedure does not smear the solution in smooth regions, even when steep gradients arise. A slight smearing can be observed for the usual filtering by mean values around the location of the arising shock at $x = 0.5$.

Figure 10.7 illustrates the results at time $t = 3$, for which the shock has fully developed and has already traveled through the whole domain once. While we observe oscillations for the DGFEM without filtering and smearing for usual filtering by mean values, the Bernstein procedure still provides sharp profiles for the discontinuous solution.

10.4.3 A concave flux function

Next, we investigate the CL

$$\partial_t u + \partial_x \left(u(1-u) \right) = 0 \tag{10.17}$$

with a concave flux function $f(u) = u(1-u)$ on $\Omega = [0,2]$, periodic BCs, and a discontinuous IC

$$u(x,0) = u_0(x) = \begin{cases} 1 & \text{if } 0.5 \leq x \leq 1.5, \\ 0 & \text{otherwise.} \end{cases}$$

For this problem a rarefaction wave develops in the solution. Figure 10.8 illustrates the results at time $t = 0.5$.

Here, the DGSEM without any filtering fails to capture the rarefaction wave around $x = 1$, yielding a wrong (physically unreasonable) weak solution. The usual filtering by mean values captures the rarefaction wave, but again smears the solution. The Bernstein procedure is also able to capture the rarefaction wave, while providing notably sharper profiles. Yet, we can observe some remaining slight oscillation for the Bernstein procedure in some cases. Remember that the Bernstein procedure is ensured to preserve bounds. Thus, the oscillations probably have been caused by the Bernstein procedure getting activated after oscillations have already been developed in the numerical solution by the DGSEM. Future work will investigate the possibility of using other shock sensors to steer the proposed Bernstein procedure in greater detail. These sensors might be activated once the TV increases over time or once the numerical solution starts to violate certain bounds.

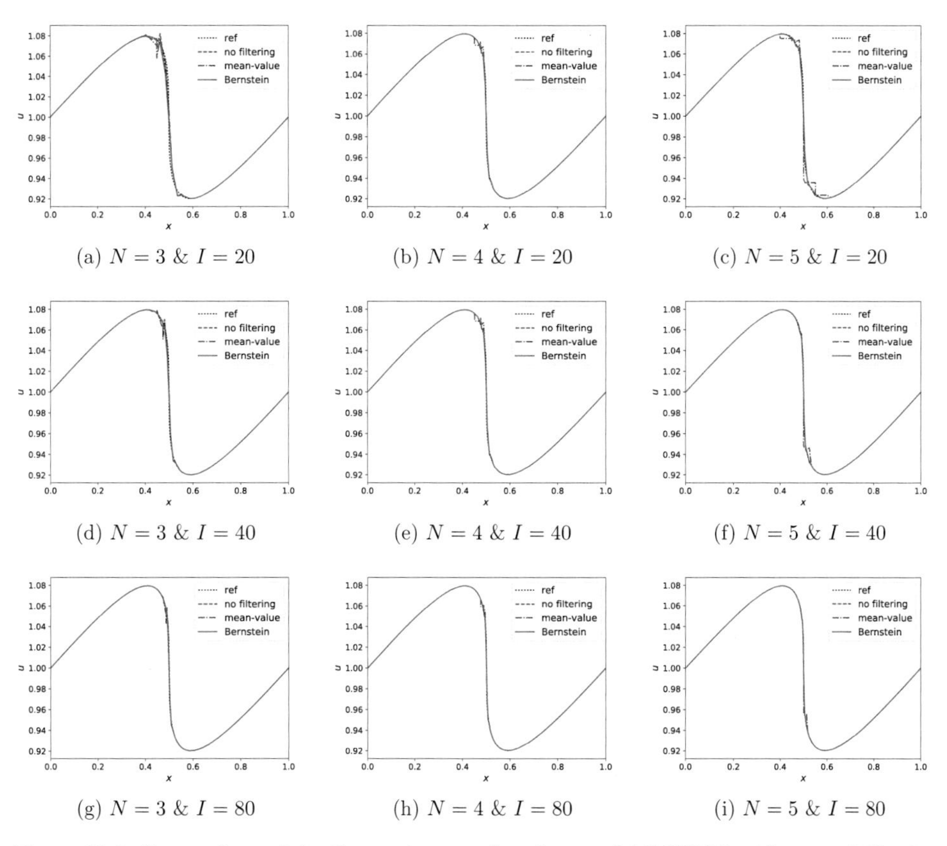

(a) $N = 3$ & $I = 20$ (b) $N = 4$ & $I = 20$ (c) $N = 5$ & $I = 20$

(d) $N = 3$ & $I = 40$ (e) $N = 4$ & $I = 40$ (f) $N = 5$ & $I = 40$

(g) $N = 3$ & $I = 80$ (h) $N = 4$ & $I = 80$ (i) $N = 5$ & $I = 80$

Figure 10.6: Comparison of the Bernstein procedure in a nodal DGSEM with a usual filtering technique and no filtering for the Inviscid Burgers equation (10.16) at time $t = 2$.

10.4.4 The Buckley–Leverett equation

Finally, we consider the Buckley–Leverett equation

$$\partial_t u + \partial_x \left(\frac{u^2}{u^2 + (1-u)^2} \right) = 0 \tag{10.18}$$

with a nonconvex flux function $f(u) = \frac{u^2}{u^2+(1-u)^2}$ on $\Omega = [0, 2]$, periodic BCs, and a discontinuous IC

$$u(x, 0) = u_0(x) := \begin{cases} 1 & \text{if } 0.5 \leq x \leq 1.5, \\ 0 & \text{otherwise.} \end{cases}$$

The Buckley–Leverett equation is often used to describe an immiscible displacement process, such as the displacement of oil by water [Ran92, Chapter 4.2]. Due to its nonconvex flux function, the Riemann solution involves a so-called *compound wave* (sometimes also referred to as a *composite curve*), which contains a shock discontinuity and a rarefaction wave at the same time. For a nonlinear system of equations this can arise in any nonlinear field that

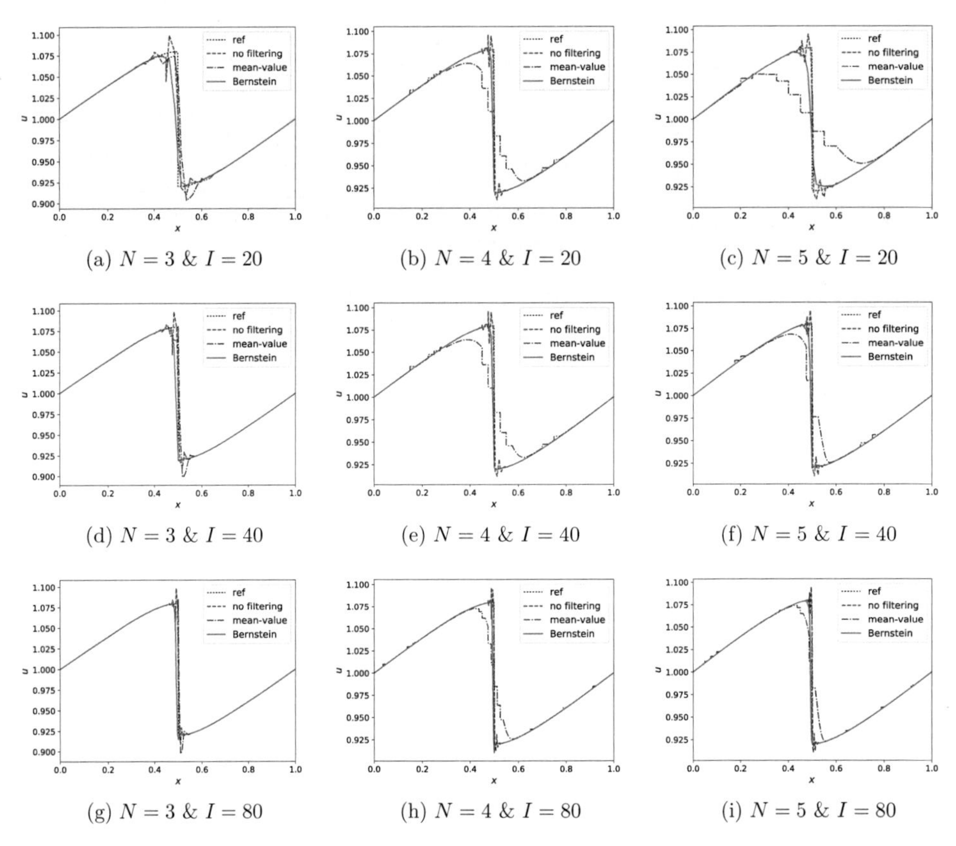

(a) $N = 3$ & $I = 20$ (b) $N = 4$ & $I = 20$ (c) $N = 5$ & $I = 20$

(d) $N = 3$ & $I = 40$ (e) $N = 4$ & $I = 40$ (f) $N = 5$ & $I = 40$

(g) $N = 3$ & $I = 80$ (h) $N = 4$ & $I = 80$ (i) $N = 5$ & $I = 80$

Figure 10.7: Comparison of the Bernstein procedure in a nodal DGSEM with a usual filtering technique and no filtering for the Inviscid Burgers equation (10.16) at time $t = 3$.

fails to be genuinely nonlinear; see [Tor13, Chapter 16.1] and [Wen72a, Wen72b, Liu75]. The reference solution has been computed using characteristic tracing again. Figure 10.9 illustrates the results at time $t = 0.5$.

For this problem, the DGSEM without filtering provides fairly poor results, polluted by heavy oscillations, in all tests. In some cases for $N = 3$ even the usual filtering by mean values shows some oscillations. This problem seems to be related to some miss-identifications of troubled elements by the PA sensor for $N = 3$. For finer meshes the problem vanishes in case of the usual filtering by mean values. Yet, we still observe the Bernstein procedure to also perform relatively poor for $N = 3$. For $N > 3$, the Bernstein procedure is again observed to provide notably better results than usual filtering by mean values. We conclude that the Bernstein procedure steered by the PA sensor (10.14) can only be recommended for higher orders $N \geq 4$. Yet, similar observations for the usual filtering by mean values indicate that the relatively poor behavior for $N = 3$ might be caused by the PA sensor, rather than by the Bernstein procedure itself.

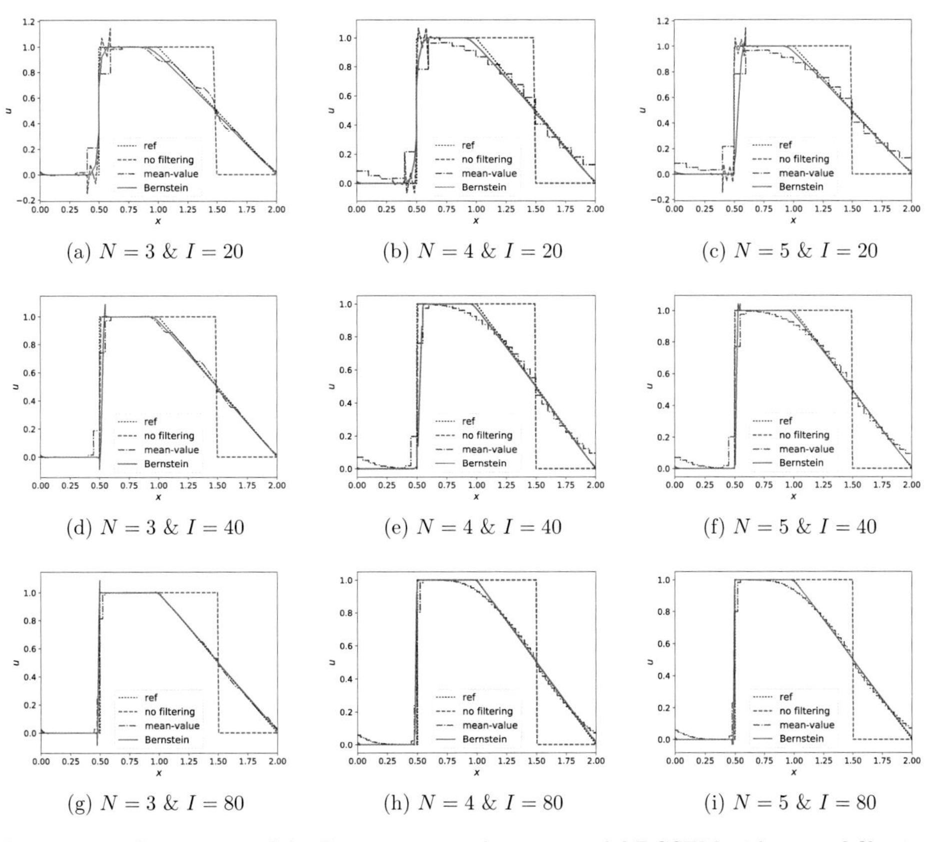

(a) $N = 3$ & $I = 20$ (b) $N = 4$ & $I = 20$ (c) $N = 5$ & $I = 20$

(d) $N = 3$ & $I = 40$ (e) $N = 4$ & $I = 40$ (f) $N = 5$ & $I = 40$

(g) $N = 3$ & $I = 80$ (h) $N = 4$ & $I = 80$ (i) $N = 5$ & $I = 80$

Figure 10.8: Comparison of the Bernstein procedure in a nodal DGSEM with a usual filtering technique and no filtering for (10.17) at time $t = 0.5$.

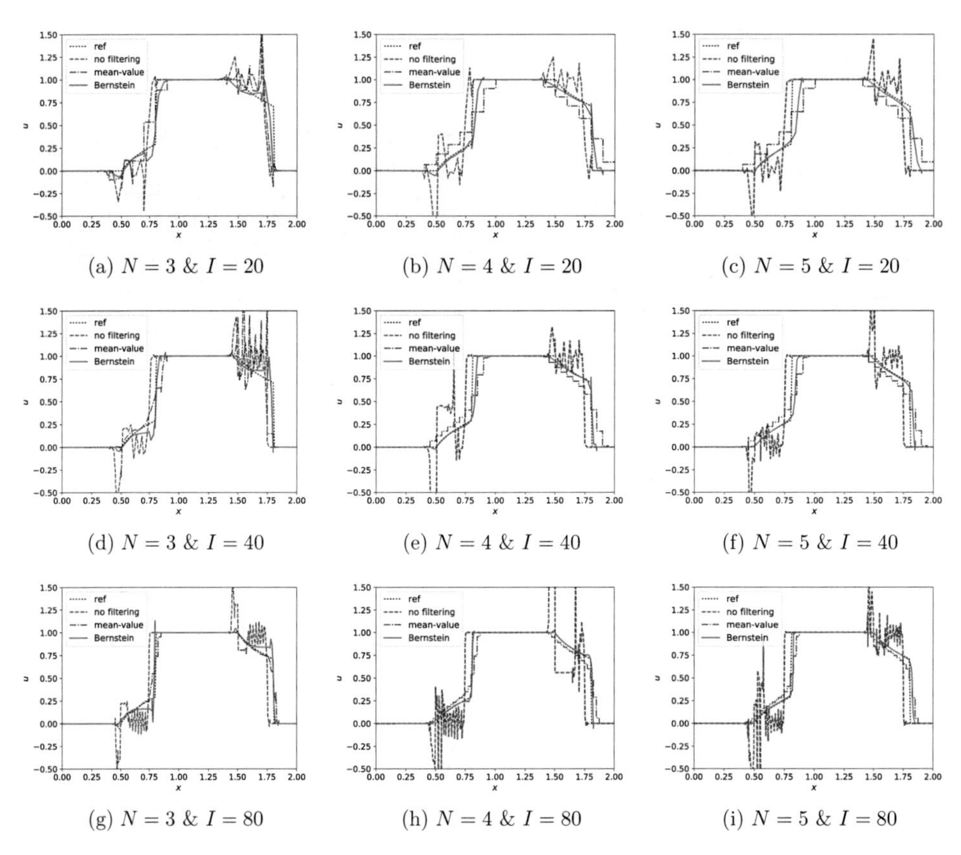

(a) $N = 3$ & $I = 20$ (b) $N = 4$ & $I = 20$ (c) $N = 5$ & $I = 20$

(d) $N = 3$ & $I = 40$ (e) $N = 4$ & $I = 40$ (f) $N = 5$ & $I = 40$

(g) $N = 3$ & $I = 80$ (h) $N = 4$ & $I = 80$ (i) $N = 5$ & $I = 80$

Figure 10.9: Comparison of the Bernstein procedure in a nodal DGSEM with a usual filtering technique and no filtering for (10.18) at time $t = 0.25$.

10.5 Concluding thoughts and outlook

In this chapter, we have proposed another novel shock capturing procedure for SE approximations for scalar hyperbolic CLs, which resulted in the publication [Gla19c]. The procedure is easy to implement and neither increases the computational complexity of the method nor adds additional time step or CFL restrictions, as other common shock capturing (AV) methods do. The procedure essentially consists of going over from the original (oscillatory) polynomial approximation to its Bernstein reconstruction in troubled elements. Thereby, the Bernstein reconstruction of a function is obtained by applying the (modified) Bernstein operator (10.10). This operator has been proved to reduce the TV of the underlying function as well as to preserve monotone (shock) profiles. Both properties distinguish the shock capturing procedure as especially suitable for scalar CLs, which come along with a TVD property for their physically reasonable solutions. We further note that the modified procedure is not just able to preserve but even to enforce certain bounds for the numerical solution, such as positivity. Finally, the proposed procedure can be calibrated to the regularity of the underlying solution by using a discontinuity sensor which, in this work, is based on comparing the values of PA operators of increasing orders. Numerical tests demonstrate that the procedure is able to significantly enhance SE approximations in the presence of shocks.

Future work will focus on the extension of the proposed Bernstein procedure to unstructured meshes in multiple dimensions and systems of CLs. Further, the investigation of other shock sensors would be of great interest. These sensors might be activated once the TV increases over time or once the numerical solution starts to violate certain bounds. Another option might be sensors based on artificial neural networks.

CHAPTER 11

High order edge sensor steered artificial viscosity operators

Finally, we have arrived at our final chapter! Here, we combine different techniques presented in previous chapters to construct novel HOES steered AV operators and discuss their discretization in the context of SBP based FD methods. These AV operators preserve stability and accuracy (in smooth regions) of the underlying method and calibrate the amount of viscosity and its distribution to the smoothness of the numerical solution and the location of discontinuities — if these are present in the solution. Building on work of Mattsson, Svärd, and Nordström [MSN04], we thoroughly justify the AV operator's design and analyze its performance on a number of benchmark problems. To the best of our knowledge, such AV operators for FD methods have not been proposed before and the material presented in this chapter resulted in the publication [Gla19b].

Outline

This chapter is organized as follows: In Chapter 11.1 we introduce and motivate generalized AV operators. Moreover, we revise the state of the art for AV operators and outline our own contribution. In Chapter 11.2, we then thoroughly justify the design of the resulting AV operators and their properties regarding conservation and (entropy) dissipation in the continuous setting. For their discretization, we focus on SBP based FD methods, which are introduced in Chapter 11.3. The discretization of the AV operators in the context of FD methods using the SBP property is investigated in Chapter 11.4. Next, Chapter 11.5 addresses how HOES can be used to calibrate the amount of viscosity and its distribution to the smoothness of the numerical solution and the location of discontinuities — if these are present in the solution. In the process, we present HOES steered AV operators for high order FD approximations. Numerical tests for the linear advection equation, the inviscid Burgers' equation, the nonlinear system of Euler equations are presented in Chapter 11.6. The tests demonstrate that we are able to better resolve discontinuous numerical solutions when HOES steered AV operators are utilized. We close this work with concluding thoughts in Chapter 11.7.

11.1 Generalized artificial viscosity operators

Once more, let us consider hyperbolic systems of CLs

$$\partial_t \boldsymbol{u} + \partial_x \boldsymbol{f}(\boldsymbol{u}) = 0, \quad t > 0, \ x \in \Omega \subset \mathbb{R}, \ \boldsymbol{u}(x,t) \in V \subset \mathbb{R}^n, \tag{11.1}$$

subject to an IC $u(x,0) = u_0$ and appropriate BCs. As we have already observed, the existence of (shock) discontinuities in the solution of hyperbolic systems of CLs pose many problems for their numerical treatment. Solving (11.1) with high accuracy is a challenging task because high order methods are known to produce spurious oscillations near (shock) discontinuities. The introduction of AV terms in (11.1), yielding the viscosity extension

$$\partial_t \boldsymbol{u} + \partial_x \boldsymbol{f}(\boldsymbol{u}) = \varepsilon \partial_x \left(a \partial_x \boldsymbol{u} \right),$$

can damp the oscillations in the computed solution and makes the convergence process mimic the vanishing viscosity method (see [BB05] or Chapter 2), which is an alternative approach to obtain physically reasonable weak solutions of (11.1). Also see Chapter 8 for a discussion of this approach in the context of (discontinuous) SE methods. Unfortunately, viscosity represents an irreversible loss of information and should therefore be introduced with great care. Thus, following [MSN04], the aim for AV operators are typically the following four properties:

(P1) In the presence of (shock) discontinuities they should efficiently reduce spurious oscillations in the numerical solution.

(P2) In smooth regions they should preserve the accuracy of the numerical solution.

(P3) They should not significantly increase the computational work of the numerical method they are incorporated in.

(P4) They should not destroy the conservation and stability properties of the numerical solution.

So far, we have just addressed (P4) in the context of SE methods; see Chapter 8. The other properties have only been mentioned briefly. For instance, computational efficiency (P3) has been shortly addressed in Chapter 9.1 and an intelligent steering (P1,P2) in Chapter 8.2. For a more detailed discussion of these properties in the context of DG methods we refer to [KWH11] and references therein. In contrast to SE methods, however, especially the clever steering of AV operators in FD methods — away from a multi element structure of the computational domain — is considerably harder to achieve.

11.1.1 State of the art

Let us shortly revisit the state of the art concerning AV techniques in the context of FD methods. As mentioned before, the idea of AV operators dates back to the pioneering work of von Neumann and Richtmyer [vNR50] and utilizes the well-known effect of dissipative mechanisms on (shock) discontinuities; when viscosity is incorporated, discontinuities in the solution are smeared out, yielding surfaces of discontinuities to be replaced by thin layers in which the solution varies (possibly rapidly but) continuously. More recently, Mattsson, Svärd, and Nordström [MSN04] have investigated general AD operators of the form $(-1)^{p-1}\partial_x^p(a(x)\partial_x^p)$ in the context of FD methods using SBP operators. In the process, they have derived conditions for the continuous AD operators as well as their discretization for which they are compatible with conservation and stability properties of hyperbolic CLs. In a subsequent work, Nordström [Nor06] also constructed AD operators especially tailored to hyperbolic CLs with variable coefficients. A similar analysis has been provided in [RGÖS18] for FR methods using SBP operators and in [GNJA+19] for DG methods. Yet, all of these works have investigated AV operators of the form

$$\varepsilon \partial_x \left(a(x) \partial_x \right),$$

with viscosity strength $\varepsilon > 0$, and where the viscosity distribution $a = a(x)$ is constructed such that conservation and stability properties are ensured, but has not been adapted to the precise location of (shock) discontinuities in the numerical solution. Often, viscosity is distributed nearly equally over the whole computational domain (or a whole element/block). Thus, especially in smooth regions, the numerical solution might get smeared unreasonably strong away from the (shock) discontinuities.

In the context of DG methods, but also in related methods such as SD and FR schemes, the concept of AV has become popular in the last decades [HH02, PP06, BD10, KWH11, GÖS18, GNJA+19, DHR19] as well. In these works also element-to-element variations of AV operators have been proposed, but their extension to FD methods is not trivial in most cases.

11.1.2 Our contribution

Here, we propose AV operators for FD methods which are steered by HOES and therefore adapt themselves to the (smoothness of the) numerical solution. As a consequence, we avoid undesired smearing of the numerical solution in smooth regions (P2), while efficiently reducing spurious oscillations near discontinuities (P1). This is possible by employing more general AV operators of the form

$$\varepsilon(\boldsymbol{u})\partial_x \left(a(x,\boldsymbol{u})\partial_x\right), \tag{11.2}$$

where the viscosity distribution $a = a(x, \boldsymbol{u})$ (nonlinearly) depends on the underlying numerical solution $\boldsymbol{u}$. Here, we propose to define a by means of HOES constructed by PA; see [AGY05, AGY08] or Chapter 9.2.2. These are able to accurately detect discontinuities in $\boldsymbol{u}$. By further enforcing the viscosity distribution a to be nonnegative and to vanish at the boundaries of the computational domain Ω, we are able to ensure the preservation of conservation and (entropy) stability properties (P4). Finally, computational efficiency (P3) is achieved by a careful scaling of the viscosity strength ε. For this scaling we propose a novel discontinuity sensor — also based on HOES — whose output is a reliably scaled smoothness estimate. For the discretization and numerical investigation of the proposed AV operators, we will focus on high order FD methods using SBP operators; see [MSN04, SN14, FHZ14]. Yet, in principle, the proposed AV operators can be incorporated in any FD method.

One particular motivation for the construction of HOES steered AV operators are the recent works of Gelb and coauthors [SGP17b, GG19a, GHL19]. Note that the work [GG19a] resulted from Chapter 9 of this thesis. There, we have utilized PA based HOES to construct a shock capturing procedure based on ℓ^1 regularization. Moreover, in [Gla19c] they were used to steer a shock capturing procedure by Bernstein polynomials and the corresponding Bernstein operator to approximate continuous functions; also see Chapter 10. Yet, to the best of our knowledge, PA based HOES have not been used to construct AV operators so far — although they seem to be especially suited for them.

11.2 Conservation and dissipation of continuous artificial viscosity operators

Let us start by providing a similar (but more general) analysis of conservation and stability properties of AV operators in the continuous setting as in Chapter 8.3. To investigate the influence of additional viscosity terms on conservation and dissipation properties of solutions $\boldsymbol{u}$ of hyperbolic systems (11.1) it is sufficient to consider the diffusion equation

$$\partial_t \boldsymbol{u} = \varepsilon \partial_x \left(A \partial_x \boldsymbol{u}\right) \tag{11.3}$$

with viscosity strength $\varepsilon \geq 0$ and a matrix-valued viscosity distribution $A = A(t, x, \boldsymbol{u})$ with respect to t, x, and $\boldsymbol{u}$. For conservation and energy dissipation, we need the viscosity distribution to vanish at the boundary $\partial\Omega$ of the computational domain Ω, i. e. $A\big|_{\partial\Omega} = 0$, and to be positive semidefinite, i. e. $\mathbf{v}^T A \mathbf{v} \geq 0$ for all $\mathbf{v} \in \mathbb{R}^n$. For entropy dissipation, it will further be convenient to replace $A : \mathbb{R}_0^+ \times \Omega \times V \to \mathbb{R}^{n \times n}$ by a scalar-valued viscosity distribution $a : \mathbb{R}_0^+ \times \Omega \times V \to \mathbb{R}$, still depending on the solution $\boldsymbol{u}$.

11.2.1 Conservation

Investigating conservation, we note that

$$\frac{\mathrm{d}}{\mathrm{d}t} \int_\Omega \boldsymbol{u} \, \mathrm{d}x = \int_\Omega \partial_t \boldsymbol{u} \, \mathrm{d}x = \varepsilon \int_\Omega \partial_x \left(A \partial_x \boldsymbol{u} \right) \mathrm{d}x = A \partial_x \boldsymbol{u}\big|_{\partial\Omega}$$

holds for a solution of (11.3), where integration is performed elementwise. Thus, for $A\big|_{\partial\Omega} = 0$, we have

$$\frac{\mathrm{d}}{\mathrm{d}t} \int_\Omega \boldsymbol{u} \, \mathrm{d}x = 0 \tag{11.4}$$

and therefore conservation.

11.2.2 Energy dissipation

Investigating the rate of change of energy, we observe that

$$\frac{1}{2} \frac{\mathrm{d}}{\mathrm{d}t} \|\boldsymbol{u}\|_{L^2(\Omega)}^2 = \frac{1}{2} \frac{\mathrm{d}}{\mathrm{d}t} \int_\Omega \boldsymbol{u} \cdot \boldsymbol{u} \, \mathrm{d}x = \int_\Omega \boldsymbol{u} \cdot \partial_t \boldsymbol{u} \, \mathrm{d}x = \varepsilon \int_\Omega \boldsymbol{u} \cdot \partial_x \left(A \partial_x \boldsymbol{u} \right) \mathrm{d}x$$

holds, where $\cdot$ denotes the usual inner product of two vectors. Next applying integration by parts, we get

$$\frac{1}{2} \frac{\mathrm{d}}{\mathrm{d}t} \|\boldsymbol{u}\|_{L^2(\Omega)}^2 = \varepsilon \boldsymbol{u} \cdot \left(A \partial_x \boldsymbol{u} \right) \big|_{\partial\Omega} - \varepsilon \int_\Omega \partial_x \boldsymbol{u} \cdot \left(A \partial_x \boldsymbol{u} \right) \mathrm{d}x. \tag{11.5}$$

Thus, for $A\big|_{\partial\Omega} = 0$ equation (11.5) reduces to

$$\frac{1}{2} \frac{\mathrm{d}}{\mathrm{d}t} \|\boldsymbol{u}\|_{L^2(\Omega)}^2 = -\varepsilon \int_\Omega \partial_x \boldsymbol{u} \cdot \left(A \partial_x \boldsymbol{u} \right) \mathrm{d}x.$$

Moreover, if A is positive semidefinite, we get

$$\frac{\mathrm{d}}{\mathrm{d}t} \|\boldsymbol{u}\|_{L^2(\Omega)}^2 \leq 0 \tag{11.6}$$

and therefore dissipation of energy.

11.2.3 Entropy dissipation

Finally, we investigate entropy dissipation. Let $\eta = \eta(\boldsymbol{u})$ be a strictly convex entropy of (11.1) and let us denote the entropy variables by $\boldsymbol{w} = \eta'(\boldsymbol{u})$. Note that for a strictly convex entropy η, the conservative variables $\boldsymbol{u}$ and the entropy variables $\boldsymbol{w}$ can be used interchangeably. Hence, the chain rule yields $\partial_x \boldsymbol{u}(\boldsymbol{w}) = \boldsymbol{u}' \partial_x \boldsymbol{w}$ with $\boldsymbol{u}' = \partial_{\boldsymbol{w}} \boldsymbol{u}$ and (11.3) can be rewritten as

$$\partial_t \boldsymbol{u} = \varepsilon \partial_x \left(A \boldsymbol{u}' \partial_x \boldsymbol{w} \right),$$

which will be convenient to investigate entropy dissipation.

As a consequence, the rate of change of the total amount of entropy is given by

$$\frac{\mathrm{d}}{\mathrm{d}t} \int_\Omega \eta(\boldsymbol{u}) \, \mathrm{d}x = \int_\Omega \boldsymbol{w} \cdot \partial_t \boldsymbol{u} \, \mathrm{d}x = \varepsilon \int_\Omega \boldsymbol{w} \cdot \partial_x \left(A \boldsymbol{u}' \partial_x \boldsymbol{w} \right) \, \mathrm{d}x. \tag{11.7}$$

Once more applying integration by parts, we find

$$\frac{\mathrm{d}}{\mathrm{d}t} \int_\Omega \eta(\boldsymbol{u}) \, \mathrm{d}x = \varepsilon \boldsymbol{w} \cdot \left(A \boldsymbol{u}' \partial_x \boldsymbol{w} \right) |_{\partial\Omega} - \varepsilon \int_\Omega \partial_x \boldsymbol{w} \cdot \left(A \boldsymbol{u}' \partial_x \boldsymbol{w} \right) \, \mathrm{d}x,$$

which reduces to

$$\frac{\mathrm{d}}{\mathrm{d}t} \int_\Omega \eta(\boldsymbol{u}) \, \mathrm{d}x = -\varepsilon \int_\Omega \partial_x \boldsymbol{w} \cdot \left(A \boldsymbol{u}' \partial_x \boldsymbol{w} \right) \, \mathrm{d}x$$

for $A|_{\partial\Omega} = 0$. Finally, if $A\boldsymbol{u}'$ is positive semidefinite, we have

$$\frac{\mathrm{d}}{\mathrm{d}t} \int_\Omega \eta(\boldsymbol{u}) \, \mathrm{d}x \leq 0 \tag{11.8}$$

and therefore dissipation of entropy.

Unfortunately, it is not trivial to ensure that the matrix product $A\boldsymbol{u}'$ is positive semidefinite. Note that $\boldsymbol{u}' = \boldsymbol{w}'(\boldsymbol{u})$ is positive definite since $\boldsymbol{w}'(\boldsymbol{u}) = \eta''(\boldsymbol{u})$ and η is a strictly convex entropy. Here, $\boldsymbol{u}'$ denotes $\partial_{\boldsymbol{w}} \boldsymbol{u}(\boldsymbol{w})$ and $\boldsymbol{w}'$ denotes $\partial_{\boldsymbol{u}} \boldsymbol{w}(\boldsymbol{u})$. Further note that there are many ways to ensure that A is positive (semi)definite as well, for instance by choosing A to be a diagonal matrix with nonnegative diagonal entries. Yet, the product $A\boldsymbol{u}'$ is only ensured to be positive semidefinite when A and $\boldsymbol{u}'$ commute, i. e. $A\boldsymbol{u}' = \boldsymbol{u}'A$; see [HJ12]. However, the only matrices which commute with all matrices are multiples of the identity matrix. Following this observation, it seems more practical to replace the matrix-valued viscosity distribution A by a scalar-valued viscosity distribution a. Thus, from here on, we restrict ourselves to AV operators of the form

$$\varepsilon \partial_x \left(a \partial_x \boldsymbol{u} \right), \tag{11.9}$$

where $a = a(t, x, \boldsymbol{u})$ is a scalar-valued function with respect to time t, space x, and the underlying solution $\boldsymbol{u}$.

11.2.4 In a nutshell

So far, we have investigated AV extensions of hyperbolic systems of CLs. Our investigation showed that conservation and energy as well as entropy dissipation is ensured for the viscosity extension

$$\partial_t \boldsymbol{u} + \partial_x \boldsymbol{f}(\boldsymbol{u}) = \varepsilon \partial_x \left(a \partial_x \boldsymbol{u} \right)$$

when a scalar-valued viscosity distribution $a = a(t, x, \boldsymbol{u})$ is chosen which is nonnegative and vanishes at the boundaries $\partial\Omega$ of the computational domain Ω. A more general matrix-valued viscosity distribution, where conservation and energy dissipation are again simple to ensure, is also possible. Yet, entropy dissipation for general entropies η seems fairly nontrivial. Henceforth, this approach is therefore not discussed in greater detail.

11.3 High order finite difference methods based on summation by parts operators

We follow the reviews [SN14, FHZ14] and provide a brief recap of SBP based FD methods — denoted by SBP-FD methods. In most cases, SBP-FD methods are used in the method of lines [LeV02], where only space is discretized and time is kept continuous first; see Chapter 3.5. Denoting the whole discrete approximation as $\mathbf{u} = (\mathbf{u}_1, \ldots, \mathbf{u}_n)$, this yields a (nonlinear) system of ordinary differential equations

$$\frac{\mathrm{d}}{\mathrm{d}t}\mathbf{u} = L(\mathbf{u}),$$

which can, for instance, be advanced in time using explicit SSP-RK method. In contrast to the previous chapters, for all tests presented here, we use a fourth order accurate explicit RK method to integrate in time: Let $\mathbf{u}^n$ be the solution at time t^n. The solution $\mathbf{u}^{n+1}$ at time t^{n+1} is computed as

$$\begin{aligned}
\mathbf{k}_1 &= \Delta t L\left(\mathbf{u}^n\right), \\
\mathbf{k}_2 &= \Delta t L\left(\mathbf{u}^n + \mathbf{k}_1/2\right), \\
\mathbf{k}_3 &= \Delta t L\left(\mathbf{u}^n + \mathbf{k}_2/2\right), \\
\mathbf{k}_4 &= \Delta t L\left(\mathbf{u}^n + \mathbf{k}_3\right), \\
\mathbf{u}^{n+1} &= \mathbf{u}^n + \left(\mathbf{k}_1 + 2\mathbf{k}_2 + 2\mathbf{k}_3 + \mathbf{k}_4\right)/6.
\end{aligned} \tag{11.10}$$

We use a fourth order accurate explicit RK method because it is a common choice for SBP-FD methods and makes the later results better comparable with existing literature and results. Addressing the spatial discretization, let us consider a mesh with N equally spaced nodes $\mathbf{x} = (x_1, \ldots, x_N)^T$ on $\Omega = [x_l, x_r]$ with $x_l = x_1 < x_2 < \cdots < x_N = x_r$. Then, at the nodes $\mathbf{x}$, the ith component of the solution $\boldsymbol{u}$ is approximated by $\mathbf{u}_i = \left(u_i^1, \ldots, u_i^N\right)^T$. At the same time, the ith component of the flux function $\boldsymbol{f}(\boldsymbol{u})$ is approximated by $\mathbf{f}_i = (f_i^1, \ldots, f_i^N)^T$, where $f_i^k = f_i(u_i^k)$ for $k = 1, \ldots, N$. Finally, first derivatives ∂_x are approximated by an SBP-FD operator satisfying

$$HD\mathbf{g} = Q\mathbf{g}, \quad \text{i.e. } \partial_x g \approx H^{-1}Q\mathbf{g},$$

where H is a diagonal positive definite matrix that defines an inner product, corresponding norm, and QR by

$$\begin{aligned}
\langle \mathbf{g}, \mathbf{h} \rangle_H &:= \mathbf{g}^T H \mathbf{h}, \\
\|\mathbf{g}\|_H^2 &= \mathbf{g}^T H \mathbf{g}, \\
\int_\Omega g h \,\mathrm{d}x &\approx \mathbf{g}^T H \mathbf{h}.
\end{aligned}$$

The restriction to a diagonal matrix H is not necessary [SN14, FHZ14], but it simplifies our subsequent investigation of conservation and energy/entropy dissipation. Note that the notation is slightly different than for the previous SE methods (see Chapter 6). Yet, the involved concepts are essentially the same, especially the concept of SBP operators.

The decisive tool in the analysis provided in Chapter 11.2 has been integration by parts,

$$\int_\Omega g\left(\partial_x h\right) \mathrm{d}x = gh|_{\partial\Omega} - \int_\Omega \left(\partial_x g\right) h \,\mathrm{d}x.$$

As in the context of SE methods, (FD) SBP operators are constructed to mimic integration by parts on a discrete level, resulting in

$$\mathbf{g}^T H D \mathbf{h} = \mathbf{g}^T E \mathbf{h} - \mathbf{g}^T D^T H \mathbf{h},$$

where $E = \operatorname{diag}(-1, 0, \ldots, 0, 1)^T$ describes the contribution at the element boundaries. Thus, with $Q = HD$, the SBP property reads

$$Q + Q^T = E.$$

Finally, accuracy of the SBP operator D is defined by its *degree* p: The degree p of an SBP operator D for the first derivative is the maximum degree for which polynomials are handled exactly, i. e. the *accuracy equations*

$$D\mathbf{x}^j = j\mathbf{x}^{j-1}, \quad j = 0, \ldots, p,$$

are satisfied. Here, $\mathbf{x}^j = (x_1^j, \ldots, x_N^j)^T$ denotes the discretization of the jth monomial x^j. All of the above requirements for SBP operators are summarized in

Definition 11.1
An operator D is an approximation of the first derivative of degree p with the SBP property if

(D1) $D = H^{-1}Q$,

(D2) H is a positive definite symmetric matrix,

(D3) $Q + Q^T = E$,

(D4) $D\mathbf{x}^j = j\mathbf{x}^{j-1}$ for $j = 0, \ldots, p$,

hold.

Explicit expressions for SBP operators can be found in [CGA93, Nor06]. It should be stressed once more that we restrict ourselves to diagonal matrices H in this chapter. Using SBP-FD operators, the hyperbolic system of CLs (11.1) can be approximated in space by

$$\frac{\mathrm{d}}{\mathrm{d}t}\mathbf{u}_i = -D\mathbf{f}_i, \quad i = 1, \ldots, n. \tag{11.11}$$

Further, to incorporate BCs, so-called SATs are usually added to the spatial approximation (11.11), yielding

$$\frac{\mathrm{d}}{\mathrm{d}t}\mathbf{u}_i = -D\mathbf{f}_i + \mathbf{SAT}_0 + \mathbf{SAT}_N, \quad i = 1, \ldots, n.$$

We refer the reader to [KL89, GKO95, Gus07, SN14, FHZ14] and references therein for a detailed description of SATs. However, for the subsequent investigation of the AV extensions of (11.1) they will not be necessary. Further extensions of SBP-FD methods include — but are not limited to — generalized SBP operators [FBZ14], skew-symmetric split forms [FCN+13, SN14, ÖGR18], and multi-block structures. These will be addressed in future works.

11.4 Conservation and dissipation of discrete artificial viscosity operators

We are now prepared to investigate the AV operator (11.2) on a discrete level. Following the concepts introduced in the previous chapter, we discretize the continuous AV operator (11.2) as εDAD and consider the system of discretized diffusion equations

$$\frac{\mathrm{d}}{\mathrm{d}t}\mathbf{u}_i = \varepsilon DAD\mathbf{u}_i, \quad i = 1, \ldots, n. \tag{11.12}$$

Here, A is a diagonal matrix given by $A = \operatorname{diag}(a(x_1), \dots, a(x_N))$, i. e. by the restriction of the viscosity distribution a to the grid points $\mathbf{x}$. Thus, we approximate second order derivatives (with variable coefficient a) by simply applying an SBP operator for the first order derivative twice, resulting in the right hand side of (11.12). Alternatively, minimum-stencil SBP operators for second order derivatives could be used [MN04, MSS08, Mat12]. Even though minimum-stencil SBP operators provide some numerical advantages, for sake of simplicity, we only consider discretizations of the simple form (11.12) in this thesis.

11.4.1 Conservation

The discrete analogue of conservation (11.4) is given by

$$\frac{\mathrm{d}}{\mathrm{d}t} \langle \mathbf{1}, \mathbf{u}_i \rangle_H = 0.$$

Revisiting the discretization (11.12), we find

$$\frac{\mathrm{d}}{\mathrm{d}t} \langle \mathbf{1}, \mathbf{u}_i \rangle_H = \mathbf{1}^T H \frac{\mathrm{d}}{\mathrm{d}t} \mathbf{u}_i = \varepsilon \mathbf{1}^T HDAD\mathbf{u}_i.$$

Next utilizing that the differentiation operator D is given by $D = H^{-1}Q$ and that the SBP property $Q + Q^T = E$ holds, this becomes

$$\begin{aligned} \frac{\mathrm{d}}{\mathrm{d}t} \langle \mathbf{1}, \mathbf{u}_i \rangle_H &= \varepsilon \mathbf{1}^T QAD\mathbf{u}_i \\ &= \varepsilon \mathbf{1}^T EAD\mathbf{u}_i - \varepsilon \mathbf{1}^T Q^T AD\mathbf{u}_i \\ &= \varepsilon \mathbf{1}^T EAD\mathbf{u}_i - \varepsilon \mathbf{1}^T D^T H^{-1} AD\mathbf{u}_i, \end{aligned} \tag{11.13}$$

where we have utilized that H is symmetric. Further note that $D\mathbf{1} = \mathbf{0}$ and (11.13) therefore reduces to

$$\frac{\mathrm{d}}{\mathrm{d}t} \langle \mathbf{1}, \mathbf{u}_i \rangle_H = \varepsilon \mathbf{1}^T EAD\mathbf{u}_i.$$

Finally, since a vanishes at the boundary $\partial\Omega$, we have $EA = 0$ and

$$\frac{\mathrm{d}}{\mathrm{d}t} \langle \mathbf{1}, \mathbf{u}_i \rangle_H = 0$$

follows, i. e. conservation.

11.4.2 Energy dissipation

The discrete analogue of continuous energy dissipation (11.6) is given by

$$\sum_{i=1}^{n} \frac{\mathrm{d}}{\mathrm{d}t} \|\mathbf{u}_i\|_H^2 \leq 0.$$

Here, the SBP property yields

$$\begin{aligned} \frac{1}{2} \frac{\mathrm{d}}{\mathrm{d}t} \|\mathbf{u}_i\|_H^2 &= \mathbf{u}_i^T H \frac{\mathrm{d}}{\mathrm{d}t} \mathbf{u}_i \\ &= \varepsilon \mathbf{u}_i^T HDAD\mathbf{u}_i \\ &= \varepsilon \mathbf{u}_i^T QAD\mathbf{u}_i \\ &= \varepsilon \mathbf{u}_i^T EAD\mathbf{u}_i - \varepsilon \mathbf{u}_i^T Q^T AD\mathbf{u}_i \\ &= \varepsilon \mathbf{u}_i^T EAD\mathbf{u}_i - \varepsilon \mathbf{u}_i^T D^T HAD\mathbf{u}_i. \end{aligned} \tag{11.14}$$

Thus, if $a|_{\partial\Omega} = 0$, we have $EA = 0$ and (11.14) reduces to

$$\frac{1}{2}\frac{\mathrm{d}}{\mathrm{d}t}\|\mathbf{u}_i\|_H^2 = -\varepsilon\,(D\mathbf{u}_i)^T HA\,(D\mathbf{u}_i)\,.$$

Note that H and A are both positive semidefinite. Further assuming a diagonal norm matrix H, both matrices are diagonal and therefore commute, resulting in their product being positive semidefinite as well. Hence, we get $(D\mathbf{u}_i)^T HA\,(D\mathbf{u}_i) \geq 0$ and therefore

$$\frac{\mathrm{d}}{\mathrm{d}t}\|\mathbf{u}_i\|_H^2 \leq 0,$$

i. e. dissipation of energy.

11.4.3 Entropy dissipation

In the continuous case, entropy dissipation has been proven by applying the chain rule to replace $\partial_x \boldsymbol{u}$ by $\boldsymbol{u}'\partial_x \boldsymbol{w}$. Unfortunately, the chain rule does not hold in the discrete case. Thus, we have to artificially replace (11.12) by

$$\frac{\mathrm{d}}{\mathrm{d}t}\mathbf{u}_i = \varepsilon \sum_{j=1}^{n} DAU'_{j,i}D\mathbf{w}_j, \tag{11.15}$$

where $U'_{j,i} = \mathrm{diag}(\boldsymbol{u}'_{j,i}(x_1), \dots, \boldsymbol{u}'_{j,i}(x_N))$. Now, let us address the discrete analogue of entropy dissipation (11.8), given by

$$\sum_{i=1}^{n}\left\langle \mathbf{w}_i, \frac{\mathrm{d}}{\mathrm{d}t}\mathbf{u}_i\right\rangle_H \leq 0.$$

Utilizing the discretization (11.15), we have

$$\left\langle \mathbf{w}_i, \frac{\mathrm{d}}{\mathrm{d}t}\mathbf{u}_i\right\rangle_H = \mathbf{w}_i^T H \frac{\mathrm{d}}{\mathrm{d}t}\mathbf{u}_i = \varepsilon \sum_{j=1}^{n} \mathbf{w}_i^T HDAU'_{j,i}D\mathbf{w}_j$$

and the SBP property yields

$$\begin{aligned}
\left\langle \mathbf{w}_i, \frac{\mathrm{d}}{\mathrm{d}t}\mathbf{u}_i\right\rangle_H &= \varepsilon \sum_{j=1}^{n} \mathbf{w}_i^T QAU'_{j,i}D\mathbf{w}_j \\
&= \varepsilon \sum_{j=1}^{n} \mathbf{w}_i^T EAU'_{j,i}D\mathbf{w}_j - \varepsilon \sum_{j=1}^{n} \mathbf{w}_i^T Q^T AU'_{j,i}D\mathbf{w}_j.
\end{aligned}$$

Thus, for $a|_{\partial\Omega} = 0$, this reduces to

$$\left\langle \mathbf{w}_i, \frac{\mathrm{d}}{\mathrm{d}t}\mathbf{u}_i\right\rangle_H = -\varepsilon \sum_{j=1}^{n} (D\mathbf{w}_i)^T HAU'_{j,i}\,(D\mathbf{w}_j)\,,$$

where we have utilized that H is symmetric. Next, let us denote $D\mathbf{w}_i$ by $\mathbf{v}_i = (v_i^1, \ldots, v_i^N)^T$. Then, summing up over all components, we get

$$
\begin{aligned}
\sum_{i=1}^{n} \left\langle \mathbf{w}_i, \frac{\mathrm{d}}{\mathrm{d}t} \mathbf{u}_i \right\rangle_H &= -\varepsilon \sum_{i,j=1}^{n} \mathbf{v}_i^T H A U'_{j,i} \mathbf{v}_j \\
&= -\varepsilon \sum_{k=1}^{N} h_k a(x_k) \sum_{i,j=1}^{k} \mathbf{v}_i^k u'_{j,i}(x_k) \mathbf{v}_j^k \\
&= -\varepsilon \sum_{k=1}^{N} \underbrace{h_k a(x_k)}_{\geq 0} \begin{pmatrix} v_1^k \\ \vdots \\ v_n^k \end{pmatrix}^T \begin{pmatrix} u'_{1,1}(x_k) & \ldots & u'_{n,1}(x_k) \\ \vdots & & \vdots \\ u'_{1,n}(x_k) & \ldots & u'_{n,n}(x_k) \end{pmatrix} \begin{pmatrix} v_1^k \\ \vdots \\ v_n^k \end{pmatrix}.
\end{aligned}
$$

Since $\boldsymbol{u}'$ is positive definite, this proves

$$
\sum_{i=1}^{n} \left\langle \mathbf{w}_i, \frac{\mathrm{d}}{\mathrm{d}t} \mathbf{u}_i \right\rangle_H \leq 0,
$$

i. e. dissipation of entropy.

Remark 11.2. In our numerical tests we have not found the entropy dissipative AV operators (11.15) to perform any better than the energy dissipative AV operators (11.12). Even for the Euler equations, see (2.5), where a pair of an entropy function and flux is given, for instance, by

$$
\eta = -\rho s, \quad \psi = -\rho s \boldsymbol{u}
$$

with specific (physical) entropy $s = \ln(p\rho^{-\gamma}) = \ln(p) - \gamma \ln(\rho)$, we observed the energy dissipative AV operators (11.12) to yield more accurate results than the entropy dissipative AV operators (11.15). Thus, in the later numerical tests, we will only focus on energy dissipative AV operators (11.12), yet comparing different viscosity distributions.

11.4.4 In a nutshell

We have investigated the discretization of AV operators. Assuming SBP operators $D = H^{-1}Q$ such that $Q + Q^T = E$ with a diagonal norm matrix H, conservation and energy dissipation carry over to the discrete case when a simple discretization $\varepsilon DAD\mathbf{u}_i$ is used on the ith component. Yet, since the proof of entropy dissipation in the continuous case involved the chain rule, we had to artificially incorporate this step and go over to the discretization $\varepsilon \sum_{j=1}^{n} DAU'_{j,i}D\mathbf{w}_j$. This discretization also ensured entropy dissipation in the discrete case and reduces to the simple discretization $\varepsilon DAD\mathbf{u}_i$ when the square entropy (energy) $\eta = \frac{1}{2}\boldsymbol{u}^2$ is used.

11.5 High order edge sensor steered artificial viscsosity operators

We now propose novel AV operators which are steered by HOES based on PA. This will enable us to calibrate the amount of viscosity and its distribution to the smoothness of the numerical solution and the location of discontinuities.

11.5.1 High order edge sensors and polynomial annihilation

Let us consider a function $s : \Omega \to \mathbb{R}$, later referred to as the *sensing variable*. For $x \in \Omega^{\circ}$, we define the *jump function* of s as

$$[s](x) := s(x^{+}) - s(x^{-}), \tag{11.16}$$

where $s(x^{-})$ and $s(x^{+})$ respectively denote the left and right limits of s. Assuming that both limits exist (which rules out the presence of essential discontinuities), we have $[s](x) \neq 0$ if and only if x is a jump discontinuity of s. Note that $[s](x) = 0$ either holds if s is continuous in x or if x is a removable discontinuity of s. Often, especially in the field of signal and image processing, jump discontinuities are also referred to as *edges*. In what follows, we are interested in constructing high order approximations to the jump function (11.16), which utilizes only the function values of s at the N grid nodes $\mathbf{x} = (x_1, \ldots, x_N)^T$ on Ω. To such approximations, we refer to as *HOES*. Examples of HOES and related methods include conjugated Fourier series [Fej13, Luk20], the concentration method of Gelb and Tadmor [GT99, GT00], estimators for local Lipschitz exponents of functions from their wavelet transform [MH92], $\lambda\tau$-representations [GJ97], and methods based on numerical optimization [Can87]. Here, we use PA, originally proposed in [AGY05], to construct HOES and to steer the viscosity distribution in our AV operators.

Even though we already introduced PA operators in Chapter 9.2.2, we briefly repeat their description in order to adapt notion to the SBP-FD methods addressed here. Let us again denote the PA operator of order m by

$$L_m[s](x) := \frac{1}{q_m(x)} \sum_{x_j \in S_x} c_j(x) s(x_j), \tag{11.17}$$

where $S_x = \{x_0(x), \ldots, x_m(x)\}$ is a local stencil of $m+1$ grid points around x. We chose the stencils S_x as subsets of the N equidistant grid points $\{x_i\}_{i=1}^N$. See Chapter 9.2.2 for the details. In particular, the normalization factor q_m is chosen such that it ensures convergence of $L_m[s]$ to the proper jump value at each jump discontinuity. At the same time, the annihilation coefficients c_j ensure mth order convergence $L_m[s] \to 0$ in regions where s has at least m continuous derivatives. This is once more summarized in

Theorem 11.3 (Theorem 3.1 in [AGY05])
Let $m \in \mathbb{N}$ and $L_m[s](x)$ be defined as in (11.17) *using a local stencil S_x with $\#S_x = m+1$. Then we have*

$$L_m[s](x) = \begin{cases} [s](\xi) + \mathcal{O}\left(h(x)\right) & \text{if } x_{j-1} \leq \xi, x \leq x_j, \\ \mathcal{O}\left(h^{\min(m,k)}(x)\right) & \text{if } s \in C^k(I_x) \text{ for } k > 0, \end{cases}$$

where $h(x) := \max\{|x_i - x_{i-1}| \mid x_{i-1}, x_i \in S_x\}$ and I_x is the smallest closed interval such that $S_x \subset I_x$.

We note that for increasing order m oscillations develop in the approximation $L_m[s]$ near jump discontinuities. These oscillations could be reduced by post-processing, e. g. by a minmod-limiter [CS91, CS89, LeV02, AGY05]. The minmod-limiter is typically used to reduce spurious oscillations in numerical solutions of hyperbolic CLs [CS91, CS89, LeV02]. Yet, in our tests, we found the PA operators without minmod-limiting to be superior. This is probably due to the fact that spurious oscillations around (but not in the location of) a jump discontinuity are not sufficiently smoothed when AV is steered by PA operators with minmod-limiting. An illustration of different PA approximations for the function

$$s : [0,1] \to \mathbb{R}, \quad s(x) := \begin{cases} e^{-300(2x-0.3)^2} & \text{if } |2x-0.3| \leq 0.25, \\ 1 & \text{if } |2x-0.9| \leq 0.2, \\ \left(1 - \left(\frac{2x-1.6}{0.2}\right)^2\right)^{\frac{1}{2}} & \text{if } |2x-1.6| \leq 0.2, \\ 0 & \text{otherwise}, \end{cases} \tag{11.18}$$

is provided by Figure 11.1.

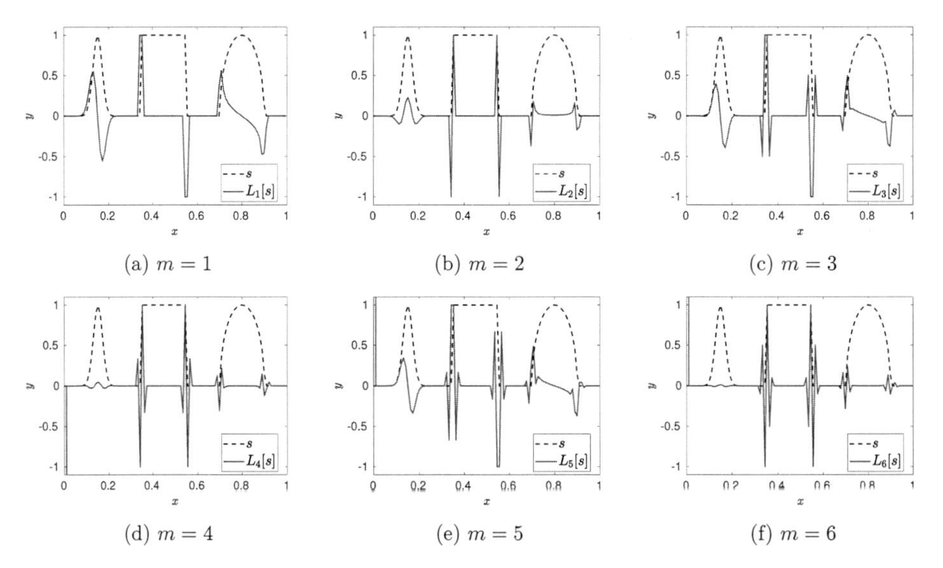

Figure 11.1: Illustration of PA approximations $L_m[s]$ for the functions s given by (11.18) and different order $m = 1, \ldots, 6$. The PA approximations are observed to successfully distinguish between jump discontinuities and steep gradients for increasing order m.

As indicated by Figure 11.1, a main advantage of PA is that for $m > 1$ it is possible to distinguish between jump discontinuities (shock waves) and steep gradients (thin continuous layers). This is of crucial advantage when steering AV operators and enables us to avoid undesired smearing of the numerical solution in smooth regions (P2), while efficiently reducing spurious oscillations near jump discontinuities (P1). To the best of our knowledge, this investigation is the first to steer AV operators by PA based HOES.

Remark 11.4. It should be noted that problems might arise for the PA operator (11.17) when evaluated at the left boundary $x_1 = x_l$ of $\Omega = [x_l, x_r]$. Often, $q_m(x) \neq 0$ is assumed for the normalization factor. Yet, when choosing $x = x_l$, we observe $q_m(x) = 0$ to hold. This observation is due to the fact that we have $S_x = S_x^+$ for $x = x_l$. Hence, we get

$$q_m(x) \overset{(9.10)}{=} \sum_{x_j \in S_x} c_j(x) \overset{(9.8)}{=} 0$$

for $m \geq 1$. Fortunately, this problem can be overcome by minmod-limiting or by excluding the values of the PA operator at the boundary points $x_1 = x_l$ and $x_N = x_r$. As our analysis in Chapter 11.2 and Chapter 11.4 has shown, the viscosity distribution a should be set to 0 at the boundaries anyway.

Remark 11.5. In this work, we only consider one dimensional CLs. It should be stressed, however, that PA can be extended to multivariate irregular data in any domain. It has been demonstrated in [AGY05] that PA is numerically cost efficient and entirely independent of any specific shape or complexity of boundaries. In particular, in [AGSX09] and [JAX11] the method has been applied to high dimensional functions that arise when solving stochastic PDEs, which reside in a high dimensional space which includes the original space and time domains as well as additional random dimensions.

11.5.2 Distributing the viscosity

We start by describing the specific PA operators which should be used to build a HOES in the context of a SBP-FD scheme. Let p be the order of accuracy of the SBP-FD method under consideration. For a diagonal norm matrix H, the scheme has interior order $2p$ and boundary order p, yielding a global order of $p+1$. For a block norm matrix H, on the other hand, the boundary order increases to $2p-1$ and the global order becomes $2p$. The edge detector should provide the same order of accuracy in order to avoid undesired smearing by the AV operator in smooth regions. Thus, we construct HOES by using PA operators of order $m = 2p$. In the interior, we utilize symmetric local stencils S_x and near boundaries we utilize nonsymmetric local stencils S_x to build the PA operators.

Let $\Omega = [x_l, x_r]$ and let us denote the equidistant grid points of the SBP-FD by

$$x_l = x_1 < x_2 < \cdots < x_N = x_r.$$

Then, for an interior grid point x_j with $p < j < N-p$, we choose the symmetric local stencil

$$S_{x_j} = \{x_{j-p}, \ldots, x_j, \ldots, x_{j+p}\} \tag{11.19}$$

around x_j, which contains x_j itself as well as the p grid points to the left and to the right hand side of x_i. For a boundary grid point x_j with $1 \leq j \leq p$ or $N-p \leq j \leq N$, we respectively choose the nonsymmetric local stencil

$$S_{x_j} = \{x_1, \ldots, x_{2p+1}\} \quad \text{or} \quad S_{x_j} = \{x_{N-2p-1}, \ldots, x_N\},$$

containing the $2p+1$ grid points nearest to the respective boundary. In both cases, the PA operator $L_{2p}[s](x_j)$ will provide order of accuracy $2p$ in smooth regions.

Remark 11.6. For an SBP-FD method with diagonal norm matrix H (as mainly considered in this work) it would be sufficient to use PA operators of order p near the boundaries to match the boundary order. Yet, the above choice allows us a direct extension to block norm matrices H and simplifies the construction of the PA operators, since all stencils have the same size.

Finally, we describe how the above constructed PA operators can be used to steer the viscosity distribution a in the AV operators (11.9) and their discretization

$$\frac{\mathrm{d}}{\mathrm{d}t}\mathbf{u}_i = \varepsilon DAD\mathbf{u}_i \tag{11.20}$$

using SBP operators.[1] Recall that $A = \operatorname{diag}(a_1, \ldots, a_N)$, where $a_j = a(x_j)$, is a diagonal matrix which contains the values of the viscosity distribution a at the grid points $\{x_j\}_{j=1}^N$. As

[1] For entropy stability, the discretization needs to be changed to (11.15) in order to mimic the chain rule.

we observed in Chapter 11.4, A needs to satisfy

$$a_1 = a_N = 0 \quad \text{and} \quad a_j \geq 0, \ j = 1, \ldots, N,$$

for conservation and energy stability to carry over to the discrete setting. Let p be the order of accuracy of the SBP-FD method under consideration. We start by defining

$$\tilde{a}_1 = \tilde{a}_N = 0, \quad \tilde{a}_j = |L_{2p}[s](x_j)|, \ j = 2, \ldots, N-1.$$

as well as

$$\tilde{a}_{\max} = \max_{1 \leq j \leq N} \tilde{a}_j. \tag{11.21}$$

Then, the discrete viscosity distribution is chosen as

$$a_j = \begin{cases} \tilde{a}_j / \tilde{a}_{\max} & \text{if } \tilde{a}_{\max} \neq 0, \\ 0 & \text{otherwise,} \end{cases} \tag{11.22}$$

for $j = 1, \ldots, N$. Assuming $\tilde{a}_{\max} \neq 0$, this yields a normalized viscosity distribution with $\|a\|_\infty = 1$. Normalizing the viscosity distribution gives us better control over the amount of viscosity introduced by the AV operator and, moreover, simplifies a suitable scaling of the viscosity strength, which will be addressed subsequently.

11.5.3 Scaling the viscosity strength

In Chapter 11.1, we have formulated four important properties every reasonable AV operator should satisfy. Chapters 11.2 and 11.4 have been devoted to property (P4), i. e. the preservation of conservation and dissipation properties of hyperbolic systems of CLs. Further, by steering the proposed AV operators by PA based HOES, we intended to ensure that spurious oscillations near (shock) discontinuities are efficiently reduced (P1), while the accuracy of the numerical method is preserved in smooth regions (P2). Therefore, we also need to guarantee a proper scaling of the viscosity strength. Moreover, it remains to ensure that the introduction of AV will not significantly increase the computational work of the underlying numerical method (P3). In particular, focus should be given to altered time step restrictions due to AV terms.

Here, we use $\tilde{a}_{\max}$ as defined in (11.21) to construct a *smoothness estimating detector* $\delta = \delta(s)$, which will be used to steer the viscosity strength ε. Consulting Theorem 11.3, the value $\tilde{a}_{\max}$ scales like

$$\tilde{a}_{\max} \sim \begin{cases} \|[s]\|_\infty & \text{if } s \text{ is discontinuous on } \Omega, \\ N^{-\min(2p,k)} & \text{if } s \in C^k(\Omega) \setminus C^{k+1}(\Omega) \text{ for } k > 0, \end{cases}$$

and, using the logarithm, we therefore have

$$\log \tilde{a}_{\max} \sim \begin{cases} \log \|[s]\|_\infty & \text{if } s \text{ is discontinuous on } \Omega, \\ -\min(2p,k) \log N & \text{if } s \in C^k(\Omega) \setminus C^{k+1}(\Omega) \text{ for } k > 0. \end{cases}$$

Let us now define the smoothness estimating detector δ as

$$\delta := -\frac{\log \tilde{a}_{\max}}{\log N}.$$

Then, δ scales like

$$\delta \sim \begin{cases} -\log \|[s]\|_{\infty} (\log N)^{-1} & \text{if } s \text{ is discontinuous on } \Omega, \\ \min(2p, k) & \text{if } s \in C^k(\Omega) \setminus C^{k+1}(\Omega) \text{ for } k > 0. \end{cases}$$

and allows us a clear interpretation. In particular, assuming $p \geq 1$ and a sufficiently large $N \in \mathbb{N}$, we expect

$$\delta \approx \begin{cases} 0 & \text{if } s \notin C^0(\Omega), \\ 1 & \text{if } s \in C^1(\Omega) \setminus C^2(\Omega), \\ 2 & \text{if } s \in C^2(\Omega) \setminus C^3(\Omega). \end{cases}$$

Even though the convergence $(\log N)^{-1} \to 0$ is fairly slow, this will not cause problems for our procedure. For the interpretation of δ, we expect $s \approx 0$ to indicate a discontinuous solution, $s \approx 1$ to indicate a C^1 (or a C^0) solution, and $s \approx 2$ to indicate a C^2 solution. A shortcoming of the above approach might be that it is not clear how to distinguish C^1 functions from functions which are only continuous. Future investigations will address the possibility to replace the above smoothness estimating detector by a data driven detector as presented in [RH18, DHR19].

Next, we address how the smoothness estimating detector δ is used to steer the viscosity strength $\varepsilon = \varepsilon(\delta)$. To overcome spurious (Gibbs–Wilbraham) oscillations in our numerical solution, AV should be fully activated in the case of a discontinuous solution ($\delta = 0$). Hence, we choose the viscosity strength $\varepsilon(0) = \varepsilon_{\max}$, where $\varepsilon_{\max}$ refers to a maximum viscosity strength which will be discussed below. Since merely continuous functions can still pose problems for a high order approximation, we aim for numerical solutions to be at least C^1. Thus, AV should start to be activated for $\delta = 1$. Here, we use a simple ramp function connecting $\varepsilon(0) = \varepsilon_{\max}$ and $\varepsilon(1) = 0$ as an activation map:

$$\varepsilon = \begin{cases} 0 & \text{if } \delta \geq 1, \\ \varepsilon_{\max}(1-\delta) & \text{if } 0 < \delta < 1, \\ \varepsilon_{\max} & \text{if } \delta \leq 0. \end{cases} \tag{11.23}$$

Finally, we address the maximum viscosity strength $\varepsilon_{\max}$. When numerically solving hyperbolic CLs with an FD method using a grid of N equidistant points, the explicit time step usually scales like

$$\Delta t \sim \frac{1}{\lambda_{\max} N},$$

where $\lambda_{\max}$ is the largest characteristic velocity. When AV terms of second order are involved, however, this scaling becomes

$$\Delta t \sim \frac{1}{\lambda_{\max} N + \varepsilon \|a\|_{L^\infty(\Omega)} N^2}.$$

Here, $\|a\|_{L^\infty(\Omega)}$ is the largest magnitude of the viscosity distribution a at time t. Let a be a normalized viscosity distribution, i. e. $\|a\|_{L^\infty(\Omega)} = 1$, and let us assume that the smoothness estimating detector δ indicates a discontinuity. In this case, the maximum viscosity strength is chosen and we have the scaling

$$\Delta t \sim \frac{1}{\lambda_{\max} N + \varepsilon_{\max} N^2}. \tag{11.24}$$

Obviously, the additional time step restriction $\Delta t \sim (\varepsilon_{\max} N^2)^{-1}$ from numerically solving a diffusion equation can be fairly damaging. Fortunately, this problem can be tackled by a clever scaling of the maximum viscosity strength $\varepsilon_{\max} \geq 0$, which has been a free parameter so far. Note that if we choose $\varepsilon_{\max} = cN^{-1}$ with a parameter $c > 0$, the scaling (11.24) reduces to

$$\Delta t \sim \frac{1}{(\lambda_{\max} + c)\, N}$$

and the time step again scales linearly with respect to the spatial resolution.

An appropriate value for the parameter $c > 0$ can be found by investigating the behavior of the diffusion term. In the subsequent analysis, we follow the work of Klöckner, Warburton, and Hesthaven [KWH11]. Let us consider the fundamental solution of the heat equation

$$\partial_t u = \varepsilon_{\max} \Delta u,$$

also known as the heat kernel [Eva10]. If we adopt the standard deviation σ as a measure of width, the heat kernel will have a width of

$$\sigma = \sqrt{2\varepsilon_{\max} \Delta t}$$

after one time step of size Δt. Hence, if we use a time step of the original size $\Delta t \approx (\lambda_{\max} N)^{-1}$ and set $\varepsilon_{\max} = cN^{-1}$, the heat kernel will have a width of

$$\sigma = \sqrt{\frac{2c}{\lambda_{\max} N^2}}.$$

Thus, we can observe that when we let $c = \frac{1}{2}\lambda_{\max} k^2$ with $k \in \mathbb{N}$, we have

$$\sigma = \frac{k}{N},$$

i. e. discontinuities are spread over k grid points at most. In our numerical tests, we have observed $k = 1$ and $k = 2$ to be too ambitious. Especially for shock discontinuities in the Euler equations, we still observed spurious oscillations in some cases. A choice of $k = 3$ has proven to be more reliable. The same observation has been made in [KWH11] for a piecewise linear AV in DG methods (see Chapter 8.2). To sum our observations up, we recommend to use

$$\varepsilon_{\max} = \frac{\lambda_{\max} k^2}{2N} \tag{11.25}$$

with $k = 3$ for the maximum viscosity strength and an adaptive time step

$$\Delta t = \frac{C}{\lambda_{\max} N + \varepsilon N^2} \tag{11.26}$$

with $C = 0.1$ and ε given by (11.23). In particular, this time step adapts to the amount of viscosity. Further, it provides the same asymptotic behavior as the original time step for the hyperbolic system of CLs without AV.

Remark 11.7. Following the discussion in [KWH11], we note that for systems of CLs, it remains to decide which characteristic velocity $\lambda_{\max}$ should be used. This choice can have far-reaching implications. For instance for the Euler equations, contact discontinuities propagate with stream velocity while shock discontinuities propagate at sonic speed. In the case of a one dimensional system, one can argue that the best course of action is to apply AV to the characteristic variables, so that each wave receives a suitable amount of dissipation; see [Rie10]. An extension of the analysis provided in chapters 11.2 and 11.4 to this setting would be of interest. Yet, doing so might work well in one spatial dimension but it is less clear how this strategy should be applied in a genuinely multidimensional system.

Here, we follow a simple strategy and choose λ_{max} the maximum characteristic velocity of the system, e. g. as

$$\lambda_{\text{max}} = \max_{x \in \Omega} a_s(x) + |u(x)|$$

for the Euler equations, where a_s is the speed of sound and u is the velocity. It should be noted, however, that this simple strategy comes with some disadvantages. For instance for the Euler equations, contact discontinuities (only propagating with stream velocity) are smeared heavier than necessary for the above choice. This is demonstrated by the subsequent numerical tests; see Sod's shock tube problem in Chapter 11.6.3. At the same time, the above choice — even though smearing contact discontinuities heavier than necessary — ensures a stable computation in all tests considered later. Future work might focus on further refinements in the scaling of the viscosity strength ε or the viscosity distribution a, for instance, by 'counting' ingoing and outgoing characteristics.

11.5.4 Choosing a sensing variable

Our analysis in Chapter 11.2 indicated that the same viscosity distribution a should be used in all components of a hyperbolic system of CLs. For a separated viscosity a_i for every component, unfortunately, we have not been able to prove entropy dissipation, since the matrix product $A\boldsymbol{u}'$ would have to be positive semidefinite, which is fairly nontrivial for general entropies. Thus, a sensing variable s should be chosen that reflects (shock) discontinuities in all components at the same time. For a scalar CL $\partial_t u + \partial_x f(u) = 0$, we can simply choose $s = u$, i. e. the solution itself. For systems of CLs, the choice for the sensing variable becomes less clear. Here, a characteristic quantity of the system should be used [PP06]. In case of the Euler equations of gas dynamics, we follow the works [KWH11, GNJA+19] and choose the fluid density ρ as the sensing variable, i. e. $s = \rho$.

11.6 Numerical results

Let us now numerically investigate the proposed HOES steered AV method proposed in Chapter 11.5. For the SBP-FD method described in Chapter 11.3, the HOES steered AV method is compared to the SBP-FD method without AV and to the SBP-FD method with a usual viscosity distribution $A = \text{diag}(0, 1, \dots, 1, 0)$. For both AV methods, the usual AV method with A as above and the HOES steered AV method with A given by (11.22), the viscosity strength and the (local) time step are respectively given by (11.25) and (11.26).

11.6.1 Linear advection equation

Let us start by considering the scalar linear advection equation

$$\partial_t u + \partial_x u = 0$$

on $\Omega = [0,1]$ with periodic BCs and discontinuous IC

$$u_0 : [0,1] \to \mathbb{R}, \quad u_0(x) := \begin{cases} e^{-300(2x-0.3)^2} & \text{if } |2x-0.3| \leq 0.25, \\ 1 & \text{if } |2x-0.9| \leq 0.2, \\ \left(1 - \left(\frac{2x-1.6}{0.2}\right)^2\right)^{\frac{1}{2}} & \text{if } |2x-1.6| \leq 0.2, \\ 0 & \text{otherwise.} \end{cases}$$

In this case the exact solution u can be determined by the method of characteristics. Periodically extending the IC u_0, the solution is given by $u(x,t) = u_0(x-t)$. The linear advection equation is the simplest PDE that can feature discontinuous solutions. Thus, the (shock capturing) method can be observed in a well-understood setting, isolated from nonlinear effects. Yet, the linear advection equation provides a fairly challenging example. Similar to contact discontinuities in Euler's equations, discontinuities are not self-steepening, i. e. once such a discontinuity is smeared by the method, it can not be recovered to its original sharp shape [KLT06].

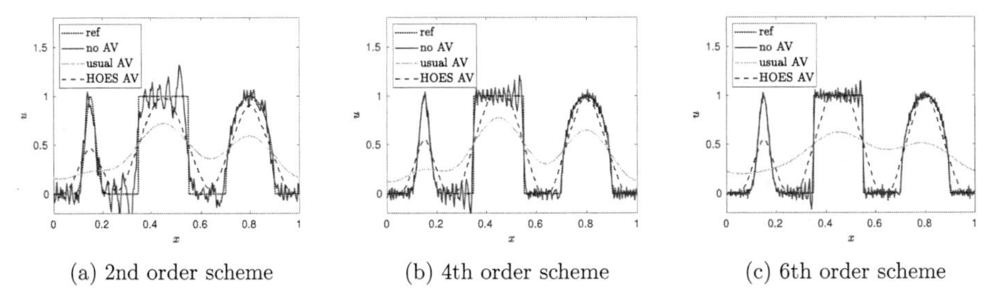

(a) 2nd order scheme (b) 4th order scheme (c) 6th order scheme

Figure 11.2: (Numerical) solutions of the linear advection equation with discontinuous IC at $t = 1$ for $N = 400$ grid points and $k = 3$.

Figure 11.2 illustrates the results of the usual and HOES steered AV method for compared to the results of the SBP-FD method without AV at time $t = 1$ and for different orders. Here, $N = 400$ grid points and $k = 3$ in (11.25) were used. While the numerical solution without AV shows spurious oscillations in all cases, both AV methods are demonstrated to smooth out these oscillations. Yet, the usual AV method is observed to smear the solution considerably stronger than the HOES steered AV method.

Accuracy of the two AV methods as well as of the SBD-FD method without AV is investigated in Table 11.1 for different orders and numbers of grid points. Table 11.1 also lists the resulting EOCs, which have been computed by performing a least squares fit for the parameters C and s in the model $y = C \cdot N^{-s}$, where y denotes the discrete $\|\cdot\|_H$-error for a fixed N. We note that the introduction of AV reduces the accuracy in nearly all cases. The usual AV method reduces accuracy a good deal more than the HOES steered AV method, however. Further, the HOES steered AV method is able to recover the EOCs of the SBP-FD method without AV.

$\|\cdot\|_H$ errors

		N				EOC
Order	Method	100	200	300	400	
2	no AV	2.8E-1	1.8E-1	1.5E-1	1.3E-1	0.5
	usual AV	4.2E-1	4.0E-1	3.6E-1	3.4E-1	0.1
	HOES AV	3.7E-1	2.7E-1	2.3E-1	2.1E-1	0.4
4	no AV	1.4E-1	9.7E-2	8.3E-2	7.5E-2	0.4
	usual AV	4.3E-1	3.7E-1	3.7E-1	3.1E-1	0.1
	HOES AV	3.1E-1	2.2E-1	2.0E-1	1.8E-1	0.4
6	no AV	9.5E-2	7.1E-2	6.4E-2	5.6E-2	0.3
	usual AV	4.3E-1	4.1E-1	4.0E-1	3.8E-1	0.1
	HOES AV	2.8E-1	2.1E-1	1.9E-1	1.8E-1	0.3

Table 11.1: Errors for the linear advection equation with discontinuous IC at time $t = 1$ and for $k = 3$.

11.6.2 Burgers' equation

Next, we investigate the proposed AV method for the nonlinear Burgers' equation

$$\partial_t u + \partial_x \left(\frac{u^2}{2} \right) = 0$$

on $\Omega = [0, 1]$ with periodical BCs and smooth IC $u_0(x) = 1 + \sin(2\pi x)/\pi$. For this problem a shock discontinuity develops in the solution at time

$$t_s = -\frac{1}{\min_{0 \leq x \leq 1} u_0'(x)} = \frac{1}{2}.$$

The exact solution can be computed using the method of characteristics by solving the implicit equation $u(x, t) = u_0(x - tu)$ in smooth regions, where u_0 is periodically extended. The jump location, separating these regions for $t > 0.5$, can be determined by the Rankine–Hugoniot condition. Here, we consider the solution at time $t = 0.4$ to investigate the influence of the AV method in smooth regions and at time $t = 1$ to investigate its effect on shock discontinuities.

Figure 11.3 demonstrates that the AV methods do not introduce any undesired smearing in smooth regions and therefore do not reduce accuracy of the high order scheme in such regions. This is due to the scaling of the viscosity strength, which has been proposed in Chapter 11.5.3 and builds up on the usage of PA based HOES. Table 11.2 reports the same observation for a vast number of different orders and numbers of grid points N. There, only for the second order method and for low resolutions ($N = 100, 200$) the solution is slightly smeared by the AV methods. The results at time $t = 1$ are reported in Table 11.2 as well. At this time, a shock discontinuity has fully developed in the solution and has already traveled through the whole domain. As a consequence, the accuracy of all three methods decreases considerably. For the SBP-FD method without AV some of the computations for sixth order even broke down before the final time could be reached. The two AV methods, on the other hand, are

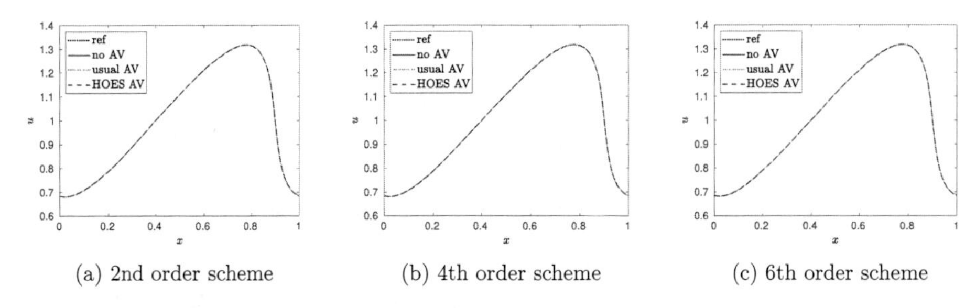

(a) 2nd order scheme (b) 4th order scheme (c) 6th order scheme

Figure 11.3: (Numerical) solutions of the Burgers' equation with a smooth IC at $t = 0.4$ for $N = 400$ grid points and $k = 3$.

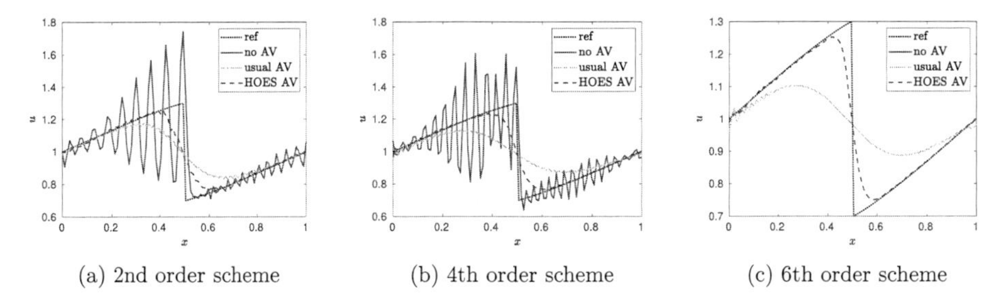

(a) 2nd order scheme (b) 4th order scheme (c) 6th order scheme

Figure 11.4: (Numerical) solutions of the Burgers' equation with a smooth IC at $t = 1$ for $N = 100$ grid points and $k = 3$.

still able to provide accurate numerical solutions, even though the EOCs decrease for them as well. Even though the EOC seems to be slightly lower, the HOES steered AV method provides more accurate solutions than the usual AV method in nearly all cases. This is also illustrated in Figure 11.4 for $N = 100$. In particular, we observe that the HOES steered AV method yields sharper shock profiles and introduces less undesired smearing away from the discontinuity.

11.6.3 Euler equations of gas dynamics

Finally, we address the extension to systems of nonlinear hyperbolic CLs. Henceforth, let us consider the one-dimensional Euler equations

$$\boldsymbol{u}_t + \boldsymbol{f}(\boldsymbol{u})_x = 0, \tag{11.27}$$

where $\boldsymbol{u}$ and $\boldsymbol{f}(\boldsymbol{u})$ respectively are the vector of conserved variables and fluxes, given by

$$\boldsymbol{u} = \begin{pmatrix} u_1 \\ u_2 \\ u_3 \end{pmatrix} = \begin{pmatrix} \rho \\ \rho u \\ E \end{pmatrix}, \quad \boldsymbol{f} = \begin{pmatrix} f_1 \\ f_2 \\ f_3 \end{pmatrix} = \begin{pmatrix} \rho u \\ \rho u^2 + p \\ u(E+p) \end{pmatrix}.$$

Here, ρ is the density, u is the velocity, p is the pressure, and E is the total energy per unit volume. Note that we have already introduced the Euler equations in Chapter 2. But since this

$\|\cdot\|_H$ errors

		$t=0.4$					$t=1$				
		N				EOC	N				EOC
Order	Method	100	200	300	400		100	200	300	400	
2	no AV	3.8E-3	1.1E-3	4.9E-4	2.8E-4	1.8	1.6E-1	1.7E-1	1.6E-1	1.7E-1	0.0
	usual AV	7.7E-3	1.5E-3	4.9E-4	2.8E-4	1.8	9.9E-2	7.2E-2	5.7E-2	4.8E-2	0.5
	HOES AV	8.9E-3	1.7E-3	4.9E-4	2.8E-4	1.8	6.9E-2	4.9E-2	4.1E-2	3.6E-2	0.5
4	no AV	5.7E-4	5.9E-5	1.3E-5	4.2E-6	2.0	1.4E-1	1.4E-1	1.4E-1	1.4E-1	0.0
	usual AV	5.7E-4	5.9E-5	1.3E-5	4.2E-6	2.0	1.1E-2	8.7E-2	7.0E-2	5.9E-2	0.5
	HOES AV	5.7E-4	5.9E-5	1.3E-5	4.2E-6	2.0	6.4E-2	5.0E-2	4.3E-2	3.7E-2	0.4
6	no AV	2.5E-4	1.3E-5	1.7E-6	3.5E-7	2.7	NaN	NaN	NaN	NaN	-
	usual AV	2.5E-4	1.3E-5	1.7E-6	3.5E-7	2.7	1.2E-1	9.6E-2	7.9E-2	6.7E-2	0.4
	HOES AV	2.5E-4	1.3E-5	1.7E-6	3.5E-7	2.7	5.5E-2	4.3E-2	3.9E-2	3.6E-2	0.3

Table 11.2: Errors for the Burgers equation with smooth IC at times $t = 0.4$ and $t = 1$ for $k = 3$.

description was back a vast number of pages we briefly recap their most important properties. The Euler equations are completed by addition of an EOS with general form

$$p = p(\rho, e),$$

where $e = E/\rho - u^2/2$ is the specific internal energy. Here, we consider ideal gases for which the EOS is given by

$$p = (\gamma - 1)\rho e$$

with γ denoting the ratio of specific heats. A pair of an entropy function and flux is given by $\eta = -\rho s$ and $\psi = -\rho s \boldsymbol{u}$ for the Euler equations, where s is the specific (physical) entropy $s = \ln(p\rho^{-\gamma}) = \ln(p) - \gamma \ln(\rho)$. There are many possible choices for the (mathematical) entropy function, e. g. $\eta = \rho g(s)$ for any strictly convex function g; see [Har83]. However, the above choice is the only one which is consistent with the entropy condition from thermodynamics [HFM86] in the presence of heat transfer.

Smooth isentropic flow

For the subsequent numerical tests, we set $\gamma = 3$ and consider a smooth isentropic flow resulting from the Euler equations on $\Omega = [-1, 1]$ with smooth IC

$$\rho_0(x) = 1 + \frac{1}{2}\sin(\pi x), \quad u_0(x) = 0, \quad p_0(x) = \rho_0^\gamma(x),$$

and periodic BCs. A similar test problem has been proposed by Cheng and Shu [CS14] as well as by Abgrall et al. [ABT19] in the context of (positivity-preserving) high order methods. Utilizing the method of characteristics, the exact density ρ and velocity u are given by

$$\rho(x,t) = \frac{1}{2}\left[\rho_0(x_1) + \rho_0(x_2)\right], \quad u(x,t) = \sqrt{3}\left[\rho(x,t) - \rho_0(x_1)\right],$$

where $x_1 = x_1(x,t)$ and $x_2 = x_2(x,t)$ are solutions of the nonlinear equations

$$x + \sqrt{3}\rho_0(x_1)t - x_1 = 0, \quad x - \sqrt{3}\rho_0(x_2)t - x_2 = 0.$$

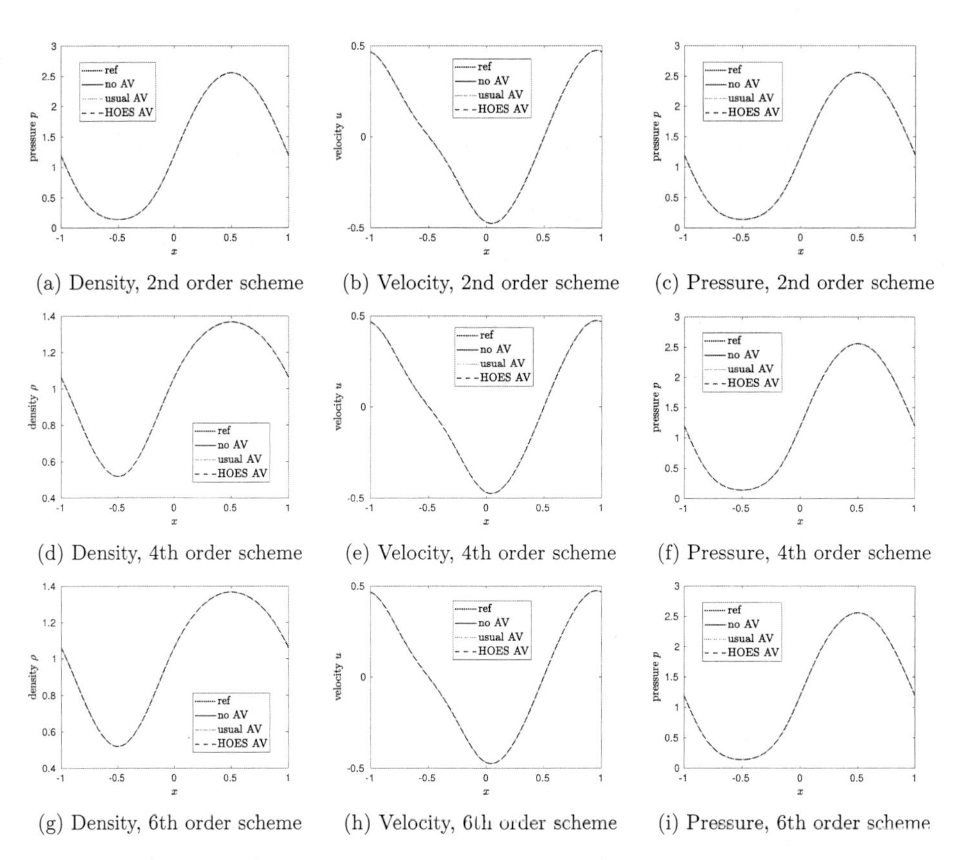

(a) Density, 2nd order scheme (b) Velocity, 2nd order scheme (c) Pressure, 2nd order scheme

(d) Density, 4th order scheme (e) Velocity, 4th order scheme (f) Pressure, 4th order scheme

(g) Density, 6th order scheme (h) Velocity, 6th order scheme (i) Pressure, 6th order scheme

Figure 11.5: (Numerical) solutions of a smooth isentropic flow for the Euler equations with a smooth IC at $t = 0.1$ and for $N = 400$ grid points.

Finally, the exact pressure p can be computed by the isentropic law $p = C\rho^\gamma$ for smooth flows; see [Tor13, Chapter 3.1].

The results at time $t = 0.1$ of the SBP-FD method without AV as well as with usual AV and HOES steered AV can be found in Figure 11.5 for $N = 400$. We observe that the AV methods do not smear the solution and therefore preserve the high accuracy of the underlying method in smooth regions. A more extensive investigation of the accuracy of the AV methods can be found in Table 11.3 (see the end of this chapter). As listed there, in all our numerical tests, we found the AV methods to not be activated for the smooth isentropic flow above.

Sod's shock tube

Next, we consider Sod's shock tube problem [Sod78] for the Euler equations on $\Omega = [-1, 1]$ with discontinuous ICs

$$\begin{cases} \rho_0(x) = 1, & u_0(x) = 1, \quad p_0(x) = 1, \quad \text{if } x < 0, \\ \rho_0(x) = 0.125, & u_0(x) = 0, \quad p_0(x) = 0.1, \quad \text{if } x > 0, \end{cases} \tag{11.28}$$

and fixed BCs $\rho(\pm 1, t) = \rho_0(\pm 1)$, $u(\pm 1, t) = u_0(\pm 1)$, and $p(\pm 1, t) = p_0(\pm 1)$. Here, $\gamma = 1.4$ is taken for the ratio of specific heat. The reference solution can be computed as described in

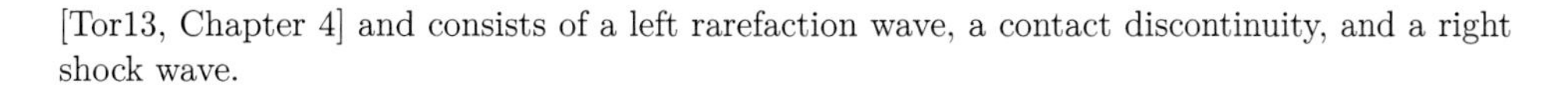
[Tor13, Chapter 4] and consists of a left rarefaction wave, a contact discontinuity, and a right shock wave.

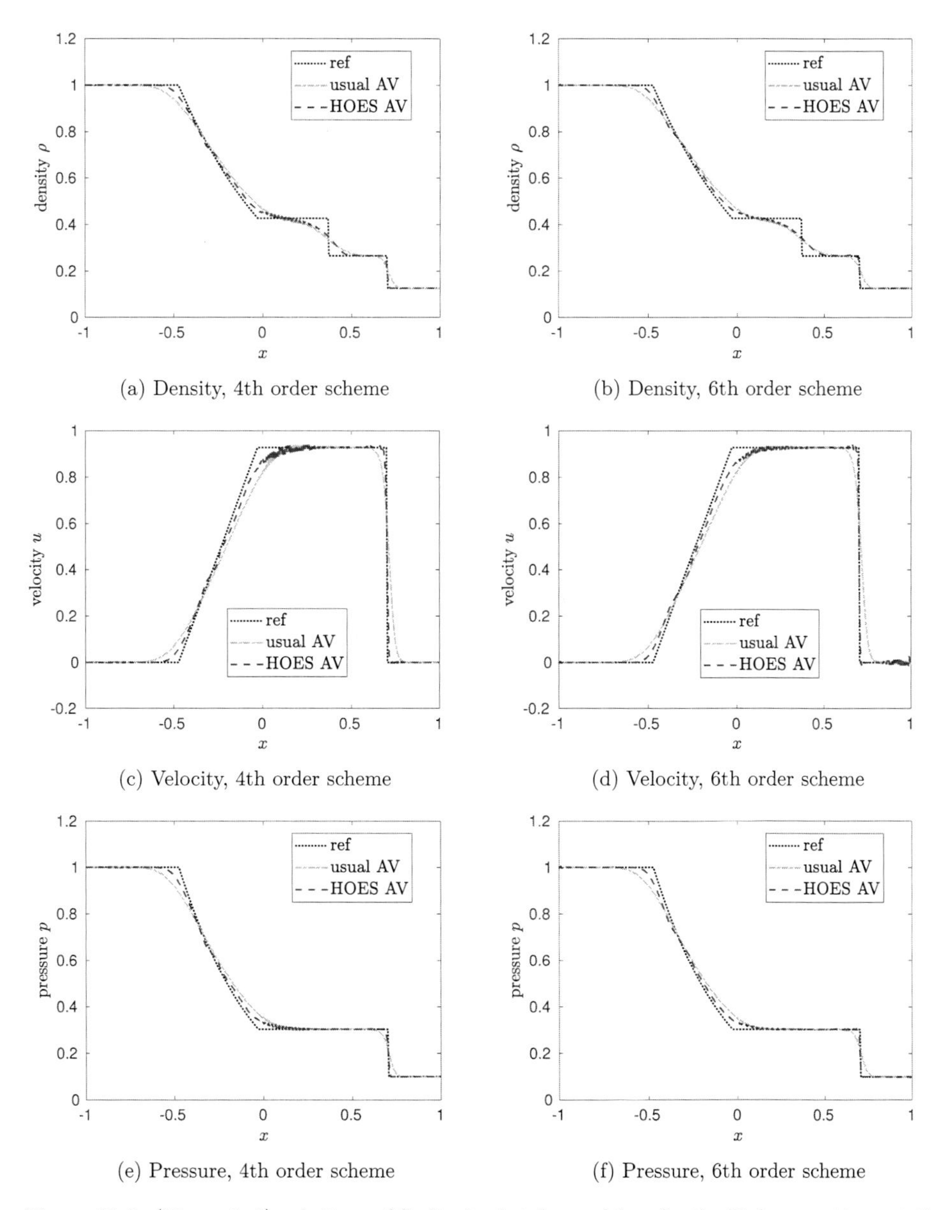

(a) Density, 4th order scheme

(b) Density, 6th order scheme

(c) Velocity, 4th order scheme

(d) Velocity, 6th order scheme

(e) Pressure, 4th order scheme

(f) Pressure, 6th order scheme

Figure 11.6: (Numerical) solutions of Sod's shock tube problem for the Euler equations at time $t = 0.4$ for $N = 400$ grid points.

Figure 11.6 illustrates the solutions at time $t = 0.4$ for $N = 400$ grid points. For this problem, the SBP-FD method without AV broke down before the final time could be reached and we only report the results for the usual AV method and the HOES steered AV method.

Moreover, for a better visual presentation, we only present the results for the fourth and sixth order SBP-FD schemes in Figure 11.6. Both AV methods are able to provide numerical solutions with only slight oscillations. The remaining minor oscillations, which can be observed in the velocity profiles, are post-shock oscillations and could easily be removed by post-processing algorithms. Further, both AV methods provide clear profiles for the rarefaction wave and the shock discontinuity, where the profiles for the HOES steered AV method are significantly sharper. Yet, both methods are observed to smear the contact discontinuity heavier than desired. This is in accordance with Remark 11.7 and could be prevented by a more clever choice of the characteristic velocity $\lambda_{\max}$ in the maximum viscosity strength (11.25). Future works will address other approaches to overcome this problem. The errors as well as EOCs for both AV methods are listed in Table 11.4 (see the end of this chapter) and show that the HOES steered AV method provides more accurate results than the usual AV method in all cases.

Shu and Osher's shock tube

The test case of Sod's shock tube (11.28) is reasonable when it is intended to demonstrate how a method handles different types of discontinuities. Yet, no small-scale features are present. A more challenging test case to observe if a method is able to capture both, discontinuities as well as small-scale smooth flows, is Shu and Osher's shock tube problem [SO89] for the Euler equations on $\Omega = [0, 4]$ with ICs

$$\begin{cases} \rho_0(x) = 3.857143, & u_0(x) = 2.629369, \quad p_0(x) = 10.333333, \quad \text{if } x \leq 2.5, \\ \rho_0(x) = 1 + 0.2 \sin(10\pi x), & u_0(x) = 0, \quad p_0(x) = 1, \quad \text{if } x > 2.5. \end{cases} \tag{11.29}$$

Again, $\gamma = 1.4$ is taken for the ratio of specific heat. For this problem the reference solution has been computed using an SSP-RK2 TVD-MUSCL scheme with 10 000 equidistant points.

Figure 11.7 illustrates the results for the usual and HOES steered AV methods at time $t = 0.2$ and $N = 400$ grid points. For a better visual presentation, we only present the results for the fourth and sixth order SBP-FD schemes again. From all profiles it is visible that both AV methods provide a sharp profile of the shock discontinuity. Yet, we further observe the HOES steered AV method to also resolve the small-scale features of the solution considerably better. Finally, we mention some minor post-shock oscillation near the left boundary. We have chosen the computational domain such that these post-shock oscillations do not interact with the physical BCs so that we can neglect resulting errors in our error analysis provided by Table 11.5 (see the end of this chapter). Again, these oscillations could be removed by post-processing algorithms. This table demonstrates that the HOES steered AV method provides more accurate solutions in all considered tests for Shu and Osher's shock tube problem.

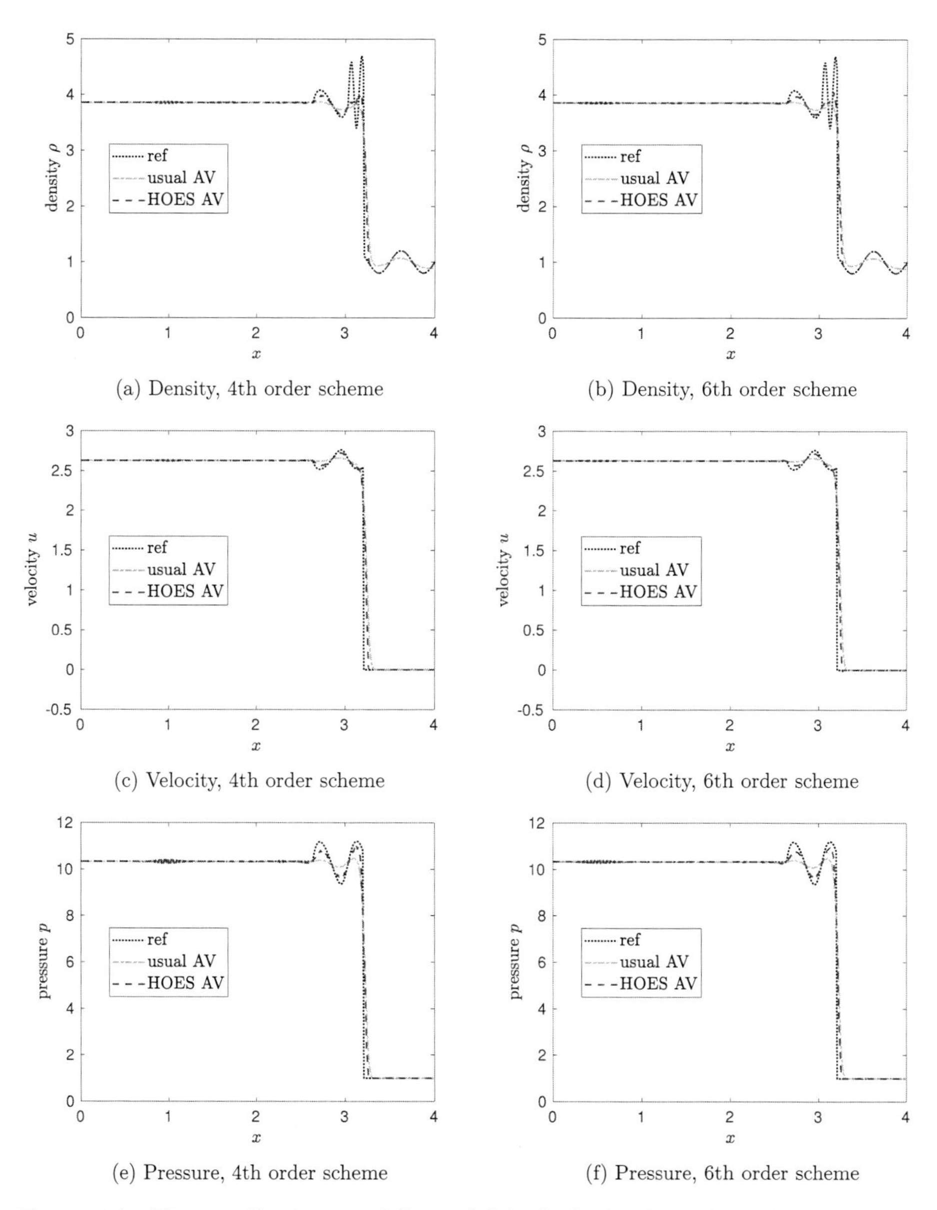

(a) Density, 4th order scheme

(b) Density, 6th order scheme

(c) Velocity, 4th order scheme

(d) Velocity, 6th order scheme

(e) Pressure, 4th order scheme

(f) Pressure, 6th order scheme

Figure 11.7: (Numerical) solutions of Shu and Osher's shock tube problem for the Euler equations at time $t = 0.2$ for $N = 400$ grid points.

		$\|\cdot\|_H$ errors														
		density					velocity					pressure				
		N				EOC	N				EOC	N				EOC
Order	Method	100	200	300	400		100	200	300	400		100	200	300	400	
2	no AV	4.2E-4	1.0E-4	4.6E-5	2.6E-5	1.7	5.7E-4	1.4E-4	6.4E-5	3.6E-5	1.7	1.6E-3	4.0E-4	1.8E-4	9.9E-5	1.7
	usual AV	4.2E-4	1.0E-4	4.6E-5	2.6E-5	1.7	5.7E-4	1.4E-4	6.4E-5	3.6E-5	1.7	1.6E-3	4.0E-4	1.8E-4	9.9E-5	1.7
	HOES AV	4.2E-4	1.0E-4	4.6E-5	2.6E-5	1.7	5.7E-4	1.4E-4	6.4E-5	3.6E-5	1.7	1.6E-3	4.0E-4	1.8E-4	9.9E-5	1.7
4	no AV	4.1E-5	6.4E-6	1.8E-6	7.1E-7	2.5	5.8E-5	7.1E-6	2.1E-6	8.9E-7	2.6	1.1E-4	1.3E-5	3.8E-6	1.6E-6	2.6
	usual AV	4.1E-5	6.4E-6	1.8E-6	7.1E-7	2.5	5.8E-5	7.1E-6	2.1E-6	8.9E-7	2.6	1.1E-4	1.3E-5	3.8E-6	1.6E-6	2.6
	HOES AV	4.1E-5	6.4E-6	1.8E-6	7.1E-7	2.5	5.8E-5	7.1E-6	2.1E-6	8.9E-7	2.6	1.1E-4	1.3E-5	3.8E-6	1.6E-6	2.6
6	no AV	4.2E-5	3.4E-6	7.3E-7	2.5E-7	2.6	3.2E-5	2.0E-6	4.1E-7	1.3E-7	2.6	6.1E-5	3.8E-6	7.4E-7	2.3E-7	2.8
	usual AV	4.2E-5	3.4E-6	7.3E-7	2.5E-7	2.6	3.2E-5	2.0E-6	4.1E-7	1.3E-7	2.6	6.1E-5	3.8E-6	7.4E-7	2.3E-7	2.8
	HOES AV	4.2E-5	3.4E-6	7.3E-7	2.5E-7	2.6	3.2E-5	2.0E-6	4.1E-7	1.3E-7	2.6	6.1E-5	3.8E-6	7.4E-7	2.3E-7	2.8

Table 11.3: Errors for a smooth isentropic flow for the Euler equations at time $t = 0.1$.

		$\|\cdot\|_H$ errors														
		density					velocity					pressure				
		N				EOC	N				EOC	N				EOC
Order	Method	100	200	300	400		100	200	300	400		100	200	300	400	
2	usual AV	7.0E-2	5.7E-2	4.9E-2	4.4E-2	0.3	2.0E-1	1.5E-1	1.3E-1	1.1E-1	0.4	8.7E-2	6.4E-2	5.3E-2	4.6E-2	0.5
	HOES AV	5.3E-2	3.7E-2	3.0E-2	2.5E-2	0.5	1.1E-1	7.5E-2	6.0E-2	5.1E-2	0.6	5.9E-2	3.8E-2	2.9E-2	2.4E-2	0.7
4	usual AV	6.8E-2	5.5E-2	4.7E-2	4.2E-2	0.3	1.9E-1	1.4E-1	1.2E-1	1.0E-1	0.4	8.4E-2	6.1E-2	5.1E-2	4.4E-2	0.5
	HOES AV	5.3E-2	4.1E-2	3.4E-2	2.9E-2	0.4	1.2E-1	6.9E-2	6.0E-2	5.0E-2	0.6	5.7E-2	3.6E-2	2.8E-2	2.2E-2	0.7
6	usual AV	7.0E-2	5.5E-2	4.7E-2	4.2E-2	0.4	2.0E-1	1.5E-1	1.2E-1	1.1E-1	0.5	8.6E-2	6.1E-2	5.0E-2	4.3E-2	0.5
	HOES AV	5.4E-2	4.2E-2	3.5E-2	3.0E-2	0.4	1.3E-1	6.6E-2	5.1E-2	4.4E-2	0.8	6.0E-2	3.8E-2	3.0E-2	2.4E-2	0.6

Table 11.4: Errors for Sod's shock tube problem for the Euler equations at time $t = 0.1$.

		$\|\cdot\|_H$ errors														
		density					velocity					pressure				
		N				EOC	N				EOC	N				EOC
2	usual AV	7.5E-1	6.0E-1	4.9E-1	4.5E-1	0.4	6.3E-1	5.2E-1	4.0E-1	3.8E-1	0.4	1.9E-0	1.5E-0	1.2E-0	1.1E-0	0.4
	HOES AV	5.4E-1	4.7E-1	3.8E-1	3.6E-1	0.3	4.4E-1	4.0E-1	2.9E-1	2.9E-1	0.3	1.4E-0	1.1E-0	8.9E-1	8.5E-1	0.4
4	usual AV	7.4E-1	5.9E-1	4.8E-1	4.5E-1	0.4	6.2E-1	5.1E-1	3.9E-1	3.8E-1	0.4	1.8E-0	1.5E-0	1.2E-0	1.1E-0	0.4
	HOES AV	6.2E-1	4.9E-1	4.0E-1	3.9E-1	0.4	4.9E-1	4.2E-1	3.1E-1	3.1E-1	0.4	1.6E-0	1.2E-0	9.4E-0	9.0E-1	0.5
6	usual AV	7.5E-1	5.9E-1	4.8E-1	4.5E-1	0.4	6.4E-1	5.2E-1	3.9E-1	3.7E-1	0.4	1.9E-0	1.5E-0	1.2E-0	1.1E-1	0.4
	HOES AV	6.0E-1	5.1E-1	4.1E-1	3.9E-1	0.3	4.7E-1	4.1E-1	3.1E-1	3.1E-1	0.3	1.5E-0	1.2E-0	9.7E-1	9.2E-1	0.4

Table 11.5: Errors for Shu and Osher's shock tube problem for the Euler equations at time $t = 0.2$.

11.7 Concluding thoughts and outlook

In this final chapter, we have proposed and investigated novel AV operators for FD methods which are steered by PA based HOES. These AV operators adapt themselves to the smoothness of the numerical solution and avoid undesired smearing in smooth regions, while efficiently reducing spurious oscillations near (shock) discontinuities. Unlike usual AV operators, which distribute viscosity almost equally over the whole computational domain, the HOES steered AV operators calibrate their viscosity distribution to the location of the discontinuities — if these are present in the solution — and therefore add dissipation only where it is needed. Moreover, we have proved that the proposed AV method preserves conservation and (entropy) stability properties of the underlying hyperbolic system of CLs. This analysis was extended to the discrete setting by using SBP operators. Yet, in principle, the proposed AV operators could be incorporated into any FD scheme. Future work will address the extension to multi-block SBP-FD methods and to SE methods as well as more refined scaling strategies for the viscosity strength, e. g. by data driven sensors and by distinguishing between shock and contact discontinuities.

CHAPTER 12

SUMMARY AND OUTLOOK

In this thesis, we have addressed different topics surrounding the numerical treatment of hyperbolic CLs. In particular, we have presented novel contributions to the numerical integration of experimental data, the construction of stable high order numerical methods for CLs, and the investigation and design of different shock capturing techniques.

Regarding numerical integration of experimental data, we first and foremost investigated stability of LS-QRs for positive weight functions, where the quadrature points were allowed to be equidistant or even scattered. Such (prescribed) sets of quadrature points often pose serious problems for usual (high order) QRs. In Chapter 4, we have been able to prove that LS-QRs are able to recover any desired degree of exactness d while being (perfectly) stable when a sufficiently large number of quadrature points N is used. For equidistant points, we found the ratio of $d \approx N^2$ to hold in all cases considered. We then extended our study of LS-QRs in Chapter 5 to the case of general (possibly nonpositive) weight functions. General weight function arise, for instance, in the numerical treatment of the Schrödinger equation and for highly oscillatory integrals. We introduced and discussed different stability concepts of QRs which arise for the case of general weight functions. Essentially, we proposed to differentiate between stable and sign-consistent QRs. Next, we developed high order representatives for both classes of QRs. For the construction of stable high order QRs, we once more utilized an LS formulation of the exactness conditions. Sign-consistent QRs, on the other hand, were constructed by adapting an existing NNLS formulation originally proposed by Huybrechs. For experimental data — including scattered quadrature points as well as measurement errors — often composite QRs with fairly low orders of degree are used, e. g. simple Monte–Carlo methods. Yet, our findings show that even for such data it is possible to construct highly accurate QRs. Future works will try to further push the boundaries in this direction: Currently we are working on an extension to cubature rules for experimental data in higher dimensions. In this context, also a comparison to commonly used Monte–Carlo would be of interest. It should be noted that in certain cases the proposed LS rules can be interpreted as high order corrections to the Monte–Carlo method and could therefore increase their accuracy. Finally, we intend to incorporate data consisting of MMVs. This might allow us to detect outliers — or in general less and more perturbed data — and to adapt the quadrature weights to the varying accuracy of the measurements, yielding a reduced data error.

In Chapter 7.1 we have used LS-QRs also to construct new stable discretizations of the DG method on equidistant and scattered collocation points. Comparable discretizations of the DG method, such as the DGSEM, usually rely on the usage of GLo or GLe points. Another new class of stable high order methods for CLs has been proposed in Chapter 7.2. In this chapter, we first revisited RBF methods which are known to be unstable for CLs in the presence of BCs. Yet, we found that this limitation can be overcome by building RBF methods from a weak form of the CL instead of the commonly used differential form. In the process, we developed and investigated conservative as well as energy stable weak RBF methods for the numerical treatment of CLs. Future work in this direction will first and foremost focus on the design of similar RBF methods — that is of RBF methods which are provable conservative as well as energy stable — from the differential form of CLs.

The main focus of this thesis was the stabilization of high order numerical methods in the presence of (shock) discontinuities, a task known as shock capturing. Investigations in this direction have been given in chapters 8, 9, 10, and 11 and can be divided into three different approaches: ℓ^1 regularization, the Bernstein procedure, and (HOES steered) AV methods.

Let us first address Chapter 9, i. e., ℓ^1 regularization. This approach was motivated by the observation that in many cases entropy solutions u of CLs are piecewise smooth and only feature a finite number of jump discontinuities. Hence, $[u](x) \neq 0$ holds for the corresponding jump function only for a finite number of arguments and we say that u has a sparse representation. The idea then was to mimic this behavior of the exact solution on a discrete level for the numerical solution. This was achieved by constructing suitable approximations to the jump function $[u]$ by PA based HOES and to apply ℓ^1 regularizations to the approximation afterwards. Similar approaches have been pursued before. However, in this thesis, the approach was applied to SE methods for the first time. We demonstrated that the SE methods provide the advantage of allowing element-to-element variations in the ℓ^1 optimization problem, enhancing its accuracy and efficiency. Finally, our numerical tests showed that the proposed procedure is able to enhance existing SE methods. Future research might include changes to the optimization problem, e. g., utilizing (re)weighted ℓ^1 or ℓ^2 optimization instead of classical (unweighted) ℓ^1 optimization. Another promising idea is to use the involved PA based HOES to also steer other shock capturing methods. In fact, this idea has already been proven to be successful in the subsequent chapters 10 and 11.

In Chapter 10, we designed and carefully investigated another new shock capturing technique for SE methods, the so-called Bernstein procedure. Again using PA operators to identify troubled elements, the corresponding numerical solution is replaced by a convex combination of the original approximation and its Bernstein reconstruction there. Building up on classical Bernstein operators, we were able to prove that the resulting Bernstein procedure is TVD and preserves monotone (shock) profiles. Moreover, the procedure can be modified to not just preserve but also to enforce certain bounds for the solution, such as positivity. Numerical results demonstrated that the proposed Bernstein procedure is able to stabilize and enhance SE methods in the presence of shock discontinuities. Future work will focus on the extension to unstructured meshes in multiple dimensions and systems of CLs. Further, the investigation of other shock sensors would be of great interest. These sensors might be activated once the TV increases over time or once the numerical solution starts to violate certain bounds.

Finally, chapters 8 and 11 addressed AV for the stabilization of high oder numerical methods for CLs. The AV approach utilizes the well-known effect of dissipative mechanisms on (shock) discontinuities; when viscosity is incorporated, discontinuities in the solution are smeared out, yielding surfaces of discontinuities to be replaced by thin layers in which the solution varies (possibly rapidly but) continuously. This approach might be the one with the strongest physical motivation. By now many researchers have adapted and refined AV in a vast number of numerical methods. In Chapter 8, among other things, we introduced and investigated conservative and stable discretizations of AV in SE methods (DG and FR) using SBP operators. Moreover, we discussed their certainly appearing but limited connection to modal filtering. At the same time, we noted some problems for usual AV methods. Often, these add dissipation nearly equally over the whole computational domain (or element) and do not always adapt adequately to the smoothness of the underlying (numerical) solution. In Chapter 11 we therefore proposed novel AV operators which are steered by PA based HOES. These AV operators are able to calibrate the amount of viscosity and its distribution to the smoothness of the numerical solution and to the location of discontinuities — if these are present in the solution. Furthermore, addressing their discretization in the context of SBP based FD methods, were are able to construct discretizations that preserve conservation, stability, and accuracy (in smooth regions) of the underlying numerical method. Still, the investigation presented in Chapter 8 for FD methods can only be considered as a first step. Future research will focus on a vast number of generalizations and refinements of the proposed HOES steered AV operators. In particular, we are already working on the extension of this approach to multi-block FD methods as well as SE and RBF methods for CLs. At the same time, we are investigating different refinements as well as alternatives for the PA based HOES, which were used to steer the viscosity strength and distribution. These alternatives could include data driven sensors build up on artificial neural networks.

Bibliography

[AAD11] M. Ainsworth, G. Andriamaro, and O. Davydov. Bernstein–Bézier finite elements of arbitrary order and optimal assembly procedures. *SIAM Journal on Scientific Computing*, 33(6):3087–3109, 2011.

[Abg17] R. Abgrall. High order schemes for hyperbolic problems using globally continuous approximation and avoiding mass matrices. *Journal of Scientific Computing*, 73(2-3):461–494, 2017.

[Abg18] R. Abgrall. A general framework to construct schemes satisfying additional conservation relations. Application to entropy conservative and entropy dissipative schemes. *Journal of Computational Physics*, 372:640–666, 2018.

[ABT16] R. Abgrall, P. Bacigaluppi, and S. Tokareva. How to avoid mass matrix for linear hyperbolic problems. In *Numerical Mathematics and Advanced Applications ENUMATH 2015*, pages 75–86. Springer, 2016.

[ABT19] R. Abgrall, P. Bacigaluppi, and S. Tokareva. High-order residual distribution scheme for the time-dependent Euler equations of fluid dynamics. *Computers & Mathematics with Applications*, 78(2):274–297, 2019.

[ADK+17] R. Anderson, V. Dobrev, T. Kolev, D. Kuzmin, M. Q. de Luna, R. Rieben, and V. Tomov. High-order local maximum principle preserving (MPP) discontinuous Galerkin finite element method for the transport equation. *Journal of Computational Physics*, 334:102–124, 2017.

[AGP16] R. Archibald, A. Gelb, and R. B. Platte. Image reconstruction from undersampled Fourier data using the polynomial annihilation transform. *Journal of Scientific Computing*, 67(2):432–452, 2016.

[AGSX09] R. Archibald, A. Gelb, R. Saxena, and D. Xiu. Discontinuity detection in multivariate space for stochastic simulations. *Journal of Computational Physics*, 228(7):2676–2689, 2009.

[AGY05] R. Archibald, A. Gelb, and J. Yoon. Polynomial fitting for edge detection in irregularly sampled signals and images. *SIAM Journal on Numerical Analysis*, 43(1):259–279, 2005.

[AGY08] R. Archibald, A. Gelb, and J. Yoon. Determining the locations and discontinuities in the derivatives of functions. *Applied Numerical Mathematics*, 58(5):577–592, 2008.

[AJ11] Y. Allaneau and A. Jameson. Connections between the filtered discontinuous Galerkin method and the flux reconstruction approach to high order discretizations. *Computer Methods in Applied Mechanics and Engineering*, 200(49-52):3628–3636, 2011.

[AMÖ18] R. Abgrall, E. l. Meledo, and P. Öffner. On the connection between residual distribution schemes and flux reconstruction. *arXiv preprint arXiv:1807.01261*, 2018.

[Bat15] H. Bateman. Some recent researches on the motion of fluids. *Monthly Weather Review*, 43(4):163–170, 1915.

[BB05] S. Bianchini and A. Bressan. Vanishing viscosity solutions of nonlinear hyperbolic systems. *Annals of Mathematics*, pages 223–342, 2005.

[BB15] N. Beisiegel and J. Behrens. Quasi-nodal third-order Bernstein polynomials in a discontinuous Galerkin model for flooding and drying. *Environmental Earth Sciences*, 74(11):7275–7284, 2015.

[BD10] G. E. Barter and D. L. Darmofal. Shock capturing with PDE-based artificial viscosity for DGFEM: Part I. Formulation. *Journal of Computational Physics*, 229(5):1810–1827, 2010.

[Ber12a] S. Bernstein. Démonstration du théoreme de Weierstrass fondée sur le calcul des probabilities. *Communications de la Société mathématique de Kharkow*, 13(1):1–2, 1912.

[Ber12b] S. Bernstein. On the best approximation of continuous functions by polynomials of a given degree. *Communications de la Société mathématique de Kharkow*, 2(13):49–194, 1912.

[Ber58] D. L. Berman. On the impossibility of constructing a linear polynomial operator furnishing an approximation within the order of the best approximation. In *Doklady Akademii Nauk*, volume 120, pages 1175–1177. Russian Academy of Sciences, 1958.

[BIG03] A. Ben-Israel and T. N. Greville. *Generalized inverses: theory and applications*, volume 15. Springer Science & Business Media, 2003.

[BM11] C. B. Boyer and U. C. Merzbach. *A history of mathematics*. John Wiley & Sons, 2011.

[BO99] C. E. Baumann and J. T. Oden. A discontinuous hp finite element method for convection-diffusion problems. *Computer Methods in Applied Mechanics and Engineering*, 175(3-4):311–341, 1999.

[Boc29] S. Bochner. Über Sturm-Liouvillesche Polynomsysteme. *Mathematische Zeitschrift*, 29(1):730–736, 1929.

[Boy01] J. P. Boyd. *Chebyshev and Fourier spectral methods*. Courier Corporation, 2001.

[BP11] H. Brass and K. Petras. *Quadrature Theory: The Theory of Numerical Integration on a Compact Interval*. Number 178. American Mathematical Soc., 2011.

[BR95] F. Bassi and S. Rebay. Accurate 2D Euler computations by means of a high order discontinuous finite element method. In *14th International Conference on Numerical Methods in Fluid Dynamics*, pages 234–240. Springer, 1995.

[Bra77] H. Brass. *Quadraturverfahren*, volume 3. Vandenhoeck + Ruprecht Gm, 1977.

[Bre00] A. Bressan. *Hyperbolic systems of conservation laws: the one-dimensional Cauchy problem*, volume 20. Oxford University Press on Demand, 2000.

[BSB01] A. Burbeau, P. Sagaut, and C.-H. Bruneau. A problem-independent limiter for high-order Runge–Kutta discontinuous Galerkin methods. *Journal of Computational Physics*, 169(1):111–150, 2001.

[Buh00] M. D. Buhmann. Radial basis functions. *Acta numerica*, 9:1–38, 2000.

[Buh03] M. D. Buhmann. *Radial basis functions: theory and implementations*, volume 12. Cambridge university press, 2003.

[Bur48] J. M. Burgers. A mathematical model illustrating the theory of turbulence. In *Advances in Applied Mechanics*, volume 1, pages 171–199. Elsevier, 1948.

[Can87] J. Canny. A computational approach to edge detection. In *Readings in Computer Vision*, pages 184–203. Elsevier, 1987.

[CDRFC19] J. Chan, D. C. Del Rey Fernandez, and M. H. Carpenter. Efficient entropy stable Gauss collocation methods. *SIAM Journal on Scientific Computing*, 41(5):A2938–A2966, 2019.

[CF99] R. Courant and K. O. Friedrichs. *Supersonic flow and shock waves*, volume 21. Springer Science & Business Media, 1999.

[CGA93] M. H. Carpenter, D. Gottlieb, and S. Abarbanel. The stability of numerical boundary treatments for compact high-order finite-difference schemes. *Journal of Computational Physics*, 108(2):272–295, 1993.

[CHQZ06] C. Canuto, M. Y. Hussaini, A. Quarteroni, and T. A. Zang. *Spectral methods*. Springer, 2006.

[Chr77] E. B. Christoffel. Sur une classe particuliere de fonctions entieres et de fractions continues. *Annali di Matematica Pura ed Applicata (1867-1897)*, 8(1):1–10, 1877.

[CHS90] B. Cockburn, S. Hou, and C.-W. Shu. The Runge–Kutta local projection discontinuous Galerkin finite element method for conservation laws. IV. The multidimensional case. *Mathematics of Computation*, 54(190):545–581, 1990.

[CLS89] B. Cockburn, S.-Y. Lin, and C.-W. Shu. TVB Runge–Kutta local projection discontinuous Galerkin finite element method for conservation laws III: one-dimensional systems. *Journal of Computational Physics*, 84(1):90–113, 1989.

[CP76] R. Cline and R. J. Plemmons. ℓ_2-solutions to underdetermined linear systems. *SIAM Review*, 18(1):92–106, 1976.

[CQ82] C. Canuto and A. Quarteroni. Error estimates for spectral and pseudospectral approximations of hyperbolic equations. *SIAM Journal on Numerical Analysis*, 19(3):629–642, 1982.

[CS89] B. Cockburn and C.-W. Shu. TVB Runge–Kutta local projection discontinuous Galerkin finite element method for conservation laws. II. General framework. *Mathematics of Computation*, 52(186):411–435, 1989.

[CS91] B. Cockburn and C.-W. Shu. The Runge–Kutta local projection P^1-discontinuous Galerkin finite element method for scalar conservation laws. *ESAIM: Mathematical Modelling and Numerical Analysis*, 25(3):337–361, 1991.

[CS98] B. Cockburn and C.-W. Shu. The Runge–Kutta discontinuous Galerkin method for conservation laws V: multidimensional systems. *Journal of Computational Physics*, 141(2):199–224, 1998.

[CS14] J. Cheng and C.-W. Shu. Positivity-preserving Lagrangian scheme for multimaterial compressible flow. *Journal of Computational Physics*, 257:143–168, 2014.

[CS17] T. Chen and C.-W. Shu. Entropy stable high order discontinuous Galerkin methods with suitable quadrature rules for hyperbolic conservation laws. *Journal of Computational Physics*, 345:427–461, 2017.

[CWB08] E. J. Candes, M. B. Wakin, and S. P. Boyd. Enhancing sparsity by reweighted ℓ_1 minimization. *Journal of Fourier Analysis and Applications*, 14(5-6):877–905, 2008.

[Daf77] C. M. Dafermos. Generalized characteristics and the structure of solutions of hyperbolic conservation laws. *Indiana University Mathematics Journal*, 26(6):1097–1119, 1977.

[Daf00] C. M. Dafermos. *Hyperbolic conservation laws in continuum physics*, volume 325. Springer, 2000.

[Det95] H. Dette. New bounds for Hahn and Krawtchouk polynomials. *SIAM Journal on Mathematical Analysis*, 26(6):1647–1659, 1995.

[DGLW16] W.-S. Don, Z. Gao, P. Li, and X. Wen. Hybrid compact-WENO finite difference scheme with conjugate Fourier shock detection algorithm for hyperbolic conservation laws. *SIAM Journal on Scientific Computing*, 38(2):A691–A711, 2016.

[DGMM+14] D. De Grazia, G. Mengaldo, D. Moxey, P. Vincent, and S. Sherwin. Connections between the discontinuous Galerkin method and high-order flux reconstruction schemes. *International Journal for Numerical Methods in Fluids*, 75(12):860–877, 2014.

[DHR19] N. Discacciati, J. S. Hesthaven, and D. Ray. Controlling oscillations in high-order Discontinuous Galerkin schemes using artificial viscosity tuned by neural networks. Technical report, 2019.

[DLGC03] A. Dervieux, D. Leservoisier, P.-L. George, and Y. Coudière. About theoretical and practical impact of mesh adaptation on approximation of functions and PDE solutions. *International Journal for Numerical Methods in Fluids*, 43(5):507–516, 2003.

[DR07] P. J. Davis and P. Rabinowitz. *Methods of numerical integration.* Courier Corporation, 2007.

[DZLD14] M. Dumbser, O. Zanotti, R. Loubère, and S. Diot. A posteriori subcell limiting of the discontinuous Galerkin finite element method for hyperbolic conservation laws. *Journal of Computational Physics*, 278:47–75, 2014.

[EB92] J. Eckstein and D. P. Bertsekas. On the Douglas–Rachford splitting method and the proximal point algorithm for maximal monotone operators. *Mathematical Programming*, 55(1-3):293–318, 1992.

[EM09] Y. C. Eldar and M. Mishali. Robust recovery of signals from a structured union of subspaces. *IEEE Transactions on Information Theory*, 55(11):5302–5316, 2009.

[Eul57] L. Euler. Principes généraux du mouvement des fluides. *Mémoires de l'Académie des Sciences de Berlin*, pages 274–315, 1757.

[Eva10] L. C. Evans. *Partial differential equations*. American Mathematical Society, 2010.

[Eva18] L. Evans. *Measure theory and fine properties of functions*. Routledge, 2018.

[Fab14] G. Faber. Über die interpolatorische Darstellung stetiger Funktionen. *Jahresbericht der Deutschen Mathematiker Vereinigung*, 23:192–210, 1914.

[Fas96] G. E. Fasshauer. Solving partial differential equations by collocation with radial basis functions. In *Proceedings of Chamonix*, volume 1997, pages 1–8. Vanderbilt University Press Nashville, TN, 1996.

[Fas07] G. E. Fasshauer. *Meshfree approximation methods with MATLAB*, volume 6. World Scientific, 2007.

[FBW16] N. Flyer, G. A. Barnett, and L. J. Wicker. Enhancing finite differences with radial basis functions: experiments on the Navier–Stokes equations. *Journal of Computational Physics*, 316:39–62, 2016.

[FBZ14] D. C. D. R. Fernández, P. D. Boom, and D. W. Zingg. A generalized framework for nodal first derivative summation-by-parts operators. *Journal of Computational Physics*, 266:214–239, 2014.

[FCN+13] T. C. Fisher, M. H. Carpenter, J. Nordström, N. K. Yamaleev, and C. Swanson. Discretely conservative finite-difference formulations for nonlinear conservation laws in split form: Theory and boundary conditions. *Journal of Computational Physics*, 234:353–375, 2013.

[FDWC02] B. Fornberg, T. A. Driscoll, G. Wright, and R. Charles. Observations on the behavior of radial basis function approximations near boundaries. *Computers & Mathematics with Applications*, 43(3-5):473–490, 2002.

[Fej13] L. Fejér. Über die Bestimmung des Sprunges der Funktion aus ihrer Fourierreihe. *Journal für die reine und angewandte Mathematik*, 142:165–188, 1913.

[FFBB16] N. Flyer, B. Fornberg, V. Bayona, and G. A. Barnett. On the role of polynomials in RBF-FD approximations: I. Interpolation and accuracy. *Journal of Computational Physics*, 321:21–38, 2016.

[FG88] D. Funaro and D. Gottlieb. A new method of imposing boundary conditions in pseudospectral approximations of hyperbolic equations. *Mathematics of Computation*, 51(184):599–613, 1988.

[FG91] D. Funaro and D. Gottlieb. Convergence results for pseudospectral approximations of hyperbolic systems by a penalty-type boundary treatment. *Mathematics of Computation*, 57(196):585–596, 1991.

[FHZ14] D. C. D. R. Fernández, J. E. Hicken, and D. W. Zingg. Review of summation-by-parts operators with simultaneous approximation terms for the numerical solution of partial differential equations. *Computers & Fluids*, 95:171–196, 2014.

[FK07] M. Feistauer and V. Kučera. On a robust discontinuous Galerkin technique for the solution of compressible flow. *Journal of Computational Physics*, 224(1):208–221, 2007.

[FZ07] B. Fornberg and J. Zuev. The Runge phenomenon and spatially variable shape parameters in RBF interpolation. *Computers & Mathematics with Applications*, 54(3):379–398, 2007.

[Gas13] G. J. Gassner. A skew-symmetric discontinuous Galerkin spectral element discretization and its relation to SBP-SAT finite difference methods. *SIAM Journal on Scientific Computing*, 35(3):A1233–A1253, 2013.

[Gau14] C. F. Gauss. Methodus nova integralium valores per approximationem inveniendi. *Commentationes Societatis Regiae Scientarium Gottingensis Recentiores 3.*, [Werke III, 163–196.], 1814.

[Gau81] W. Gautschi. A survey of Gauss–Christoffel quadrature formulae. In *EB Christoffel*, pages 72–147. Springer, 1981.

[Gau97] W. Gautschi. *Numerical analysis.* Springer Science & Business Media, 1997.

[Gau04] W. Gautschi. *Orthogonal polynomials.* Oxford university press Oxford, 2004.

[GC08] A. Gelb and D. Cates. Detection of edges in spectral data III. Refinement of the concentration method. *Journal of Scientific Computing*, 36(1):1–43, 2008.

[GG19a] J. Glaubitz and A. Gelb. High order edge sensors with ℓ^1 regularization for enhanced discontinuous Galerkin methods. *SIAM Journal on Scientific Computing*, 41(2):A1304–A1330, 2019.

[GG19b] J. Glaubitz and A. Gelb. Stability of radial basis function methods for one dimensional scalar conservation laws in weak form. 2019. Submitted.

[GH01] D. Gottlieb and J. S. Hesthaven. Spectral methods for hyperbolic problems. *Journal of Computational and Applied Mathematics*, 128(1):83–131, 2001.

[GHL19] A. Gelb, X. Hou, and Q. Li. Numerical analysis for conservation laws using ℓ_1 minimization. *Journal of Scientific Computing*, pages 1–26, 2019.

[GHS93] K. Guo, S. Hu, and X. Sun. Conditionally positive definite functions and Laplace–Stieltjes integrals. *Journal of Approximation Theory*, 74(3):249–265, 1993.

[GHW02] F. X. Giraldo, J. S. Hesthaven, and T. Warburton. Nodal high-order discontinuous Galerkin methods for the spherical shallow water equations. *Journal of Computational Physics*, 181(2):499–525, 2002.

[Gib98] J. W. Gibbs. Fourier's series. *Nature*, 59(1522):200, 1898.

[Gib99] J. W. Gibbs. Fourier's series. *Nature*, 59(1539):606, 1899.

[GJ97] M. Gokmen and A. K. Jain. $\lambda\tau$-space representation of images and generalized edge detector. *IEEE Transactions on Pattern Analysis and Machine Intelligence*, 19(6):545–563, 1997.

[GKO95] B. Gustafsson, H.-O. Kreiss, and J. Oliger. *Time dependent problems and difference methods*, volume 24. John Wiley & Sons, 1995.

[GKS11] S. Gottlieb, D. I. Ketcheson, and C.-W. Shu. *Strong stability preserving Runge–Kutta and multistep time discretizations.* World Scientific, 2011.

[Gla19a] J. Glaubitz. Discrete least squares quadrature rules on equidistant and scattered points. 2019. Submitted.

[Gla19b] J. Glaubitz. High order edge sensor steered artificial viscosity operators. 2019. Submitted.

[Gla19c] J. Glaubitz. Shock capturing by Bernstein polynomials for scalar conservation laws. *Applied Mathematics and Computation*, 363:124593, 2019.

[Gla19d] J. Glaubitz. Stable high order quadrature rules for scattered data and general weight functions. 2019. Submitted.

[GLT89] R. Glowinski and P. Le Tallec. *Augmented Lagrangian and operator-splitting methods in nonlinear mechanics*, volume 9. SIAM, 1989.

[GM75] R. Glowinski and A. Marroco. Sur l'approximation, par éléments finis d'ordre un, et la résolution, par pénalisation-dualité d'une classe de problèmes de Dirichlet non linéaires. *Revue française d'automatique, informatique, recherche opérationnelle. Analyse numérique*, 9(R2):41–76, 1975.

[GM76] D. Gabay and B. Mercier. A dual algorithm for the solution of nonlinear variational problems via finite element approximation. *Computers & Mathematics with Applications*, 2(1):17–40, 1976.

[GMP+08] J.-L. Guermond, F. Marpeau, B. Popov, et al. A fast algorithm for solving first-order PDEs by L1-minimization. *Communications in Mathematical Sciences*, 6(1):199–216, 2008.

[GNJA+19] J. Glaubitz, A. Nogueira Jr, J. Almeida, R. Cantão, and C. Silva. Smooth and Compactly Supported Viscous Sub-cell Shock Capturing for Discontinuous Galerkin Methods. *Journal of Scientific Computing*, 79(1):249–272, 2019.

[GO09] T. Goldstein and S. Osher. The split Bregman method for L1-regularized problems. *SIAM Journal on Imaging Sciences*, 2(2):323–343, 2009.

[GÖ20] J. Glaubitz and P. Öffner. Stable discretisations of high-order discontinuous Galerkin methods on equidistant and scattered points. *Applied Numerical Mathematics*, 2020.

[GÖRS16] J. Glaubitz, P. Öffner, H. Ranocha, and T. Sonar. Artificial viscosity for correction procedure via reconstruction using summation-by-parts operators. In *XVI International Conference on Hyperbolic Problems: Theory, Numerics, Applications*, pages 363–375. Springer, 2016.

[GÖS18] J. Glaubitz, P. Öffner, and T. Sonar. Application of modal filtering to a spectral difference method. *Mathematics of Computation*, 87(309):175–207, 2018.

[GP03] H. Gzyl and J. L. Palacios. On the approximation properties of Bernstein polynomials via probabilistic tools. *Boletın de la Asociación Matemática Venezolana*, 10(1):5–13, 2003.

[GP08] J.-L. Guermond and R. Pasquetti. Entropy-based nonlinear viscosity for Fourier approximations of conservation laws. *Comptes Rendus Mathematique*, 346(13-14):801–806, 2008.

[GPP11] J.-L. Guermond, R. Pasquetti, and B. Popov. Entropy viscosity method for nonlinear conservation laws. *Journal of Computational Physics*, 230(11):4248–4267, 2011.

[GPR08] A. Gelb, R. B. Platte, and W. S. Rosenthal. The discrete orthogonal polynomial least squares method for approximation and solving partial differential equations. *Communications in Computational Physics*, 3(3):734–758, 2008.

[GR91] E. Godlewski and P.-A. Raviart. *Hyperbolic systems of conservation laws.* Ellipses, 1991.

[GRS19] J. Glaubitz, D. Rademacher, and T. Sonar. *Lernbuch Analysis 1: Das Wichtigste ausführlich für Bachelor und Lehramt.* Springer-Verlag, 2019.

[GS83] I. Grotrian-Steinweg. *Die Quadraturverfahren vom Newton-Cotes-Typ.* Technische Universität Carolo-Wilhelmina zu Braunschweig, 1983.

[GS97] D. Gottlieb and C.-W. Shu. On the Gibbs phenomenon and its resolution. *SIAM Review*, 39(4):644–668, 1997.

[GS98] S. Gottlieb and C.-W. Shu. Total variation diminishing Runge-Kutta schemes. *Mathematics of Computation of the American Mathematical Society*, 67(221):73–85, 1998.

[GS19] A. Gelb and T. Scarnati. Reducing effects of bad data using variance based joint sparsity recovery. *Journal of Scientific Computing*, 78(1):94–120, 2019.

[GST01] S. Gottlieb, C.-W. Shu, and E. Tadmor. Strong stability-preserving high-order time discretization methods. *SIAM Review*, 43(1):89–112, 2001.

[GT99] A. Gelb and E. Tadmor. Detection of edges in spectral data. *Applied and Computational Harmonic Analysis*, 7(1):101–135, 1999.

[GT00] A. Gelb and E. Tadmor. Detection of edges in spectral data II. Nonlinear enhancement. *SIAM Journal on Numerical Analysis*, 38(4):1389–1408, 2000.

[GT06a] A. Gelb and E. Tadmor. Adaptive edge detectors for piecewise smooth data based on the minmod limiter. *Journal of Scientific Computing*, 28(2):279–306, 2006.

[GT06b] A. Gelb and J. Tanner. Robust reprojection methods for the resolution of the Gibbs phenomenon. *Applied and Computational Harmonic Analysis*, 20(1):3–25, 2006.

[Gus07] B. Gustafsson. *High order difference methods for time dependent PDE*, volume 38. Springer Science & Business Media, 2007.

[GVL12] G. H. Golub and C. F. Van Loan. *Matrix computations*. JHU Press, 2012.

[GWK16a] G. J. Gassner, A. R. Winters, and D. A. Kopriva. Split form nodal discontinuous Galerkin schemes with summation-by-parts property for the compressible Euler equations. *Journal of Computational Physics*, 327:39–66, 2016.

[GWK16b] G. J. Gassner, A. R. Winters, and D. A. Kopriva. A well balanced and entropy conservative discontinuous Galerkin spectral element method for the shallow water equations. *Applied Mathematics and Computation*, 272:291–308, 2016.

[Hah49] W. Hahn. Über Orthogonalpolynome, die q-Differenzengleichungen genügen. *Mathematische Nachrichten*, 2(1-2):4–34, 1949.

[Har71] R. L. Hardy. Multiquadric equations of topography and other irregular surfaces. *Journal of Geophysical Research*, 76(8):1905–1915, 1971.

[Har83] A. Harten. On the symmetric form of systems of conservation laws with entropy. *Journal of Computational Physics*, 49(1):151–164, 1983.

[Har84] A. Harten. On a class of high resolution total-variation-stable finite-difference schemes. *SIAM Journal on Numerical Analysis*, 21(1):1–23, 1984.

[Har06] R. Hartmann. Adaptive discontinuous Galerkin methods with shock-capturing for the compressible Navier–Stokes equations. *International Journal for Numerical Methods in Fluids*, 51(9-10):1131–1156, 2006.

[HCP12] A. Huerta, E. Casoni, and J. Peraire. A simple shock-capturing technique for high-order discontinuous Galerkin methods. *International Journal for Numerical Methods in Fluids*, 69(10):1614–1632, 2012.

[HFM86] T. J. Hughes, L. Franca, and M. Mallet. A new finite element formulation for computational fluid dynamics: I. Symmetric forms of the compressible Euler and Navier-Stokes equations and the second law of thermodynamics. *Computer Methods in Applied Mechanics and Engineering*, 54(2):223–234, 1986.

[HH79] E. Hewitt and R. E. Hewitt. The Gibbs–Wilbraham phenomenon: An episode in Fourier analysis. *Archive for History of Exact Sciences*, 21(2):129–160, 1979.

[HH02] R. Hartmann and P. Houston. Adaptive discontinuous Galerkin finite element methods for the compressible Euler equations. *Journal of Computational Physics*, 183(2):508–532, 2002.

[HJ12] R. A. Horn and C. R. Johnson. *Matrix analysis*. Cambridge university press, 2012.

[HK08] J. Hesthaven and R. Kirby. Filtering in Legendre spectral methods. *Mathematics of Computation*, 77(263):1425–1452, 2008.

[HLS15] T. Y. Hou, Q. Li, and H. Schaeffer. Sparse+ low-energy decomposition for viscous conservation laws. *Journal of Computational Physics*, 288:150–166, 2015.

[HLW06] E. Hairer, C. Lubich, and G. Wanner. *Geometric numerical integration: structure-preserving algorithms for ordinary differential equations*, volume 31. Springer Science & Business Media, 2006.

[HM98] Y. Hon and X. Mao. An efficient numerical scheme for Burgers' equation. *Applied Mathematics and Computation*, 95(1):37–50, 1998.

[HNW91] E. Hairer, S. P. Nørsett, and G. Wanner. *Solving ordinary differential equations I*. Springer-Vlg, 1991.

[HO09] D. Huybrechs and S. Olver. Highly oscillatory quadrature. *Highly Oscillatory Problems*, (366):25–50, 2009.

[Hu96] F. Q. Hu. On absorbing boundary conditions for linearized Euler equations by a perfectly matched layer. *Journal of Computational Physics*, 129(1):201–219, 1996.

[Huy07] H. T. Huynh. A flux reconstruction approach to high-order schemes including discontinuous Galerkin methods. *AIAA paper*, 4079:2007, 2007.

[Huy09] D. Huybrechs. Stable high-order quadrature rules with equidistant points. *Journal of Computational and Applied Mathematics*, 231(2):933–947, 2009.

[HW07] J. S. Hesthaven and T. Warburton. *Nodal discontinuous Galerkin methods: algorithms, analysis, and applications.* Springer Science & Business Media, 2007.

[HWV14] H. T. Huynh, Z. J. Wang, and P. E. Vincent. High-order methods for computational fluid dynamics: A brief review of compact differential formulations on unstructured grids. *Computers & Fluids*, 98:209–220, 2014.

[IN04] A. Iserles and S. P. Nørsett. On quadrature methods for highly oscillatory integrals and their implementation. *BIT Numerical Mathematics*, 44(4):755–772, 2004.

[INO06] A. Iserles, S. Nørsett, and S. Olver. Highly oscillatory quadrature: The story so far. In *Numerical Mathematics and Advanced Applications*, pages 97–118. Springer, 2006.

[Isk03] A. Iske. Radial basis functions: basics, advanced topics and meshfree methods for transport problems. *Rendiconti del Seminario Matematico, Universitae Politecnico di Torino*, 61(3):247–285, 2003.

[JAX11] J. D. Jakeman, R. Archibald, and D. Xiu. Characterization of discontinuities in high-dimensional stochastic problems on adaptive sparse grids. *Journal of Computational Physics*, 230(10):3977–3997, 2011.

[JJS95] J. Jaffre, C. Johnson, and A. Szepessy. Convergence of the discontinuous Galerkin finite element method for hyperbolic conservation laws. *Mathematical Models and Methods in Applied Sciences*, 5(03):367–386, 1995.

[JS94] G. S. Jiang and C.-W. Shu. On a cell entropy inequality for discontinuous Galerkin methods. *Mathematics of Computation*, 62(206):531–538, 1994.

[Kan90] E. J. Kansa. Multiquadrics — A scattered data approximation scheme with applications to computational fluid-dynamics — II solutions to parabolic, hyperbolic and elliptic partial differential equations. *Computers & Mathematics with Applications*, 19(8-9):147–161, 1990.

[KCH06] A. Kanevsky, M. H. Carpenter, and J. S. Hesthaven. Idempotent filtering in spectral and spectral element methods. *Journal of Computational Physics*, 220(1):41–58, 2006.

[Ket08] D. I. Ketcheson. Highly efficient strong stability-preserving Runge–Kutta methods with low-storage implementations. *SIAM Journal on Scientific Computing*, 30(4):2113–2136, 2008.

[KG14] D. A. Kopriva and G. J. Gassner. An energy stable discontinuous Galerkin spectral element discretization for variable coefficient advection problems. *SIAM Journal on Scientific Computing*, 36(4):A2076–A2099, 2014.

[KH00] E. Kansa and Y. Hon. Circumventing the ill-conditioning problem with multiquadric radial basis functions: applications to elliptic partial differential equations. *Computers and Mathematics with Applications*, 39(7-8):123–138, 2000.

[Kir11] R. C. Kirby. Fast simplicial finite element algorithms using Bernstein polynomials. *Numerische Mathematik*, 117(4):631–652, 2011.

[KK03] R. M. Kirby and G. E. Karniadakis. De-aliasing on non-uniform grids: algorithms and applications. *Journal of Computational Physics*, 191(1):249–264, 2003.

[KL89] H.-O. Kreiss and J. Lorenz. *Initial-boundary value problems and the Navier-Stokes equations*, volume 47. Siam, 1989.

[KLT06] D. Kuzmin, R. Löhner, and S. Turek. *Flux-corrected transport: Principles, algorithms, and applications*. Scientific Computation. Springer Berlin Heidelberg, 2006.

[KO79] H.-O. Kreiss and J. Oliger. Stability of the Fourier method. *SIAM Journal on Numerical Analysis*, 16(3):421–433, 1979.

[Kop09] D. A. Kopriva. *Implementing spectral methods for partial differential equations: Algorithms for scientists and engineers*. Springer Science & Business Media, 2009.

[KS74] H.-O. Kreiss and G. Scherer. Finite element and finite difference methods for hyperbolic partial differential equations. In *Mathematical Aspects of Finite Elements in Partial Differential Equations*, pages 195–212. Elsevier, 1974.

[KS77] H.-O. Kreiss and G. Scherer. On the existence of energy estimates for difference approximations for hyperbolic systems. Technical report, Department of Scientific Computing, Uppsala University, 1977.

[KS06] V. I. Krylov and A. H. Stroud. *Approximate calculation of integrals*. Courier Corporation, 2006.

[KWH11] A. Klöckner, T. Warburton, and J. S. Hesthaven. Viscous shock capturing in a time-explicit discontinuous Galerkin method. *Mathematical Modelling of Natural Phenomena*, 6(3):57–83, 2011.

[KXR+04] L. Krivodonova, J. Xin, J.-F. Remacle, N. Chevaugeon, and J. E. Flaherty. Shock detection and limiting with discontinuous Galerkin methods for hyperbolic conservation laws. *Applied Numerical Mathematics*, 48(3-4):323–338, 2004.

[Lav89] J. Lavery. Solution of steady-state one-dimensional conservation laws by mathematical programming. *SIAM Journal on Numerical Analysis*, 26(5):1081–1089, 1989.

[Lav91] J. E. Lavery. Solution of steady-state, two-dimensional conservation laws by mathematical programming. *SIAM Journal on Numerical Analysis*, 28(1):141–155, 1991.

[Lax57] P. D. Lax. Hyperbolic systems of conservation laws II. *Communications on Pure and Applied Mathematics*, 10(4):537–566, 1957.

[Lax73] P. D. Lax. *Hyperbolic systems of conservation laws and the mathematical theory of shock waves*, volume 11. SIAM, 1973.

[LeV92] R. J. LeVeque. *Numerical methods for conservation laws*, volume 132. Springer, 1992.

[LeV02] R. J. LeVeque. *Finite volume methods for hyperbolic problems*, volume 31. Cambridge university press, 2002.

[LF03] E. Larsson and B. Fornberg. A numerical study of some radial basis function based solution methods for elliptic PDEs. *Computers & Mathematics with Applications*, 46(5-6):891–902, 2003.

[LH95] C. L. Lawson and R. J. Hanson. *Solving least squares problems.* Siam, 1995.

[Liu75] T.-P. Liu. The Riemann problem for general systems of conservation laws. *Journal of Differential Equations*, 18(1):218–234, 1975.

[LKSM17] C. Lohmann, D. Kuzmin, J. N. Shadid, and S. Mabuza. Flux-corrected transport algorithms for continuous Galerkin methods based on high order Bernstein finite elements. *Journal of Computational Physics*, 344:151–186, 2017.

[Lob52] R. Lobatto. *Lessen over de differentiaal-en integraal-rekening*, volume 2. De Gebroeders Van Cleef, 1852.

[Lor12] G. G. Lorentz. *Bernstein polynomials.* American Mathematical Soc., 2012.

[LT98] D. Levy and E. Tadmor. From semidiscrete to fully discrete: Stability of Runge–Kutta schemes by the energy method. *SIAM Review*, 40(1):40–73, 1998.

[Luk20] F. Lukács. Über die Bestimmung des Sprunges der Funktion aus ihrer Fourierreihe. *Journal für die reine und angewandte Mathematik*, 150:107–112, 1920.

[LYJZ13] C. Li, W. Yin, H. Jiang, and Y. Zhang. An efficient augmented Lagrangian method with applications to total variation minimization. *Computational Optimization and Applications*, 56(3):507–530, 2013.

[Mat12] K. Mattsson. Summation by parts operators for finite difference approximations of second-derivatives with variable coefficients. *Journal of Scientific Computing*, 51(3):650–682, 2012.

[MH92] S. Mallat and W. L. Hwang. Singularity detection and processing with wavelets. *IEEE Transactions on Information Theory*, 38(2):617–643, 1992.

[Mic84] C. A. Micchelli. Interpolation of scattered data: distance matrices and conditionally positive definite functions. In *Approximation Theory and Spline Functions*, pages 143–145. Springer, 1984.

[MMO78] A. Majda, J. McDonough, and S. Osher. The Fourier method for nonsmooth initial data. *Mathematics of Computation*, 32(144):1041–1081, 1978.

[MN04] K. Mattsson and J. Nordström. Summation by parts operators for finite difference approximations of second derivatives. *Journal of Computational Physics*, 199(2):503–540, 2004.

[MO16] A. Meister and S. Ortleb. A positivity preserving and well-balanced DG scheme using finite volume subcells in almost dry regions. *Applied Mathematics and Computation*, 272:259–273, 2016.

[MOS12] A. Meister, S. Ortleb, and T. Sonar. Application of spectral filtering to discontinuous Galerkin methods on triangulations. *Numerical Methods for Partial Differential Equations*, 28(6):1840–1868, 2012.

[MOSW13] A. Meister, S. Ortleb, T. Sonar, and M. Wirz. An extended discontinuous Galerkin and spectral difference method with modal filtering. *ZAMM-Journal of Applied Mathematics and Mechanics/Zeitschrift für Angewandte Mathematik und Mechanik*, 93(6-7):459–464, 2013.

[MP16] J. M. Martel and R. B. Platte. Stability of radial basis function methods for convection problems on the circle and sphere. *Journal of Scientific Computing*, 69(2):487–505, 2016.

[MSN04] K. Mattsson, M. Svärd, and J. Nordström. Stable and accurate artificial dissipation. *Journal of Scientific Computing*, 21(1):57–79, 2004.

[MSS08] K. Mattsson, M. Svärd, and M. Shoeybi. Stable and accurate schemes for the compressible Navier–Stokes equations. *Journal of Computational Physics*, 227(4):2293–2316, 2008.

[NC99] J. Nordström and M. H. Carpenter. Boundary and interface conditions for high-order finite-difference methods applied to the Euler and Navier–Stokes equations. *Journal of Computational Physics*, 148(2):621–645, 1999.

[Nor06] J. Nordström. Conservative finite difference formulations, variable coefficients, energy estimates and artificial dissipation. *Journal of Scientific Computing*, 29(3):375–404, 2006.

[ÖGR18] P. Öffner, J. Glaubitz, and H. Ranocha. Stability of correction procedure via reconstruction with summation-by-parts operators for Burgers' equation using a polynomial chaos approach. *ESAIM: Mathematical Modelling and Numerical Analysis*, 52(6):2215–2245, 2018.

[ÖGR19] P. Öffner, J. Glaubitz, and H. Ranocha. Analysis of artificial dissipation of explicit and implicit time-integration methods. 2019. Accepted in International Journal of Numerical Analysis and Modeling.

[Ole57] O. A. Oleinik. Discontinuous solutions of non-linear differential equations. *Uspekhi Matematicheskikh Nauk*, 12(3):3–73, 1957.

[Ole64] O. Oleinik. The Cauchy problem for nonlinear equations in a class of discontinuous functions. *American Mathematical Society Translations*, 42:7–12, 1964.

[Ols95a] P. Olsson. Summation by parts, projections, and stability. I. *Mathematics of Computation*, 64(211):1035–1065, 1995.

[Ols95b] P. Olsson. Summation by parts, projections, and stability. II. *Mathematics of Computation*, 64(212):1473–1493, 1995.

[Osh84] S. Osher. Riemann solvers, the entropy condition, and difference. *SIAM Journal on Numerical Analysis*, 21(2):217–235, 1984.

[ÖSW13] P. Öffner, T. Sonar, and M. Wirz. Detecting strength and location of jump discontinuities in numerical data. *Applied Mathematics*, 4(12):1, 2013.

[PD04] R. B. Platte and T. A. Driscoll. Computing eigenmodes of elliptic operators using radial basis functions. *Computers and Mathematics with Applications*, 48(3-4):561–576, 2004.

[PD05] R. B. Platte and T. A. Driscoll. Polynomials and potential theory for Gaussian radial basis function interpolation. *SIAM Journal on Numerical Analysis*, 43(2):750–766, 2005.

[PD06] R. B. Platte and T. A. Driscoll. Eigenvalue stability of radial basis function discretizations for time-dependent problems. *Computers & Mathematics with Applications*, 51(8):1251–1268, 2006.

[Pet90] K. Petras. On the minimal norms of polynomial projections. *Journal of Approximation Theory*, 62(2):206–212, 1990.

[Phi03] G. M. Phillips. *Interpolation and approximation by polynomials*, volume 14. Springer Science & Business Media, 2003.

[PIN09] P. Pettersson, G. Iaccarino, and J. Nordström. Numerical analysis of the Burgers' equation in the presence of uncertainty. *Journal of Computational Physics*, 228(22):8394–8412, 2009.

[PIN15] M. P. Pettersson, G. Iaccarino, and J. Nordström. *Polynomial chaos methods for hyperbolic partial differential equations: Numerical techniques for fluid dynamics problems in the presence of uncertainties.* Springer, 2015.

[PP06] P.-O. Persson and J. Peraire. Sub-cell shock capturing for discontinuous Galerkin methods. In *44th AIAA Aerospace Sciences Meeting and Exhibit*, page 112, 2006.

[PTK11] R. B. Platte, L. N. Trefethen, and A. B. Kuijlaars. Impossibility of fast stable approximation of analytic functions from equispaced samples. *SIAM Review*, 53(2):308–318, 2011.

[QS05] J. Qiu and C.-W. Shu. A Comparison of troubled-cell indicators for Runge–Kutta discontinuous Galerkin methods using weighted essentially nonoscillatory limiters. *SIAM Journal on Scientific Computing*, 27(3):995–1013, 2005.

[Ran92] J. L. Randall. Numerical methods for conservation laws. *Lectures in Mathematics ETH Zürich*, 1992.

[RGÖS16] H. Ranocha, J. Glaubitz, P. Öffner, and T. Sonar. Time discretisation and L_2 stability of polynomial summation-by-parts schemes with Runge–Kutta methods. *arXiv preprint arXiv:1609.02393*, 2016.

[RGÖS18] H. Ranocha, J. Glaubitz, P. Öffner, and T. Sonar. Stability of artificial dissipation and modal filtering for flux reconstruction schemes using summation-by-parts operators. *Applied Numerical Mathematics*, 128:1–23, 2018.

[RH73] W. H. Reed and T. Hill. Triangular mesh methods for the neutron transport equation. Technical report, Los Alamos Scientific Laboratory, New Mexico, USA, 1973.

[RH18] D. Ray and J. S. Hesthaven. An artificial neural network as a troubled-cell indicator. *Journal of Computational Physics*, 367:166–191, 2018.

[Ric91] F. Richards. A Gibbs phenomenon for spline functions. *Journal of Approximation Theory*, 66(3):334–351, 1991.

[Rie60] B. Riemann. *Über die Fortpflanzung ebener Luftwellen von endlicher Schwingungsweite*. Verlag der Dieterichschen Buchhandlung, 1860.

[Rie10] F. Rieper. On the dissipation mechanism of upwind-schemes in the low Mach number regime: A comparison between Roe and HLL. *Journal of Computational Physics*, 229(2):221–232, 2010.

[RÖ18] H. Ranocha and P. Öffner. L_2 stability of explicit Runge–Kutta schemes. *Journal of Scientific Computing*, 75(2):1040–1056, 2018.

[RÖS16] H. Ranocha, P. Öffner, and T. Sonar. Summation-by-parts operators for correction procedure via reconstruction. *Journal of Computational Physics*, 311:299–328, 2016.

[Rou84] E. J. Routh. On some properties of certain solutions of a differential equation of the second order. *Proceedings of the London Mathematical Society*, 1(1):245–262, 1884.

[Run01] C. Runge. Über empirische Funktionen und die Interpolation zwischen äquidistanten Ordinaten. *Zeitschrift für Mathematik und Physik*, 46(224-243):20, 1901.

[San16] T. Sanders. Matlab imaging algorithms: Image reconstruction, restoration, and alignment, with a focus in tomography, 2016.

[Sch38] I. J. Schoenberg. Metric spaces and completely monotone functions. *Annals of Mathematics*, pages 811–841, 1938.

[Sch73] D. G. Schaeffer. A regularity theorem for conservation laws. *Advances in Mathematics*, 11(3):368–386, 1973.

[Sch95a] R. Schaback. Creating surfaces from scattered data using radial basis functions. *Mathematical Methods for Curves and Surfaces*, 477, 1995.

[Sch95b] R. Schaback. Error estimates and condition numbers for radial basis function interpolation. *Advances in Computational Mathematics*, 3(3):251–264, 1995.

[Sch05] R. Schaback. Multivariate interpolation by polynomials and radial basis functions. *Constructive Approximation*, 21(3):293–317, 2005.

[Sch11] A. Schönhage. *Approximationstheorie.* Walter de Gruyter, 2011.

[SCHO13] H. Schaeffer, R. Caflisch, C. D. Hauck, and S. Osher. Sparse dynamics for partial differential equations. *Proceedings of the National Academy of Sciences*, 110(17):6634–6639, 2013.

[Ser91] D. Serre. Richness and the classification of quasilinear hyperbolic systems. In *Multidimensional Hyperbolic Problems and Computations*, pages 315–333. Springer, 1991.

[SGP17a] T. Sanders, A. Gelb, and R. B. Platte. Composite SAR imaging using sequential joint sparsity. *Journal of Computational Physics*, 338:357–370, 2017.

[SGP17b] T. Scarnati, A. Gelb, and R. B. Platte. Using ℓ_1 regularization to improve numerical partial differential equation solvers. *Journal of Scientific Computing*, pages 1–28, 2017.

[Shu88] C.-W. Shu. Total-variation-diminishing time discretizations. *SIAM Journal on Scientific and Statistical Computing*, 9(6):1073–1084, 1988.

[SJ14] A. Sheshadri and A. Jameson. Shock detection and capturing methods for high order discontinuous-galerkin finite element methods. In *32nd AIAA Applied Aerodynamics Conference*, page 2688, 2014.

[SM14] M. Sonntag and C.-D. Munz. Shock capturing for discontinuous Galerkin methods using finite volume subcells. In *Finite Volumes for Complex Applications VII-Elliptic, Parabolic and Hyperbolic Problems*, pages 945–953. Springer, 2014.

[Smo12] J. Smoller. *Shock waves and reaction-diffusion equations*, volume 258. Springer Science & Business Media, 2012.

[SN14] M. Svärd and J. Nordström. Review of summation-by-parts schemes for initial–boundary-value problems. *Journal of Computational Physics*, 268:17–38, 2014.

[SO88] C.-W. Shu and S. Osher. Efficient implementation of essentially non-oscillatory shock-capturing schemes. *Journal of Computational Physics*, 77(2):439–471, 1988.

[SO89] C.-W. Shu and S. Osher. Efficient implementation of essentially non-oscillatory shock-capturing schemes, II. *Journal of Computational Physics*, 83(1):32–78, 1989.

[Sod78] G. A. Sod. A survey of several finite difference methods for systems of nonlinear hyperbolic conservation laws. *Journal of Computational Physics*, 27(1):1–31, 1978.

[Son16] T. Sonar. *3000 Jahre Analysis: Geschichte-Kulturen-Menschen.* Springer-Verlag, 2016.

[SRG10] W. Stefan, R. A. Renaut, and A. Gelb. Improved total variation-type regularization using higher order edge detectors. *SIAM Journal on Imaging Sciences*, 3(2):232–251, 2010.

[Str94] B. Strand. Summation by parts for finite difference approximations for d/dx. *Journal of Computational Physics*, 110(1):47–67, 1994.

[Sze39] G. Szegö. *Orthogonal polynomials*, volume 23. American Mathematical Soc., 1939.

[Tad90] E. Tadmor. Shock capturing by the spectral viscosity method. *Computer Methods in Applied Mechanics and Engineering*, 80(1-3):197–208, 1990.

[Tan06] J. Tanner. Optimal filter and mollifier for piecewise smooth spectral data. *Mathematics of Computation*, 75(254):767–790, 2006.

[TBI97] L. N. Trefethen and D. Bau III. *Numerical linear algebra*. Siam, 1997.

[Tor13] E. F. Toro. *Riemann solvers and numerical methods for fluid dynamics: a practical introduction*. Springer Science & Business Media, 2013.

[Tra09] J. A. Trangenstein. *Numerical solution of hyperbolic partial differential equations*. Cambridge University Press, 2009.

[Tre11] L. N. Trefethen. Six myths of polynomial interpolation and quadrature. 2011.

[Tre13] L. N. Trefethen. *Approximation theory and approximation practice*, volume 128. Siam, 2013.

[TT93] E. Tadmor and T. Tassa. On the piecewise smoothness of entropy solutions to scalar conservation laws. *Communications in Partial Differential Equations*, 18(9-10):1631–1652, 1993.

[TT02] E. Tadmor and J. Tanner. Adaptive mollifiers for high resolution recovery of piecewise smooth data from its spectral information. *Foundations of Computational Mathematics*, 2(2):155–189, 2002.

[TW12] E. Tadmor and K. Waagan. Adaptive spectral viscosity for hyperbolic conservation laws. *SIAM Journal on Scientific Computing*, 34(2):A993–A1009, 2012.

[Van91] H. Vandeven. Family of spectral filters for discontinuous problems. *Journal of Scientific Computing*, 6(2):159–192, 1991.

[VCJ11] P. E. Vincent, P. Castonguay, and A. Jameson. A new class of high-order energy stable flux reconstruction schemes. *Journal of Scientific Computing*, 47(1):50–72, 2011.

[VFWJ15] P. E. Vincent, A. M. Farrington, F. D. Witherden, and A. Jameson. An extended range of stable-symmetric-conservative Flux Reconstruction correction functions. *Computer Methods in Applied Mechanics and Engineering*, 296:248–272, 2015.

[vNR50] J. von Neumann and R. D. Richtmyer. A method for the numerical calculation of hydrodynamic shocks. *Journal of Applied Physics*, 21(3):232–237, 1950.

[WAG15] G. Wasserman, R. Archibald, and A. Gelb. Image reconstruction from Fourier data using sparsity of edges. *Journal of Scientific Computing*, 65(2):533–552, 2015.

[Wei85] K. Weierstrass. Über die analytische Darstellbarkeit sogenannter willkürlicher Functionen einer reellen Veränderlichen. *Sitzungsberichte der Königlich Preußischen Akademie der Wissenschaften zu Berlin*, 2:633–639, 1885.

[Wen72a] B. Wendroff. The Riemann problem for materials with nonconvex equations of state I: Isentropic flow. *Journal of Mathematical Analysis and Applications*, 38(2):454–466, 1972.

[Wen72b] B. Wendroff. The Riemann problem for materials with nonconvex equations of state: II: General Flow. *Journal of Mathematical Analysis and Applications*, 38(3):640–658, 1972.

[Wen04] H. Wendland. *Scattered data approximation*, volume 17. Cambridge university press, 2004.

[WH96] G. Wanner and E. Hairer. *Solving ordinary differential equations II.* Springer Berlin Heidelberg, 1996.

[Whi11] G. B. Whitham. *Linear and nonlinear waves*, volume 42. John Wiley & Sons, 2011.

[Wil48] H. Wilbraham. On a certain periodic function. *The Cambridge and Dublin Mathematical Journal*, 3:198–201, 1848.

[Wil70a] M. W. Wilson. Discrete least squares and quadrature formulas. *Mathematics of Computation*, 24(110):271–282, 1970.

[Wil70b] M. W. Wilson. Necessary and sufficient conditions for equidistant quadrature formula. *SIAM Journal on Numerical Analysis*, 7(1):134–141, 1970.

[YGH97] B. Yang, D. Gottlieb, and J. S. Hesthaven. Spectral simulations of electromagnetic wave scattering. *Journal of Computational Physics*, 134(2):216–230, 1997.

[YW13] M. Yu and Z. Wang. On the connection between the correction and weighting functions in the correction procedure via reconstruction method. *Journal of Scientific Computing*, 54(1):227–244, 2013.

Index

Glossary

AD artificial dissipation. 138

AF active flux. 91

AV artificial viscosity. 3

BC boundary condition. 3

CG continuous Galerkin. 91

CL conservation law. 1

DG discontinuous Galerkin. 2

DGDLS discontinuous Galerkin discrete least squares. 100

DGSEM discontinuous Galerkin spectral element method. 3

DLS discrete least squares. 2

DOP discrete orthonormal polynomials. 42

EOC experimental order of convergence. 108

EOS equation of state. 9

FD finite difference. 3

FE finite element. 91

FR flux reconstruction. 2

FV finite volume. 91

GLe Gauss–Legendre. 2

GLo Gauss–Lobatto. 2

HOES high order edge sensor(s). 4

IC initial condition. 10

ID initial data. 10

IVP initial value problem. 10

LS least squares. 2

MMV multiple measurement vectors. 69

NNLS nonnegative least squares. 71

PA polynomial annihilation. 151

PDE partial differential equation. 1

QR quadrature rule. 1

RBF radial basis function. 3

RD residual distribution. 91

RK Runge–Kutta. 27

SAT simultaneous approximation term. 97

SBP summation by parts. 3

SD spectral difference. 91

SE spectral element. 2

SSP strong stability preserving. 27

TV total variation. 23

TVB total variation bounded. 92

TVD total variation diminishing. 4

WENO weighted essentially nonoscillatory. 91